Progress in Mathematics

Volume 215

Categorical Decomposition Techniques in Algebraic Topology

International Conference in Algebraic Topology, Isle of Skye, Scotland, June 2001

Gregory Arone
John Hubbuck
Ran Levi
Michael Weiss
Editors

Springer Basel AG

Editors:

Gregory Arone
Department of Mathematics
University of Virginia
P.O. Box 400137
Charlottesville, VA 22904
USA
e-mail: zga2m@virginia.edu

John Hubbuck
Ran Levi
Michael Weiss
Department of Mathematical Sciences
University of Aberdeen
Aberdeen AB24 3UE
UK
e-mail: j.hubbuck@maths.abdn.ac.uk
ran@maths.abdn.ac.uk
weiss@maths.abdn.ac.uk

2000 Mathematics Subject Classification 55-xx

A CIP catalogue record for this book is available from the Library of Congress, Washington D.C., USA

Bibliographic information published by Die Deutsche Bibliothek
Die Deutsche Bibliothek lists this publication in the Deutsche Nationalbibliografie; detailed bibliographic data is available in the Internet at <http://dnb.ddb.de>.

ISBN 978-3-0348-9601-6

Originally published by Birkhäuser Verlag, Basel in 2004
Softcover reprint of the hardcover 1st edition 2004

Printed on acid-free paper produced of chlorine-free pulp. TCF ∞

ISBN 978-3-0348-9601-6 ISBN 978-3-0348-7863-0 (eBook)
DOI 10.1007/978-3-0348-7863-0

www.birkhauser-science.com

9 8 7 6 5 4 3 2 1

Contents

Preface

The current volume constitutes the proceedings of the International Conference in Algebraic Topology, held on the Isle of Skye, Scotland in June 2001.

Many of the the talks at the conference focused on categorical decomposition techniques currently used in algebraic topology, such as Goodwillie's "calculus of functors" and the various approximation techniques that have proved so useful for the study of classifying spaces.

The contents represent these, and other themes in algebraic toplogy, as they are being developed by experts in the field. For instance, the homotopy theory of classifying spaces is represented by the articles of Aguade-Broto-Saumell, Davis and Iwase-Mimura. The papers by Betley, Kuhn and Panov-Ray-Vogt deal with general categorical decomposition techniques. The papers of Anton, Goerss-Henn-Mahowald and Hodgkin-Ostvaer bring us to the forefront of computational homotopy theory. Other papers deal with assorted topics of current interest in algebraic topology.

Progress in Mathematics, Vol. 215
Algebraic Topology: Categorical Decomposition Techniques, 1–20

The Functor T and the Cohomology of Mapping Spaces

Jaume Aguadé, Carles Broto, and Laia Saumell

1. Introduction

In his fundamental work [15] Lannes has introduced a functor T defined in the category $\mathcal{K}$ (resp. $\mathcal{U}$) of unstable algebras (resp. modules) over the Steenrod algebra which has many important applications in homotopy theory. This functor is, in some sense, the algebraic analogue of the mapping space functor $\mathrm{Map}(BV,-)$ for an elementary abelian group V. More precisely, there is a functor $T = T_V : \mathcal{K} \to \mathcal{K}$ which is adjoint to the functor $-\otimes H^*(BV)\colon \mathcal{K} \to \mathcal{K}$, and which computes, under some hypothesis, $H^*(\mathrm{Map}(BV, -^{\wedge}_p))$. Here, and throughout this paper, $H^*(-)$ denotes $H^*(-;\mathbb{F}_p)$ for some fixed prime p and $Y^{\wedge}_p$ denotes the p-completion of Y in the sense of Bousfield-Kan ([7]). We also assume that all constructions are done simplicially. For practical purposes, one would like to work on a component of the mapping space containing a particular map $f\colon BV \to X$. This is done by considering an algebraic "connected component" of the functor T which we denote T_f. Hence, we will assume that we have fixed a map $f\colon BV \to X$ so that we have a natural transformation

$$\Gamma\colon T_f(H^*(X)) = T(H^*(X)) \otimes_{(T(H^*X))^0} \mathbb{F}_p{}^{\{f\}} \to H^*(\mathrm{Map}(BV, X^{\wedge}_p)_f)$$

which is, under quite general conditions, an isomorphism. Here $\mathbb{F}_p{}^{\{f\}}$ means $\mathbb{F}_p$ considered as a module over $T(H^*X)^0$ via the restriction to degree zero of the adjoint of the map $f^*\colon H^*X \to H^*BV$ and $\mathrm{Map}(Y,Z)_g$ denotes the subspace of the mapping space $\mathrm{Map}(Y,Z)$ which contains all maps $g\colon Y \to Z$ with $g^* = f^*\colon H^*(Z) \to H^*(Y)$. (In the cases which we deal with in this paper, this subspace is just the connected component which contains f.)

It is clear that some hypothesis on X are needed for Γ to be an isomorphism. For instance, if we take $V = 0$, the spaces X such that Γ is an isomorphism are called $\mathbb{F}_p$-good (cf. [7]) and it is known that there are spaces which are not $\mathbb{F}_p$-good (cf. [6]). It is more difficult to find an example in which X is a 1-connected p-complete space with $H^*(X)$ of finite type and, nevertheless, T does not compute the cohomology of the mapping space. We provide an example of this kind in the last section of this paper.

The authors acknowledge support from DGES grant PB97-0203.

The relationship between $T_f H^*(X)$ and $H^*(\mathrm{Map}(BV, X_p^\wedge)_f)$ was made clear in the work of Dror-Smith ([10]) and Morel ([17], [18]). It turns out that $T_f H^*(X)$ is isomorphic to the continuous cohomology of the profinite space $\mathrm{Map}(BV, Y)$ where Y is the p-completion of the profinite completion of X. On the other side, there are some well-known conditions on the low-dimensional behavior of $T_f H^*(X)$ which imply that Γ is an isomorphism. Lannes ([15]) proves that Γ is an isomorphism if $H^*(X)$ and $T_f H^*(X)$ are of finite type and $T_f H^*(X)$ vanishes in degree 1 or, more in general, if $T_f H^*(X)$ is *free in degrees* ≤ 2 (see [15], 3.2). The purpose of this paper is to find more general conditions which imply also that Γ is an isomorphism.

In [1] it was necessary to compute the cohomology of some mapping spaces in cases in which the functor T does not vanish in degree one and where Lannes' *freeness* condition does not hold. The problem was solved there by some *ad hoc* method adapted to the specific value of $T_f H^*(X)$. The same kind of problem was encountered in [8] and again an *ad hoc* method was devised to solve it. Once again, along the investigation of the classifying spaces of Kac-Moody groups from a homotopy point of view ([2], see also [3]) it was necessary to compute the cohomology of some mapping spaces in cases when Lannes' condition does not hold.

What we want to do here is to put all these *ad hoc* methods into a quite general framework in such a way that we can deduce from a single theorem all computations of the cohomology of mapping spaces that we have just mentioned as well as some other computations which could arise. In Section 3 we introduce a condition – which looks rather technical but has many direct applications – called *T-representability* or *finite T-representability* and then Theorem 3, our main theorem, states that finite T-representability together with the usual finiteness conditions is a sufficient condition for the homomorphism

$$\Gamma\colon T_f(H^*(X)) \to H^*(\mathrm{Map}(BV, X_p^\wedge)_f)$$

to be an isomorphism.

This condition of T-representability looks much more complex than Lannes' freeness condition. However, we show in Sections 4 and 5 of this paper how this condition of T-representability can be deduced from the low-dimensional behavior of $T_f(H^*(X))$. This provides us with several concrete examples of cases in which the functor T effectively computes the cohomology of the mapping space. We recover Lannes' criterion, as well as the cases which have been studied in [1], [8] and [2]. Our arguments here heavily depend on group cohomology calculations. The main idea in Sections 4 and 5 is to reproduce the low-dimensional behavior of $T_f H^*(X)$ by the cohomology of a finite p-group P.

Given a p-complete space X and a map $f\colon BV \to X$, the structure in low dimensions of $T_f(H^*(X))$ is formally captured by an auxiliary algebra L that we define as follows. Let W_1^* denote the $\mathbb{F}_p$-vector space of elements of degree one of $T_f(H^*(X))$. Adjoint to the inclusion it is defined a map of unstable algebras over the Steenrod algebra $U(W_1^*) \to T_f(H^*(X))$, where U is Steenrod's free unstable algebra functor. If Q_2^* denotes the kernel of this map in degree two, we obtain a factorization $U(W_1^*)//U(Q_2^*) \to T_f(H^*(X))$ which is an isomorphism in degree one and a monomorphism in degree two. We define $L = U(W_1^*)//U(Q_2^*)$.

To one such algebra L, we attach a system of finite p-groups $\mathcal{C}(L)$ by the property that their mod p cohomology algebras mimic the behavior of L in low dimensions; more precisely, a finite p-group belongs to $\mathcal{C}(L)$ if there is a homomorphism of unstable algebras over the Steenrod algebra $\rho\colon L \to H^*(P)$ which is an isomorphism in degree one and a monomorphism in degree two. The particular nature of the system $\mathcal{C}(L)$ would imply T-representability and even finite T-representability. The case in which the system consists of only one finite p-group behaves particularly well and is developed independently in Section 4. In cases in which we can deduce finite T-representability, we also show that the fundamental group of $\mathrm{Map}(BV, X_p^\wedge)_f$ is either a finite p-group of the system or a pro-finite p-group obtained as a limit of a chain in the system.

As said before, the paper ends with an example where T does not compute the cohomology of $\mathrm{Map}(BV, X_p^\wedge)_f$ for some 1-connected space X of finite type. This example helps in understanding the scope of the main theorem.

We are grateful to F.-X. Dehon for his comments on a previous version of this paper.

2. Finiteness conditions and the geometric interpretation of T

To simplify the notation, let us say that a connected space Y is of finite $\widehat{\mathbb{Z}}_p$-type if $\pi_1(Y)$ is a finite p-group and $\pi_i(Y)$ is a finitely generated $\widehat{\mathbb{Z}}_p$-module for all $i > 1$. Let us say that Y is of finite $\mathbb{F}_p$-type if $H^i(Y)$ is a finite-dimensional $\mathbb{F}_p$-vector space for all i. For 1-connected, p-complete spaces, both conditions are equivalent (see [1], 5.7). On the other side, if Y is of finite $\widehat{\mathbb{Z}}_p$-type then Y is p-complete and of finite $\mathbb{F}_p$-type. To prove this, consider the universal cover of Y, $\widetilde{Y} \to Y \to K(\pi_1(Y), 1)$. $\widetilde{Y}$ is of finite $\mathbb{F}_p$-type and $\pi_1(Y)$ is a finite p-group. Hence, by [7] II.5.2 the fibration is nilpotent in the sense of [7], II.4 and Y is p-complete. Then, the Serre spectral sequence shows that Y has also finite $\mathbb{F}_p$-type.

We would like to remark that a finitely generated $\widehat{\mathbb{Z}}_p$-module is an Ext-p-complete abelian group in the sense of [7], VI.3. Conversely, an Ext-p-complete abelian group A has a canonical structure of $\widehat{\mathbb{Z}}_p$-module given by a natural isomorphism $A \cong \mathrm{Ext}(\mathbb{Z}_{p^\infty}, A)$. This implies that the category of finitely generated $\widehat{\mathbb{Z}}_p$-modules is a *full* subcategory of the category of abelian groups. Moreover, these same ideas can be used to prove that the category of finitely generated $\widehat{\mathbb{Z}}_p$-modules is closed under abelian group extensions.

Let us recall also the well-known fact that $\varprojlim$ is an exact functor in the category of towers of finite groups and in the category of towers of finitely generated $\widehat{\mathbb{Z}}_p$-modules (Jensen, [14]). Also, a homomorphism of towers of finite groups $\{G_s\} \to \{H_s\}$ is a pro-monomorphism (pro-epimorphism) if and only if it induces a monomorphism (epimorphism) $\varprojlim G_s \to \varprojlim H_s$.

Finally, we recall the geometric interpretation of the functor T developed by Dror Farjoun-Smith in [10]. Let X be a space such that H^*X is of finite type and vanishes in degree one. Then $\pi_1 X$ is $\mathbb{F}_p$-perfect and we get that $X_p^\wedge$ is simply connected, p-complete and $H^*X \cong H^*X_p^\wedge$. Also, the homotopy groups of $X_p^\wedge$ are finitely generated $\widehat{\mathbb{Z}}_p$-modules. Let $f\colon BV \to X$ be a map and consider $T_f H^*X$. Let $\{P_s X_p^\wedge\}$ be the Postnikov tower of $X_p^\wedge$ and denote $E_s = \mathrm{Map}(BV, P_s X_p^\wedge)_{f_s}$ (f_s is the map induced by f). Then Dror Farjoun-Smith prove ([10]) that

$$T_f H^*X \cong \varinjlim H^*(E_s),$$

the isomorphism being induced by the natural map. Hence, the homomorphism

$$\Gamma\colon T_f(H^*(X)) \to H^*(\mathrm{Map}(BV, X_p^\wedge)_f)$$

is an isomorphism if and only if $H^*(\mathrm{holim}\, E_s) \cong \varinjlim H^*(E_s)$ and so the problem that we are investigating in this note is, in a certain sense, a problem about the commutation of homology and inverse limits.

We claim that the spaces E_s are of finite $\widehat{\mathbb{Z}}_p$-type. We prove this by induction on s. Recall that the spaces E_s are defined by $E_s = \mathrm{Map}(BV, P_s X_p^\wedge)_{f_s}$. Hence, we have principal fibrations $E_s \to E_{s-1}$ whose fibre is a union of components from $\mathrm{Map}(BV, K(\pi_s X_p^\wedge, s))$. Each component is a product of Eilenberg-MacLane spaces $K(H^{s-j}(BV; \pi_s X_p^\wedge), j)$, $1 \le j \le s$. Notice that $X_p^\wedge$ is simply connected. Hence, if we take into account the remarks on the category of finitely generated $\widehat{\mathbb{Z}}_p$-modules that we have made above, the homotopy long exact sequence of these fibrations yields the result by induction.

3. T-representability

Here is our condition on the low-dimensional behavior of $T_f H^*X$:

Definition 1. *We say that $f\colon BV \to X$ is T-representable if there is an increasing sequence $\alpha(s)$ and a map of towers*

$$g\colon \{E_{\alpha(s)}\} \to \{B_s\}$$

such that

1. *$B_s = K(G_s, 1)$ and G_s is a finite p-group for all s.*
2. *g induces a pro-isomorphism in H_1 and a pro-epimorphism in H_2.*
3. *$H^*(\varprojlim G_s) \cong \varinjlim H^*(G_s)$, induced by the natural map.*

If we have a T-representation $g\colon \{E_{\alpha(s)}\} \to \{B_s\}$, then we denote $B_\infty = \mathrm{holim}\, B_s$ and $G_\infty = \varprojlim G_s$. Then, since the groups G_s are finite, we have $B_\infty = K(G_\infty, 1)$.

Here are some straightforward examples of T-representability:

1. If $T_f H^*X$ vanishes in degree one then we get T-representability in a trivial way, by taking $B_s = *$.

2. If $T_f H^* X$ is *free in degrees* ≤ 2 (see [15], 3.2) we also get T-representability in the following way. Let W be the $\mathbb{F}_p$-vector space $T_f^1 H^* X$. We can take $\{B_s\}$ to be the constant tower $\{K(W, 1)\}$. Then, conditions 1 and 3 are trivially satisfied and 2 is just the condition of being free in degrees ≤ 2.

In the next sections, we present some other examples of T-representability.

In order to prove that Γ is an isomorphism we need that the T-representation satisfies some finiteness condition. Notice that a T-representation $\{E_{\alpha(s)}\} \to \{B_s\}$ produces a structure of $H^*(G_\infty)$-module on $T_f H^* X = \varinjlim H^*(E_s)$.

Definition 2. *We say that $f\colon BV \to X$ is finitely T-representable if $T_f H^* X$ is of finite type and there is a T-representation $\{E_{\alpha(s)}\} \to \{B_s\}$ such that either of these conditions holds true:*

1. *G_∞ is a finite p-group.*
2. *$H^*(G_\infty)$ is of finite type and $\mathrm{Tor}^{*,*}_{H^*(G_\infty)}(T_f H^* X, \mathbb{F}_p)$ is finite-dimensional in each total degree.*

In most practical cases, it happens that $T_f H^* X$ is free over $H^*(G_\infty)$ and the above finiteness condition on Tor holds trivially.

We are now ready to state the main result of this paper which says that finite T-representability implies that the functor T computes the cohomology of the appropriate mapping space.

Theorem 3. *If X is a space of finite $\mathbb{F}_p$-type such that $H^1 X = 0$ and $f\colon BV \to X$ is finitely T-representable, then*

$$\Gamma\colon T_f(H^*(X)) \to H^*(\mathrm{Map}(BV, X_p^\wedge)_f)$$

is an isomorphism. Furthermore, $\mathrm{Map}(BV, X_p^\wedge)_f$ is p-complete and if the tower of p-groups $\{G_s\}$ provides a T-representation and we write $G_\infty = \varprojlim G_s$, then we have $\pi_1 \mathrm{Map}(BV, X_p^\wedge)_f \cong G_\infty$.

Proof. We want to show that $H^*(\mathrm{holim}\, E_s) \cong \varinjlim H^*(E_s)$. We will work in homology. Notice that if we assume that both $T_f H^* X = \varinjlim H^*(E_s)$ and $H^*(B_\infty)$ are of finite type then we have that $H_*(\varprojlim G_s) \cong \varprojlim H_*(G_s)$ and we need to prove that $H_*(\mathrm{holim}\, E_s) \cong \varprojlim H_*(E_s)$.

We start with a T-representation $\{E_s\} \to \{B_s\}$ (we take $\alpha(s) = s$ without loss of generality) and we denote by F_s the homotopy fibre of the map $E_s \to B_s$. Then, we split the proof in several steps.

(1) *Without loss of generality, we can assume that F_s is connected.*

To see this, consider $\phi_s\colon \pi_1 E_s \to \pi_1 B_s$ and define $B'_s = K(\mathrm{Im}\, \phi_s, 1)$. We have a map of towers $\{E_s\} \to \{B'_s\}$ and if we could prove that $\{B'_s\} \to \{B_s\}$ is a weak pro-homotopy equivalence ([7], III.3) then it would be clear that $\{E_s\} \to \{B'_s\}$ is also a (finite) T-representation with connected fibers.

Recall now that the maximal subgroups of a finite p-group are normal and of index p. Then, define $H_s \triangleleft G_s$ to be equal to G_s if $\phi_s\colon \pi_1 E_s \to \pi_1 B_s$ is onto and to be a maximal subgroup of G_s containing the image of $\pi_1 E_s$ if ϕ_s is not onto. We want to show that $\{H_s\} \to \{G_s\}$ is a pro-epimorphism i.e., we need to show that $\varprojlim G_s/H_s = 1$. By hypothesis, $\{H_1 E_s\} \to \{H_1 G_s\}$ is a pro-epimorphism and so it is $\{H_1 H_s\} \to \{H_1 G_s\}$. But the cokernel of $H_1 H_s \to H_1 G_s$ contains a copy of G_s/H_s.

(2) *The tower $\{H_1 F_s\}$ is pro-trivial.*

We have for each s a fibration $F_s \to E_s \to B_s$ with connected fibre. Notice that the spaces F_s are of finite $\widehat{\mathbb{Z}}_p$-type and so they are p-complete and of finite $\mathbb{F}_p$-type. Hence, to prove that $\{H_1 F_s\}$ is pro-trivial we can prove that the (ascending) tower of finite $\mathbb{F}_p$-vector spaces $\{H^1 F_s\}$ is pro-trivial.

Let us consider the cohomology Serre spectral sequences $E_*^{*,*}(s)$ of the fibrations $F_s \to E_s \to B_s$. Let $\alpha \in E_2^{0,1}(s) = H^0(B_s; H^1 F_s) = (H^1 F_s)^{G_s}$ and let us consider $d\alpha \in H^2 B_s$. Since $\{H^2 B_s\} \to \{H^2 E_s\}$ is a pro-monomorphism, if we take s large enough we can assume $d\alpha = 0$. Hence, α represents an element in $H^1 E_s$ not in the image of $H^1 B_s$. Since $\{H^1 B_s\} \to \{H^1 E_s\}$ is a pro-epimorphism, again if we take s large enough we see that $\alpha = 0$.

Thus, we have seen that for any s there is h such that

$$(H^1 F_s)^{G_s} \to (H^1 F_{s+h})^{G_{s+h}}$$

is trivial. We finish now with a lemma:

Lemma 4. *Let $\{V_i\}$ be an ascending tower of $\mathbb{F}_p$-vector spaces. Let $\{G_i\}$ be a descending tower of finite p-groups. Assume that each V_i is a G_i-module and that the obvious coherence condition is satisfied. Then, if the tower of invariants $\{V_i^{G_i}\}$ is pro-trivial then the tower $\{V_i\}$ is also pro-trivial.*

Proof. Recall that when a finite p-group G acts on an $\mathbb{F}_p$-vector space V, then there is a finite filtration of V by G-submodules, $0 = V^{(0)} \subset V^{(1)} \subset \cdots \subset V^{(k)} = V$ such that G acts trivially on each quotient $V^{(i)}/V^{(i-1)}$.

In our case, we want to prove that for any s there is N such that $V_s \to V_{s+N}$ is the trivial map. We find N by induction along the filtration of V_s. Assume we have found h such that all elements of $V_s^{(i)}$ die in V_{s+h}. By hypothesis, all elements in $(V_{s+h})^{G_{s+h}}$ die in some V_{s+h+k}. Since $V_s^{(i+1)} \to V_{s+h}$ factors through $V_s^{(i+1)}/V_s^{(i)}$, it factors also through $(V_{s+h})^{G_{s+h}}$ and so $V_s^{(i+1)} \to V_{s+h+k}$ is trivial and the induction goes on. □

(3) *The tower $\{H_i F_s\}$ is pro-constant for any i.*

This statement is the point where the finiteness assumption is needed. To prove that the tower $\{H_i F_s\}$ is pro-constant it is enough to show that $\varprojlim H_i F_s$ is finite. This is proven using the Serre spectral sequence or the Eilenberg-Moore spectral sequence, depending on which finiteness condition we assume.

If $\varprojlim G_s$ is a finite p-group then we use the (homology) Serre spectral sequence in the following way. The fibrations $F_s \to E_s \to B_s$ produce an inverse system of Serre spectral sequences of finite-dimensional $\mathbb{F}_p$-vector spaces

$$H_*(B_s; H_* F_s) \Longrightarrow H_* E_s.$$

Since $\varprojlim$ is exact in this context, we get a spectral sequence

$$\varprojlim H_*(B_s; H_*F_s) \implies \varprojlim H_*E_s.$$

By hypothesis, this spectral sequence converges to a graded $\mathbb{F}_p$-vector space of finite type. Let us investigate the E^2-term. First of all, we have an isomorphism (use the spectral sequences of Theorem 4.4 in [14])

$$H_*(\varprojlim G_s; \varprojlim H_*F_s) \cong \varprojlim H_*(\varprojlim G_s; H_*F_s).$$

Now, H_iF_s is a $\varprojlim G_s$-module via $\varprojlim G_s \to G_s$ and G_s is a finite p-group. Hence, H_iF_s has a finite filtration whose factors have trivial $\varprojlim G_s$-action. By the T-representability hypothesis, $H_*(\varprojlim G_s; N) \cong \varprojlim H_*(G_s; N)$ if N is a trivial $\mathbb{F}_pG_s$-module. Then, by induction along the filtration of H_iF_s and using the exactness of $\varprojlim$ on finite $\mathbb{F}_p$-vector spaces, we get isomorphisms

$$\varprojlim H_*(\varprojlim G_s; H_*(F_s)) \cong \varprojlim_r \varprojlim_s H_*(G_s; H_*F_r) \cong \varprojlim_s H_*(G_s; H_*F_s).$$

Hence, we have a spectral sequence starting at $H_*(\varprojlim G_s; \varprojlim H_*(F_s))$ and converging to something of finite type. By hypothesis, $\varprojlim G_s$ is a finite p-group. Then, we can show inductively that $\varprojlim H_*(F_s)$ is of finite type, using the following easy lemma:

Lemma 5. *If G is a finite p-group and M is an $\mathbb{F}_pG$-module such that $H_0(G; M)$ is finite, then M is finite.*

Proof. Let I be the augmentation ideal of $\mathbb{F}_pG$. Since G is a finite p-group, it is well known that $I^n = 0$ for some n and so it suffices to prove inductively that $I^rM/I^{r+1}M$ is finite. $M/IM = H_0(G; M)$ is finite by hypothesis. The homology exact sequence

$$H_1(G; I^rM/I^{r+1}M) \to H_0(G; I^{r+1}M) \to H_0(G; I^rM)$$

yields the inductive step. □

If the finiteness condition 2 (involving Tor) is assumed then we use the cohomology Eilenberg-Moore spectral sequence. It converges strongly by [11] because the fundamental group of the base space is G_s, a finite p-group, thus acting nilpotently on the mod p cohomology of the fibre. In the limit, the exactness of $\varinjlim$ produces a spectral sequence

$$\mathrm{Tor}^{*,*}_{\varinjlim H^*(G_s)}(\varinjlim H^*E_s, \mathbb{F}_p) \implies \varinjlim H^*F_s.$$

The assumption on Tor implies that $\varinjlim H^*F_s$ is of finite type.

(4) holim F_s *is p-complete, simply connected, and* $H_*(\operatorname{holim} F_s) \cong \varprojlim H_*F_s$.

This is proved as follows. For a space X let us denote by $\{R_sX\}$ the Bousfield-Kan $\mathbb{F}_p$-tower of X ([7]). Each R_sF_s is $\mathbb{F}_p$-nilpotent, and therefore the coaugmentation of R_s induces a weak pro-homotopy equivalence $\{R_sF_s\} \to \{R_sR_sF_s\}$. Hence, also the map $\{R_sF_s\} \to \{R_sR_sF_s\}$ obtained by applying the functor R_s to $\phi\colon F_s \to R_sF_s$ is a weak pro-homotopy equivalence and then, the map of towers

$\{F_s\} \to \{R_sF_s\}$ induces pro-isomorphisms $\{H_i(F_s)\} \cong \{H_i(R_sF_s)\}$ for all i (see [7], III.6.6). Now the tower of fibrations of $\mathbb{F}_p$-nilpotent spaces $\{R_sF_s\}$ satisfies the hypothesis of Lemma 9.3 in [5]. We can conclude that holim R_sF_s is p-complete and simply connected and there are pro-isomorphisms $\{H_i(\text{holim}\, R_sF_s)\} \cong \{H_i(R_sF_s)\}$ for all i. Since each F_s is p-complete, holim R_sF_s = holim F_s. We deduce that there are pro-isomorphisms $\{H_i(\text{holim}\, F_s)\} \cong \{H_i(F_s)\}$ for all i and the claim follows.

(5) *Final step.*

Consider the inverse system of fibrations $\{F_s \to E_s \to B_s\}$. For simplicity, let us denote $E_\infty = \text{holim}\, E_s$, $F_\infty = \text{holim}\, F_s$. Since the homotopy fibre of a map is a particular kind of homotopy limit, we have a fibration $F_\infty \to E_\infty \to B_\infty$.

As discussed above, the fibrations $F_s \to E_s \to B_s$ produce an inverse system of Serre spectral sequences of finite $\mathbb{F}_p$-vector spaces

$$H_*(B_s; H_*F_s) \quad \Longrightarrow \quad H_*E_s$$

and a limit spectral sequence

$$\varprojlim H_*(B_s; H_*F_s) \quad \Longrightarrow \quad \varprojlim H_*E_s.$$

Let us analyze the E^2-term of this second spectral sequence. He have seen that the tower $\{H_*F_\infty\}$ is pro-isomorphic to the tower $\{H_*F_s\}$. Notice now that it is easy to show that a homomorphism of towers is a pro-isomorphism if and only if it is a pro-isomorphism of pointed sets. We mean that $\{H_*F_\infty\}$ and $\{H_*F_s\}$ are also pro-isomorphic as towers of $\mathbb{F}_pG_\infty$-modules. Hence, the towers $\{H_*(B_\infty; H_*F_\infty)\}$ and $\{H_*(B_\infty; H_*F_s)\}$ are pro-isomorphic too and this implies that

$$H_*(B_\infty; H_*F_\infty) \cong \varprojlim H_*(B_\infty; H_*F_s).$$

Now, the same argument as in step 3 of the proof shows that

$$\varprojlim H_*(B_\infty; H_*F_s) \cong \varprojlim H_*(B_s; H_*F_s)$$

and the proof of the theorem is complete. □

4. Examples I

In this section and in the next one we want to show how we can obtain T-representability for cohomology algebras, assuming some low-dimensional conditions on $T_f(H^*(X))$. As we have already pointed out, if we assume that $T_fH^*(X)$ is free in degrees ≤ 2 then we have T-representability in a trivial way, by a constant tower BW where W is an $\mathbb{F}_p$-vector space, and we want to deal now with a more general situation.

Given two algebras over the Steenrod algebra A and B, we will say that a $\mathcal{K}$-map $f\colon B \to A$ is an *n-equivalence* if it is bijective in degrees $r < n$ and injective in degree n. A sequence of connected algebras over the Steenrod algebra $C \to B \to A$ is called an *n-approximation of* A if the composition is trivial in positive degrees and induces an n-equivalence $B//C \to A$. T-representability will be seen to be related to n-approximations for small values of n.

If A is a finite type connected unstable algebra over the Steenrod algebra, we can obtain a 2-approximation of A in the following way. Let W_1 be the $\mathbb{F}_p$-vector space dual to A^1. The $\mathbb{F}_p$-linear map $W_1^* \to A$ extends to a $\mathcal{K}$-algebra homomorphism

$$\varphi\colon U(W_1^*) \to A,$$

where U is Steenrod's free unstable algebra functor. Let now Q_2 be the $\mathbb{F}_p$-vector space dual to the kernel of φ in degree 2. Then, we have a canonical 2-approximation

$$U(Q_2^*) \xrightarrow{\psi} U(W_1^*) \xrightarrow{\varphi} A.$$

This 2-approximation produces a central extension of finite p-groups in the following way. Let $H_2(W_1) \to Q_2$ be dual to the inclusion $Q_2^* \subset U^2(W_1^*) \cong H^2(W_1)$. This defines an element $\omega \in H^2(W_1; Q_2)$ and therefore a central extension of finite p-groups

(1) $$1 \longrightarrow Q_2 \xrightarrow{i} P \xrightarrow{\pi} W_1 \longrightarrow 1.$$

Then, one can check by a spectral sequence argument that π induces a 2-approximation

(2) $$U(Q_2^*) \xrightarrow{\psi} U(W_1^*) \xrightarrow{B\pi^*} H^*(P).$$

Notice that this 2-approximation depends only on the structure of the initial algebra A in degrees one and two.

As always throughout this paper, let X be a connected space of finite $\mathbb{F}_p$-type with $H^1(X) = 0$ and let $f\colon BV \to X$ be a map. Assume that $T_f(H^*(X))$ is of finite type and let us perform the above construction starting with $A = T_f(H^*(X))$. We obtain a 2-approximation like (2).

Theorem 6. *If the 2-approximation* (2) *associated to $T_f(H^*X)$ is a 3-approximation, then f is finitely T-representable.*

Hence the question of finite T-representability is essentially reduced to one about group cohomology. Notice also that if $T_f(H^*(X))$ is free in degrees ≤ 2 then this theorem applies trivially ($Q_2^* = 0$) and so this theorem generalizes Lannes' criterion.

Proof of Theorem 6. The idea here is to prove that the constant tower $\{BP\}$ can be used to prove T-representability. Let $\{E_s\}$ be the tower of fibrations with $E_s = \mathrm{Map}(BV, P_s X_p^\wedge)_{f_s}$ as in Section 2. It provides in mod p cohomology a direct system $\{H^*(E_s)\}$ with limit $T_f(H^*(X))$. On the other side, we have the canonical 2-approximation

$$U(Q_2^*) \xrightarrow{\psi} U(W_1^*) \xrightarrow{\varphi} T_f(H^*(X)).$$

Then, for some large index s, there is a factorisation

$$\begin{array}{ccc} U(W_1^*)//U(Q_2^*) & & \\ \Big\downarrow \hat{g}_s^* & \searrow^{\hat{\varphi}} & \\ H^*(E_s) & \xrightarrow[j_s]{} & T_f(H^*(X)) \end{array}$$

where $g_s^* \colon U(W_1^*) \to H^*(E_s)$ is induced by a map $g_s \colon E_s \to BW_1$ which should lift to BP. Hence, we have a diagram

$$\begin{array}{ccccc} & & H^*(BP) & & \\ & {}^{B\pi^*}\nearrow & & \searrow^{\bar{g}_s^*} \quad \searrow^{\bar{g}^*} & \\ U(W_1^*)//U(Q_2^*) & \xrightarrow[\hat{g}_s^*]{} & H^*(E_s) & \xrightarrow[j_s]{} & T_f(H^*(X)) \end{array}$$

where we defined $\bar{g}^* = j_s \circ \bar{g}_s^*$. $B\pi^*$ is a 3-equivalence by hypothesis and therefore $\bar{g}^* \colon H^*(BP) \to T_f(H^*(X))$ is a 2-equivalence. It follows that the obvious map of towers $\{E_s\} \to \{BP\}$ induces a pro-isomorphism in H_1 and a pro-epimorphism in H_2 and so $f \colon BV \to X$ is T-representable. □

The question is now to decide when the sequence

$$U(Q_2^*) \xrightarrow{\psi} U(W_1^*) \xrightarrow{\pi^*} H^*(P) \tag{3}$$

associated to a group extension $Q_2 \rightarrowtail P \twoheadrightarrow W_1$ is a 3-approximation. This question can be investigated by means of the Serre spectral sequence of the extension. The first differential $d_2 \colon E_2^{0,1} \to E_2^{2,0}$ is identified with the map ψ in (3) in degree 2. The next differential $d_3 \colon E_3^{0,2} \to E_3^{3,0}$ is the transgression

$$\tau \colon \beta Q_2^* \longrightarrow U^3(W_1^*)/W_1^* \cdot Q_2^*$$

that satisfies $\tau(\beta q) = \beta(\psi(q))$ modulo $\operatorname{Im}\psi$.

We thus have $\mathbb{F}_p$-vector spaces isomorphisms $E_\infty^{1,0} \cong W_1^* \cong H^1(P)$, and $E_\infty^{2,0} \cong U^2(W_1^*)/Q_2^* \subset H^2(P)$, with complement isomorphic to $E_\infty^{1,1} \oplus E_\infty^{0,2}$; and furthermore, there are exact sequences

$$0 \longrightarrow E_\infty^{1,1} \longrightarrow W_1^* \otimes Q_2^* \xrightarrow{m} U^3(W_1^*) \longrightarrow U^3(W_1^*)/W_1^* \cdot Q_2^* \longrightarrow 0$$

$$0 \longrightarrow E_\infty^{0,2} \longrightarrow \beta Q_2^* \xrightarrow{\tau} U^3(W_1^*)/W_1^* \cdot Q_2^* \longrightarrow E_\infty^{3,0} \subset H^3(P),$$

where m is given by multiplication in $U(W_1^*)$: $m(\omega \otimes q) = \omega \cdot \psi(q)$.

It follows that the necessary and sufficient condition for (3) to be a 3-approximation is that m and τ are injective. If $p = 2$, then m is always injective because $U(W_1^*)$ does not contain zero divisors but for an odd p this is not necessarily so. We have thus proved

Proposition 7. (3) *is a 3-equivalence if and only if either $p = 2$ and the transgression map $\tau \colon Sq^1 Q_2^* \to U^3(W_1^*)/W_1^* \cdot Q_2^*$ is injective; or p is odd and both the transgression $\tau \colon \beta Q_2^* \to U^3(W_1^*)/W_1^* \cdot Q_2^*$ and $m \colon Q_2^* \otimes W_1^* \to U^3(W_1^*)$ are injective.* □

We will now look more closely at the particular case in which Q_2^* is 1-dimensional generated by a quadratic form $q \in H^2(W_1)$. In this case the extension $\mathbb{Z}/p \rightarrowtail P \twoheadrightarrow W_1$ is an extraspecial group and the cohomology of these groups has been largely studied. The mod 2 cohomology of extraspecial 2-groups was described by Quillen ([19]). At odd primes our knowledge is not complete. Diethelm ([9]) and Leary ([16]) computed the mod p cohomology of extraspecial groups or order p^3. A general reference is a beautiful paper by Benson and Carlson [4].

For $p = 2$, if n is the dimension of W_1^* and h is the codimension of a maximal isotropic subspace of q in W_1^*, we have

$$H^*(P; \mathbb{F}_2) \cong P[x_1, x_2, \ldots, x_n]/(q_o, q_1, \ldots, q_{h-1}) \otimes P[\zeta]$$

where x_i are all of degree 1, $q_0 = q$ has degree 2, $q_{i+1} = Sq^{2^i} q_i$, and $\deg \zeta = 2^h$.

Corollary 8. *Assume that $p = 2$ and $T_f(H^*(X))$ has a 2-approximation*

$$U(Q_2^*) \xrightarrow{\psi} U(W_1^*) \xrightarrow{\varphi} T_f(H^*(X))$$

where Q_2^* is generated by a single quadratic form $q \in U^2(W_1^*)$ with the property that a maximal isotropic subspace of q has codimension $h \geq 2$. Then $f\colon BV \to X$ is finitely T-representable. □

Example 9. If $p = 2$ and $T_f(H^*(X))$ has two generators x, y in degree one which satisfy a single relation q in degree 2, there are three possible cases: $q = x^2$, $q = xy$ or $q = x^2 + xy + y^2$. In the first two cases, the codimension of a maximal isotropic subspace is 1 and Corollary 8 does not apply. Here the group P is either $\mathbb{Z}/4\mathbb{Z} \times \mathbb{Z}/2\mathbb{Z}$ or the dihedral group of 8 elements. Those cases will be considered in the next section. In the case in which $q = x^2 + xy + y^2$, P is the quaternion group of order 8, $h = 2$ and Corollary 8 applies. □

If p is odd, $q \in H^2(W_1)$ is not really a quadratic form. We have

$$H^2(W_1; \mathbb{F}_p) \cong \Lambda^2(W_1^*) \otimes \beta W_1^*,$$

thus q can be written as $q = s + \lambda$, where $s \in \Lambda^2(E_1^*)$ is a symplectic form and $\lambda \in \beta W_1^*$ a linear form.

Corollary 10. *Assume that p is odd and $T_f(H^*(X))$ has a 2-approximation*

$$U(Q_2^*) \xrightarrow{\psi} U(W_1^*) \xrightarrow{\varphi} T_f(H^*(X))$$

where Q_2^* is generated by a single form $q = s + \lambda \in U^2(W_1^*)$. Then, if both s and λ are non-trivial, $f\colon BV \to X$ is finitely T-representable.

Proof. Let $\mathbb{Z}/p \rightarrowtail P \twoheadrightarrow W_1$ be the extension classified by $q = s + \lambda$. If we write $H^*(\mathbb{Z}/p) \cong P[z] \otimes \Lambda(t)$, the differential d_2 in the Serre spectral sequence of the extension hits the extension class $d_2(t) = q$. If $\lambda \neq 0$, then q is not a zero divisor and $E_3^{*,1} = 0$. The next differential is $d_3(z) = \beta(q) = \beta(s)$, and if $s \neq 0$, then it is not divisible by λ. Hence $d_3(z) \neq 0$ in $E_3^{*,*}$ and z does not survive to $E_\infty^{*,*}$. It follows that $U(E_1^*)//U(Q_2^*) \longrightarrow H^*(P; \mathbb{F}_p)$ is a 3-equivalence. □

Example 11. If p is odd and $T_f(H^*(X))$ has two 1-dimensional generators y_1, y_2, that satisfy a single relation $q \in \langle y_1y_2, x_1, x_2\rangle$, where we write $x_i = \beta y_i$, then there are three possibilities.

1. If $q = x_1$ then $P \cong \mathbb{Z}/p^2 \times \mathbb{Z}/p$ and Corollary 10 does not apply.
2. If $q = y_1y_2$ then P is the non-abelian group of order p^3 and exponent p. Its mod p cohomology is described in [16]. Again Corollary 10 does not apply.
3. If $q = y_1y_2 + x_1$ then P is now a non-abelian split metacyclic group of order p^3. The cohomology ring was computed in [9]. Now, Corollary 10 applies. □

5. Examples II: Systems of p-groups with fixed low-dimensional cohomology

This section is devoted to a further investigation of cases in which Theorem 6 does not apply. That is, cases in which the 2-approximation of $T_f(H^*(X))$ determines an extension $Q_2 \to P \to W_1$ for which $U(W_1^*)//U(Q_2^*) \to H^*(P;\mathbb{F}_p)$ is not a 3-equivalence (cf. Example 9 for $q = x^2$, xy).

Fix the algebra $L = U(W_1^*)//U(Q_2^*)$. As a first step we attach to L a system of finite p-groups, or, more precisely, a system of isomorphism classes of finite p-groups:

$$\mathcal{C}(L) = \{\, P \text{ finite } p\text{-group} \,|\, \text{there is a 2-equivalence } \rho\colon L \to H^*(P) \,\}$$

where we just write P for the isomorphism class that it represents. $\mathcal{C}(L)$ is a poset with the relation $P \lessdot P'$ if there is an epimorphism $\pi\colon P' \twoheadrightarrow P$. It is useful to observe that this epimorphism induces a commutative diagram in cohomology

$$\begin{array}{ccc} & L & \\ {}^{\rho_1}\swarrow & & \searrow^{\rho_2} \\ H^*(P) & \xrightarrow[\pi^*]{} & H^*(P') \end{array}$$

where ρ_1 and ρ_2 are 2-equivalences. In fact, if K is the kernel of π, we have an exact sequence $0 \to H^1(P) \to H^1(P') \to H^1(K)^P$, so that π^* is a monomorphism in degree 1, but $H^1(P)$ and $H^1(P')$ have the same finite dimension. This shows the commutativity of the diagram in degree 1. In degree 2 the commutativity of the diagram follows because L is generated by degree one elements.

Now we state some properties of the system $\mathcal{C}(L)$.

(A) If $P \in \mathcal{C}(L)$ and $1 \to V \to P' \to P \to 1$ is a central extension, with V elementary abelian p-group, determined by an inclusion $V^* \subset H^2(P)$, then, $P' \in \mathcal{C}(L)$ if and only if $V^* \cap \rho(L^2) = \{0\}$.

(B) A relation $P \lessdot P'$ can be refined to a chain $P = P_1 \lessdot P_2 \lessdot P_3 \lessdot \cdots \lessdot P_r = P'$ in $\mathcal{C}(L)$ such that, for each k there is a central extension $1 \to V_k \to P_k \to P_{k-1} \to 1$, with V_k elementary abelian p-group.

(C) There is an initial element. Namely, the extension given by the 2-approximation $U(Q_2^*) \to U(W_1^*) \to L$.

(D) An element P in $\mathcal{C}(L)$ is maximal if and only if the 2-equivalence $\rho\colon L \to H^*(P)$ is actually an isomorphism in degree 2.

We will say that a system $\mathcal{C}(L)$ is of *finite type* if for any infinite chain

$$P_{\lambda_1} \lessdot P_{\lambda_2} \lessdot \cdots \lessdot P_{\lambda_s} \lessdot \cdots$$

the mod p cohomology of the p pro-finite group $P_{\lambda_\infty} = \varprojlim P_{\lambda_s}$ is of finite type and $H^*(P_{\lambda_\infty}) \cong \varinjlim H^*(P_{\lambda_s})$.

Before providing some examples of system of p-groups we show how to prove the above properties. For (A), we use the Serre spectral sequence for the extension $1 \to V \to P' \to P \to 1$. In low dimensions it provides an exact sequence

$$\text{(4)} \qquad 0 \to H^1(P) \to H^1(P') \to H^1(V) \xrightarrow{d_2} H^2(P) \to H^2(P').$$

But the differential d_2 is given by the extension class, so, in our case it is identified with the inclusion $V^* \subset H^2(P)$ and the exact sequence breaks to an isomorphism $H^1(P) \cong H^1(P')$ and a monomorphism $H^2(P)/V^* \rightarrowtail H^2(P')$. The composition $L \xrightarrow{\rho} H^*(P) \to H^*(P')$ in degree 2 factors through this monomorphism and this proves the statement.

For (B) we start with an epimorphism $P' \twoheadrightarrow P$. If it is an isomorphism we are done, otherwise, let K be the kernel. Using the action of P' on K by conjugation we can find an elementary abelian subgroup of K, V, which is central in P', and then we can form the factorization $P' \twoheadrightarrow P'/V \twoheadrightarrow P$, and repeat the same argument inductively until we obtain a sequence of epimorphisms $P' = P_r \twoheadrightarrow P_{r-1} \twoheadrightarrow \cdots \twoheadrightarrow P_2 \twoheadrightarrow P_1 = P$ where each step is a central extension. We can compose the restrictions with the 2-equivalence $\rho\colon L \to H^*(P)$ and get a diagram

$$\begin{array}{ccccccc} L & & & & & & \\ \downarrow{\scriptstyle\rho} & \searrow & & \cdots & \overset{\rho'}{\searrow} & & \\ H^*(P_1) & \longrightarrow & H^*(P_2) & \longrightarrow & \cdots & \longrightarrow & H^*(P_r)\,. \end{array}$$

In degree one ρ and ρ' are isomorphisms, while the exact sequence (4) applies to each step $V_k \rightarrowtail P_k \twoheadrightarrow P_{k-1}$ and then the morphisms in the bottom row are all monomorphisms, hence isomorphisms. In degree two (4) now implies that $H^2(P_{k-1}) \to H^2(P_k)$ factors as $H^2(P_{k-1}) \twoheadrightarrow H^2(P_{k-1})/V_k^* \rightarrowtail H^2(P_k)$ and since both ρ and ρ' are injective in degree two, it follows that at each step $V_k^* \cap \rho_k(L^2) = \{0\}$ and then property (A) applies and $P_k \in \mathcal{C}(L)$.

We will now show that the central extension $Q_2 \to P \to W_1$ determined by $L = U(W_1^*)//U(Q_2^*)$ is an initial object in $\mathcal{C}(L)$, thus proving property (C). Assume that $P' \in \mathcal{C}(L)$. The 2-equivalence $\rho\colon L \to H^*(P')$ gives $W_1^* \cong H^1(P')$, and then it gives an epimorphism $P' \twoheadrightarrow W_1 \cong H_1(P')$. Let K be the kernel. The Serre spectral sequence for the extension $1 \to K \to P' \to W_1 \to 1$ gives an exact sequence $0 \to H^1(K)^{W_1} \xrightarrow{d_2} H^2(W_1) \to H^2(P')$. The last homomorphism factors through L in degree 2 $H^2(W_1) \twoheadrightarrow L^2 \rightarrowtail H^2(P')$ and so, therefore, $Q_2^* \cong$

$H^1(K)^{W_1} \subset H^1(K)$. The dual of this inclusion provides an epimorphism $K \twoheadrightarrow Q_2$ and a map of extensions

$$\begin{array}{ccccccccc} 0 & \longrightarrow & K & \longrightarrow & P' & \longrightarrow & W_1 & \longrightarrow & 0 \\ & & \downarrow & & \downarrow & & \| & & \\ 0 & \longrightarrow & Q_2 & \longrightarrow & P & \longrightarrow & W_1 & \longrightarrow & 0 \end{array}$$

where the middle vertical homomorphism will also be an epimorphism, thus proving $P \lessdot P'$.

Finally we prove (D). Assume first that $\rho(L^2) \cong H^2(P)$ and that P' is another group in $\mathcal{C}(L)$ with $P \lessdot P'$. According to (B) there is another group $P'' \in \mathcal{C}(L)$ and a central extension $V \to P'' \to P$. This extension is classified by an inclusion $V^* \subset H^2(P)$, that according to (A) satisfies $V^* \cap \rho(L^2) = 0$. But this can only be true if $V = 0$, that is $P'' \cong P$ and then, also, $P' \cong P$.

Conversely, if $\rho(L^2) \ncong H^2(P)$ there is a non-trivial complement V^* of $\rho(L^2)$ in $H^2(P)$, defining a central extension $V \to P' \to P$ with $P' \in \mathcal{C}(L)$, by (A), so that $P \lessdot P'$ with $P \ncong P'$ and P is not maximal in $\mathcal{C}(L)$.

Example 12 (The cyclic system). We call cyclic system to the system The system attached to $\Lambda(x)$, a single exterior generator of degree 1, is called the *cyclic system.* Clearly, $\mathcal{C}(\Lambda(x))$ consists of an infinite chain of cyclic groups $P_i \cong \mathbb{Z}/p^{i+1}$ with limit $P_\infty \cong \widehat{\mathbb{Z}}_p$. so, this is a system of finite type.

Example 13 (The dihedral system). The system attached to $P[x,y]/(x^2+xy)$ at the prime 2 is called the *dihedral system.* This system consists of dihedral, generalized quaternion and semidihedral 2-groups. The respective cohomology rings are described in [12, 13] (see also [20]). For dihedral groups, $n \geq 1$,

$$H^*(D_{2^{n+2}}) \cong P[x,y,w]/(x^2+xy)$$

with x and y of degree 1 and w of degree 2. Clearly, $\rho_n \colon P[x,y]/(x^2+xy) \to H^*(D_{2^{n+2}})$ is a 2-equivalence and has codimension 1 in degree 2. A possible complement q for $\operatorname{Im}\rho_n$ in degree 2 is either w, $w+y^2$, $w+x^2$ or $w+x^2+y^2$, and the respective central extensions are either the dihedral group $D_{2^{n+3}}$, for $q=w$, or equivalently for $q = w+x^2+y^2$, the generalized quaternion group $Q_{2^{n+3}}$, for $q = w+y^2$, that gives the cohomology ring

$$H^*(Q_{2^{n+3}}) \cong P[x,y,v]/(x^2+xy,y^3)$$

where $\deg v = 4$, or the semidihedral group $SD_{2^{n+3}}$, for $q = w+x^2$, with cohomology ring

$$H^*(SD_{2^{n+3}}) \cong P[x,y,u,t]/(x^2+xy,xu,x^3,u^2+(x^2+y^2)t)$$

where $\deg u = 3$ and $\deg t = 4$.

Since the given maps

$$\rho'_{n+1} \colon P[x,y]/(x^2+xy) \to H^*(Q_{2^{n+3}})$$

or

$$\rho''_{n+1}\colon P[x,y]/(x^2+xy) \to H^*(SD_{2^{n+3}})$$

are 3-equivalences, the generalized quaternion groups and the semidihedral groups are maximal elements in the system.

Therefore, we have only one infinite chain in the system. The one consisting of the dihedral groups

$$\cdots \longrightarrow BD_{2^{i+2}} \xrightarrow{B\pi_{i+1}} BD_{2^{i+1}} \xrightarrow{B\pi_i} \cdots \xrightarrow{B\pi_3} BD_8.$$

Notice that the group D_{2^n} is a semidirect product $\mathbb{Z}/2^{n-1} \rtimes \mathbb{Z}/2$. We see that $\varprojlim D_{2^n} = \widehat{\mathbb{Z}}_p \rtimes \mathbb{Z}/2$ with action given by sign change. Let D_∞ be the infinite dihedral group $D_\infty = \mathbb{Z} \rtimes \mathbb{Z}/2 \cong \mathbb{Z}/2 * \mathbb{Z}/2$. Notice now that the fibration $B\mathbb{Z} \to BD_\infty \to B\mathbb{Z}/2$ is homologically nilpotent. This implies that $\varprojlim D_{2^n}$ is the p-completion of D_∞ and that BD_∞ is $\mathbb{F}_p$-good. Hence

$$H^*(\varprojlim D_{2^n}) \cong H^*((BD_\infty)^\wedge_p) \cong H^*(BD_\infty) \cong \mathbb{F}_2[u,v]/u^2+uv \cong \varinjlim H^*(D_{2^n}).$$

So, $\mathcal{C}\big(P[x,y]/(x^2+xy)\big)$ is a system of finite type.

We indicate one last example. The system associated to the remaining case of Example 9; that is $L = P[x,y]/(x^2)$ for $p=2$. It is worth mentioning that, although L splits $L \cong \Lambda[x] \otimes P[y]$, the system is not only composed of products of cyclic groups.

Example 14 (The system of $L = P[x,y]/(x^2)$ at the prime 2). For $L = P[x,y]/(x^2)$, $p=2$, the system $\mathcal{C}(L)$ consists of the groups $\mathbb{Z}/2^n \times \mathbb{Z}/2$, with

$$H^*(\mathbb{Z}/2^n \times \mathbb{Z}/2) \cong P[x,y,v]/(x^2); \quad \deg v = 2$$

and the non-abelian metacyclic groups $M_{2^{n+2}}$, $n \geq 2$, defined as the central extension

$$\mathbb{Z}/2 \to M_{2^{n+2}} \to \mathbb{Z}/2^n \times \mathbb{Z}/2$$

classified by the extension class $v + xy \in H^2(\mathbb{Z}/2^n \times \mathbb{Z}/2; \mathbb{Z}/2)$. Their cohomology, as computed by Rusin [20, Lemma 10] is

$$H^*(M_{2^{n+2}}) \cong P[x,y,z,t]/(x^2, xy^2, xz, z^2); \quad \deg(z)=3,\ \deg(t)=4\,.$$

In particular these are maximal elements in the system. Then, the possible infinite chains are chains of products of finite cyclic 2-groups and $\mathbb{Z}/2$. The limit of such chains is $\hat{\mathbb{Z}}_2 \times Z/2$, and therefore the system is of finite type.

Assume that we have a space X of finite $\mathbb{F}_p$-type with $H^1(X)=0$ and a map $f\colon BV \to X$. If $T_f(H^*(X))$ is of finite type, then we have a 2-approximation $U(Q_2^*) \to U(W_1^*) \to T_f(H^*(X))$. Write $L = U(W_1^*)//U(Q_2^*)$, and let $\mathcal{C}(L)$ the associated system of finite p-groups.

Lemma 15. *There is a chain of groups in $\mathcal{C}(L)$, $P_1 \lessdot P_2 \lessdot \cdots \lessdot P_i \cdots$, and a map of towers*

$$g\colon \{E_{s(i)}\} \longrightarrow \{BP_i\}$$

such that

1. *For each* i, $g_i\colon E_{s(i)} \to BP_i$ *induces a commutative diagram*

$$\begin{array}{ccccc} L & & & & \\ \downarrow{\scriptstyle \rho_i} & & \searrow{\scriptstyle \varphi} & & \\ H^*(P_i) & \xrightarrow[g_i^*]{} & H^*(E_{s(i)}) & \xrightarrow[j_{s_i}]{} & T_f(H^*(X)) \end{array}$$

where $\rho_i\colon L \to H^*(P_i)$ *is a* 2*-equivalence.*

2. *For each* i, *the epimorphism* $\pi_i\colon P_{i+1} \twoheadrightarrow P_i$ *satisfies* $\pi_i^*(\ker(j_{s_i} \circ g_i^*)) = 0$.

Proof. We construct the tower $\{BP_i\}$ and maps $g_i\colon E_{s(i)} \to BP_i$ inductively, starting at $g_1\colon E_{s(1)} \to BP_1$, where P_1 is the initial object of $\mathcal{C}(L)$ and g_1 is constructed like in the proof of Theorem 6.

Assume now that $g_i\colon E_{s(i)} \to BP_i$ has been constructed. Denote $V_{i+1}^* = \ker(j_{s_i} \circ g_i^*) \subset H^2(BP_i)$. Since ρ_i and φ are 2-equivalences, $V_{i+1}^* \cap \rho_i(L^2) = 0$. Then, if we define P_{i+1} by the central extension $V_{i+1} \to P_{i+1} \xrightarrow{\pi_i} P_i$ classified by the inclusion $V_{i+1}^* \subset H^2(BP_i)$, P_{i+1} belongs to the system $\mathcal{C}(L)$ by the property (A). Furthermore, there is a larger index $s(i+1)$ such that the composition $V_{i+1}^* \subset H^2(BP_i) \to H^2(E_{s(i)}) \to H^2(E_{s(i+1)})$ is trivial, so that there is a lifting $g_{i+1}\colon E_{s(i+1)} \to BP_{i+1}$

$$\begin{array}{ccccc} E_{s(i+1)} & \xrightarrow{g_{i+1}} & BP_{i+1} & & \\ \downarrow & & \downarrow{\scriptstyle B\pi_i} & & \\ E_{s(i)} & \xrightarrow{g_i} & BP_i & \longrightarrow & K(V_{i+1}, 2) \end{array}$$

and, by construction, $\pi_i^*(\ker(j_{s_i} \circ g_i^*)) = \pi_i^*(V_{i+1}^*) = 0$. □

Theorem 16. *Let* X *be a space of finite* $\mathbb{F}_p$*-type such that* $H^1(X;\mathbb{F}_p) = 0$ *and* $f\colon BV \to X$ *a map. Assume that*

$$U(Q_2^*) \xrightarrow{\psi} U(W_1^*) \xrightarrow{\varphi} T_f(H^*(X;\mathbb{F}_p)) \tag{5}$$

is a 2*-approximation of* $T_f(H^*(X;\mathbb{F}_p))$ *and* $L = U(W_1^*)//U(Q_2^*)$. *If the system* $\mathcal{C}(L)$ *is of finite type, then* $f\colon BV \to X$ *is* T*-representable.*

Proof. According to Lemma 15 we can build a tower $\{BP_i\}$, with $P_i \in \mathcal{C}(L)$ and a map $g\colon \{E_{s(i)}\} \to \{BP_i\}$.

Eventually, the tower $\{BP_i\}$ might become constant. This would happen at a stage $g_i\colon E_{s(i)} \to BP_i$ where $V_{i+1}^* = \ker(j_{s_i} \circ g_i^*) = 0$ (see the proof of Lemma 15); that is, $j_{s_i} \circ g_i^*\colon H^*(P_i) \to T_f(H^*(X))$ is a 2-equivalence.

Otherwise we obtain an infinite chain $P_1 \lessdot P_2 \lessdot \cdots \lessdot P_i \lessdot \cdots$ in $\mathcal{C}(L)$. We have assumed that the system $\mathcal{C}(L)$ is of finite type, hence, if we write $P_\infty = \varprojlim P_i$, then $H^*(P_\infty) = \varinjlim H^*(P_i)$, and this is an $\mathbb{F}_p$-algebra of finite type. Furthermore we obtain an induced map

$$H^*(P_\infty) \to T_f(H^*(X))$$

which will be an isomorphism in degree one. Also condition (2) of the tower $\{BP_i\}$ stated in Lemma 15 implies that this is injective, hence a 2-equivalence.

So, in either case, the map of towers $g\colon \{E_{s(i)}\} \to \{BP_i\}$ is a T-representation for $f\colon BV \to X$. □

Remark 17. Notice that our conditions for a finite type system of p-groups $\mathcal{C}(L)$ includes already the fact that the pro-finite groups P_∞ obtained as limits of infinite chains in $\mathcal{C}(L)$ has finite type mod p-cohomology, so the only remaining condition for showing that $f\colon BV \to X$ is finitely T-representable is the Tor condition. The counterexamples in the last section will show that this condition cannot be removed.

Remark 18. The cases analyzed in Section 4 correspond to the situation in which a system $\mathcal{C}(L)$ consists of a single element. Fix an algebra $L = U(W_1*)//U(Q_2^*)$ and let P be the initial object of the system $\mathcal{C}(L)$; that is, P is the extension $1 \to Q_2 \to P \to W_1 \to 1$ determined by the homomorphism $U(Q_2^*) \to U(W_1^*)$. If the induced homomorphism $\rho\colon L \to H^*(P)$ is a 3-equivalence, then P is also a maximal element of $\mathcal{C}(L)$, thus P is the only element in $\mathcal{C}(L)$.

Corollary 19. *Let X be a space of finite $\mathbb{F}_p$-type such that $H^1(X) = 0$ and $f\colon BV \to X$ a map. If there is a 2-equivalence $\Lambda(x) \to T_f(H^*(X))$, where $\deg x = 1$, then $f\colon BV \to X$ is T-representable.*

This is the special case of T-representability that was studied in [1]

Corollary 20. *Let X be a space of finite $\mathbb{F}_p$-type such that $H^1(X) = 0$ and $f\colon BV \to X$ a map. If there is a 2-equivalence $P[x,y]/(x^2+xy) \to T_f(H^*(X))$, where $\deg x = \deg y = 1$, then $f\colon BV \to X$ is T-representable.*

The situation encountered in this corollary is significant in the investigation of the homotopy type of the classifying spaces of rank two Kac-Moody groups ([2], [3]).

Corollary 21. *Let X be a space of finite $\mathbb{F}_p$-type such that $H^1(X) = 0$ and $f\colon BV \to X$ a map. If there is a 2-equivalence $P[x,y]/(x^2) \to T_f(H^*(X))$, where $\deg x = \deg y = 1$, then $f\colon BV \to X$ is T-representable.*

Remark 22. As a final remark, let us note that all our above calculations assume only very little knowledge of the structure of $T_f H^* X$. This has the advantage that this structure can be recognized very easily. However, in many applications a richer structure can be successfully applied. These include the algebra structure in higher degrees, and also Steenrod operations in low degrees. A very important tool in many cases might be the Bockstein spectral sequence. This makes sense, since $T_f H^* X$ inherits a Bockstein spectral sequence from the isomorphism $T_f H^* X \cong \varinjlim H^*(E_s)$. If we have information about the higher Bockstein operations in $T_f H^* X$ we can obtain finite T-representability in cases where the fundamental group of the mapping space $\mathrm{Map}(BV, X_p^\wedge)_f$ is a finite abelian p-group, not necessarily elementary abelian.

6. A counterexample

In this section we show a family of examples in which a particular component of a mapping space with source $B\mathbb{Z}/p$ and target a 1-connected p-complete space of finite $\mathbb{F}_p$-type is not p-good, hence T does not compute its cohomology.

The idea is simple. Choose a space X which is 1-connected and of finite $\mathbb{F}_p$-type, a non trivial map $f\colon B\mathbb{Z}/p \to X$, and a finite complex B which is p-bad. The space $W = X \wedge B_+$ will also be 1-connected and of finite $\mathbb{F}_p$-type, however for the composition $w\colon B\mathbb{Z}/p \to X \to W$, the mapping space $\mathrm{Map}(B\mathbb{Z}/p, W_p^\wedge)_w$ is likely to be p-bad. We will work out the particular case $X = \mathbb{C}P^\infty$, $f\colon B\mathbb{Z}/p \to \mathbb{C}P^\infty$ a non-trivial map, and $B = S^1 \vee S^m$.

Define $Y = \mathbb{C}P^\infty \times (S^1 \vee S^m)$, and W as the homotopy cofibre of $S^1 \vee S^m \to Y$; that is $W = \mathbb{C}P^\infty \wedge (S^1 \vee S^m)_+$. We can easily obtain the mod p cohomology algebras of these spaces. We choose an odd prime p, although similar considerations can be made for $p = 2$:

$$H^*(Y) \cong P[e] \otimes \Lambda[a,b]/(ab)\,, \quad \deg e = 2,\ \deg a = 1,\ \deg b = m$$
$$H^*(W) \cong P[e] \otimes \Lambda[x,y]/(xy)\,, \quad \deg x = 3\,, \deg y = m+2\,.$$

Moreover, the map $g\colon Y \to W$, from Y to the homotopy cofibre, induces in cohomology $g^*(e) = e$, $g^*(x) = ea$, and $g^*(y) = eb$. It follows that $\mathcal{P}^1 x = xe^{p-1}$ and $\mathcal{P}^1 y = ye^{p-1}$. Notice that W is 1-connected of finite $\mathbb{F}_p$-type.

The map $B\mathbb{Z}/p \times \mathbb{C}P^\infty \to \mathbb{C}P^\infty$ induced by the multiplication $\mathbb{Z}/p \times S^1 \to S^1$, extends to a map

$$\mu\colon B\mathbb{Z}/p \times \mathbb{C}P^\infty \times (S^1 \vee S^m) \longrightarrow \mathbb{C}P^\infty \times (S^1 \vee S^m) \longrightarrow \mathbb{C}P^\infty \wedge (S^1 \vee S^m)_+\,;$$

that is, $\mu\colon B\mathbb{Z}/p \times Y \to W$. If we write $H^*(B\mathbb{Z}/p) = \Lambda[u] \otimes P[v]$, the map in cohomology induced by μ is $\mu^*\colon H^*(W) \to H^*(B\mathbb{Z}/p) \otimes H^*(Y)$, determined by $\mu^*(e) = v \otimes 1 + 1 \otimes e$, $\mu^*(x) = 1 \otimes ae$, and $\mu^*(y) = 1 \otimes be$.

Lemma 23. *Let $w\colon B\mathbb{Z}/p \to W$ the restriction of μ to $B\mathbb{Z}/p$ at the base point of Y. The adjoint of μ^* is an isomorphism $\tilde{\mu}^*\colon T_w(H^*(W)) \to H^*(Y)$.*

Proof. We sketch the proof. It follows the methods of [1, §3]. $H^*(W)$ fits in an exact sequence of unstable $H^*(W)$-$\mathcal{U}$-modules that can also be seen as $P[e]$-$\mathcal{U}$-modules

$$0 \longrightarrow xP[e] \otimes yP[e] \longrightarrow H^*(W) \longrightarrow P[e] \longrightarrow 0\,.$$

Since $xP[e] \cong yP[e] \cong \Sigma eP[e]$ as unstable $P[e]$-$\mathcal{U}$-modules, where Σ denotes the suspension functor, we obtain $T_w(xP[e]) \cong T_w(yP[e]) \cong \Sigma T_w(eP[e]) \cong \Sigma P[e] = zP[e]$, with trivial Steenrod algebra action on z. Hence, there is an exact sequence

$$0 \longrightarrow aP[e] \otimes bP[e] \longrightarrow T_w(H^*(W)) \longrightarrow P[e] \longrightarrow 0$$

and the induced homomorphism $\varepsilon\colon H^*(W) \to T_w(H^*(W))$ maps $\varepsilon(e) = e$, $\varepsilon(x) = ae$, and $\varepsilon(y) = be$. Since the composition $H^*(W) \xrightarrow{\varepsilon} T_w(H^*(W)) \xrightarrow{\tilde{\mu}^*} H^*(Y)$ should coincide with g^*, it turns out that $\tilde{\mu}^*\colon T_w(H^*(W)) \to H^*(Y)$ is an isomorphism. □

Corollary 24. *μ induces a homotopy equivalence $Y_p^\wedge \xrightarrow{\simeq} \mathrm{Map}(B\mathbb{Z}/p, W_p^\wedge)_w$.*

Proof. It follows directly from Lemma 23 above and [15, 3.3.2]. □

According to [6], $S^1 \vee S^m$ is $\mathbb{F}_p$-bad, hence, so is $Y = \mathbb{C}P^\infty \times (S^1 \vee S^m)$. More precisely, Bousfield shows that, for $m \geq 2$, $H_{2m}((S^1 \vee S^m)_p^\wedge)$ is uncountable, and this implies that $H_{2m}(Y_p^\wedge)$ is uncountable, as well. Also, in the case $m = 1$, either $H_2((S^1 \vee S^1)_p^\wedge)$ or $H_3((S^1 \vee S^1)_p^\wedge)$ is uncountable, and the same will be true for $H_*(Y_p^\wedge)$. This means that, definitely, $T_w(H^*(W))$ is not isomorphic to $H^*(\mathrm{Map}(B\mathbb{Z}/p, W_p^\wedge)_w)$.

For $m = 1$, $T_w(H^*(W))$ is not T-representable. $\Lambda[a,b]/(ab) \to T_w(H^*(W))$ is a 2-approximation, but the possible towers of groups $\{G_s\}$ associated to this 2-approximation would give $G_\infty = \varprojlim G_s = (\mathbb{Z}*\mathbb{Z})_p^\wedge$, the p-profinite completion of $\mathbb{Z}*\mathbb{Z}$, which is the fundamental group of $(S^1 \vee S^1)_p^\wedge$, but in this case $\varinjlim H^*(G_s) \neq H^*(G_\infty)$.

For $m \geq 2$, $T_w(H^*(W))$ is T-representable according to Example 12, but it is not finitely T-representable because $\mathrm{Tor}^{*,*}_{\Lambda(a)}(T_w(H^*(W)), \mathbb{F}_p)$ is not finite-dimensional in each total degree. We can decompose

$$T_w(H^*(W)) \cong P[e] \otimes \Lambda[a,b]/(ab) \cong (P[e] \otimes \Lambda[a]) \oplus bP[e]$$

as $\Lambda[a]$-modules, and then

$$\begin{aligned}\mathrm{Tor}^{*,*}_{\Lambda(a)}(T_w(H^*(W), \mathbb{F}_p) &\cong \mathrm{Tor}^{*,*}_{\Lambda(a)}((P[e] \otimes \Lambda[a]) \oplus bP[e], \mathbb{F}_p)\\ &\cong \mathrm{Tor}^{*,*}_{\Lambda(a)}(P[e] \otimes \Lambda[a], \mathbb{F}_p) \oplus \mathrm{Tor}^{*,*}_{\Lambda(a)}(bP[e], \mathbb{F}_p) \cong P[e] \oplus (\Gamma[\gamma_0] \otimes bP[e])\end{aligned}$$

where $\Gamma[\gamma_0]$ is an algebra of divided powers on a generator of total degree zero, thus making $\mathrm{Tor}^{*,*}_{\Lambda(a)}(T_w(H^*(W)), \mathbb{F}_p)$ infinite-dimensional in total degree zero.

This last example shows also that the condition on Tor cannot be removed from the hypothesis of the main theorem.

References

[1] J. Aguadé, C. Broto, D. Notbohm, *Homotopy classification of spaces with interesting cohomology and a conjecture of Cooke. Part I.* Topology 33 (1994), 455–492.

[2] J. Aguadé, C. Broto, N. Kitchloo, L. Saumell, *Homotopy type of classifying spaces of rank two Kac-Moody groups.* (In preparation.)

[3] J. Aguadé, C. Broto, N. Kitchloo, L. Saumell, *Cohomology of classifying spaces of central quotients of rank two Kac-Moody groups.* Preprint.

[4] D.J. Benson, J.F. Carlson, *The cohomology of extraspecial groups,* Bull. London Math. Soc. 24 (1992), 209–235. *Erratum to: "The cohomology of extraspecial groups",* Bull. London Math. Soc. 25 (1993), 498.

[5] A.K. Bousfield, *On the homology spectral sequence of a cosimplicial space.* Amer. J. Math. 109 (1987), 361–394.

[6] A.K. Bousfield, *On the p-adic completions of nonnilpotent spaces.* Trans. Amer. Math. Soc. 331 (1992), 335–359.

[7] A.K. Bousfield, D.M. Kan, *Homotopy limits, completions and localizations.* Lecture Notes in Math., 304. Springer-Verlag, Berlin-New York, 1972.

[8] C. Broto, R. Levi, *Loop structures on homotopy fibres of self maps of spheres.* Amer. J. Math. 122 (2000), 547–580.

[9] T. Diethelm, *The mod p cohomology rings of the nonabelian split metacyclic p-groups.* Arch. Math. 44 (1985), 29–38.

[10] E. Dror Farjoun, J. Smith, *A geometric interpretation of Lannes' functor T.* Astérisque, 191 (1990), 87–95.

[11] W.G. Dwyer, *Strong convergence of the Eilenberg-Moore spectral sequence.* Topology 13 (1974), 255–265.

[12] L. Evens, S. Priddy, *The cohomology of the semidihedral group.* Conference on algebraic topology in honor of Peter Hilton (Saint John's, Nfld., 1983), 61–72, Contemp. Math., 37, Amer. Math. Soc., Providence, RI, 1985.

[13] Z. Fiedorowicz, S. Priddy, *Homology of classical groups over finite fields and their associated infinite loop spaces.* Lecture Notes in Math., 674. Springer, Berlin, 1978.

[14] C.U. Jensen, *Les foncteurs dérivés de* $\varprojlim$ *et leurs applications en théorie des modules.* Lecture Notes in Math., Vol. 254. Springer-Verlag, Berlin-New York, 1972.

[15] J. Lannes, *Sur les espaces fonctionnels dont la source est le classifiant d'un p-groupe abélien élémentaire.* Publ. Math. I.H.E.S., 75 (1992), 135–244.

[16] I.J. Leary, *The mod-p cohomology rings of some p-groups*, Math. Proc. Cambridge Philos. Soc. 112 (1992), 63–75.

[17] F. Morel, *Quelques remarques sur la cohomologie modulo p continue des pro-p-espaces et les résultats de J. Lannes concernant les espaces fonctionnels* $hom(BV, X)$. Ann. Sci. École Norm. Sup., 26 (1993), 309–360.

[18] F. Morel, *Ensembles profinis simpliciaux et interprétation géométrique du foncteur* T. Bull. Soc. math. France, 124 (1996), 347–373.

[19] D. Quillen, *The* mod 2 *cohomology rings of extra-special* 2*-groups and the spinor groups*, Math. Ann. 194 (1971) 197–212.

[20] D.J. Rusin, *The* mod 2 *cohomology of metacyclic* 2*-groups*, J. Pure Appl. Algebra 44 (1987), 315–327.

Jaume Aguadé, Carles Broto, Laia Saumell
Dep. de Matemàtiques
Universitat Autònoma de Barcelona
E-08193 Bellaterra, Spain
e-mail: `aguade@mat.uab.es`
e-mail: `broto@mat.uab.es`
e-mail: `laia@mat.uab.es`

Progress in Mathematics, Vol. 215
Algebraic Topology: Categorical Decomposition Techniques, 21–30

On Morava K-theories of an S-arithmetic Group

Marian F. Anton

Abstract. We completely describe the Morava K-theories with respect to the prime p for the étale model of the classifying space of $GL_m(\mathbb{Z}[\sqrt[p]{1}, 1/p])$ when p is an odd regular prime. For $p = 3$ and $m = 2$ (and conjecturally for $m = \infty$) these cohomologies are the same as those of the classifying space itself.

1. Introduction

By using an Eilenberg-Moore type spectral sequence, Tanabe calculated the Morava K-theories for the classifying spaces of certain Chevalley groups. In particular, if $K(n)$ is the n-th Morava K-theory with the ring of coefficients

$$K(n)^*(pt) = \mathbb{F}_p[v_n, v_n^{-1}]$$

where p is a prime and v_n has degree $2(p^n - 1)$, and if q is a power of a prime different from p, then [12]

$$K(n)^*BGL_m(\mathbb{F}_q) \approx K(n)^*BGL_m(\bar{\mathbb{F}}_q)_{\psi^q} \approx \frac{K(n)^*(pt)[[c_1, \ldots, c_m]]}{(c_1 - \psi^q c_1, \ldots, c_m - \psi^q c_m)} \tag{1.1}$$

i.e., a ring of formal power series in certain "Chern classes" $c_1, \ldots, c_m$ modulo an ideal given in terms of generators. Here ψ^q is the "Adams operation" induced from the Frobenius automorphism $x \mapsto x^q$ of the algebraic closure $\bar{\mathbb{F}}_q$ of the field $\mathbb{F}_q$ with q elements. The same formula (1.1) holds for the p-adic version $\hat{K}(n)$ of $K(n)$ obtained by replacing $K(n)^*(pt)$ with $\hat{K}(n)^*(pt) = \mathbb{Z}_p[v_n, v_n^{-1}]$ where $\mathbb{Z}_p$ denotes the ring of p-adic integers [12].

On the other hand, if $A = \mathbb{Z}[\sqrt[p]{1}, 1/p]$ and p is a *regular* prime in the sense of number theory, then Dwyer and Friedlander [5, 6] calculated the mod p cohomology of a space $BGL_m(A_{\text{ét}})$ which is naturally associated to the classifying space $BGL_m(A)$ of the S-arithmetic group $GL_m(A)$. We call the space $BGL_m(A_{\text{ét}})$ *the étale model at p* for the classifying space $BGL_m(A)$ and recall that it is endowed with a natural map [4, 2.5]

$$f_A : BGL_m(A) \to BGL_m(A_{\text{ét}}) \tag{1.2}$$

2000 *Mathematics Subject Classification.* 55N20,19F27,11F75.
Key words and phrases. Morava K-theory, étale homotopy, unitary bundles.

The goal of this article is to show how we can use these two calculations in order to completely describe the Morava K-theories with respect to the prime p of the étale model above. The main result is

Theorem 1.1. *If $A = \mathbb{Z}[\sqrt[p]{1}, 1/p]$ with p an odd regular prime, then the n-th Morava K-theory with respect to the prime p of the étale model $BGL_m(A_{\text{ét}})$ is an exterior algebra given by the formula*

$$K(n)^*BGL_m(A_{\text{ét}}) \approx K(n)^*BGL_m(\mathbb{F}_q)\langle\sigma_1, \dots, \sigma_m\rangle^{\otimes(p-1)/2}$$

where q is a prime $\equiv 1 \mod p$ but $\not\equiv 1 \mod p^2$, the tensor product is over the ring $K(n)^*BGL_m(\mathbb{F}_q)$, and σ_i has degree $2i-3$ $(1 \le i \le m)$. Moreover, the same formula holds for the p-adic version $\hat{K}(n)$.

In particular, if $m = \infty$ and $n = 1$ (and conjecturally for $n > 1$) the above theorem and (1.1) give $K(n)$ and $\hat{K}(n)$ theories of the classifying space $BGL_\infty(A)$ itself for p odd and regular, according to [7]. Here GL_∞ denotes the union of all GL_n for $n \ge 1$ with respect to the block inclusions.

Also, if $p = 3$ and $m = 2$, then we showed that the natural map (1.2) is a mod p equivalence [1]. Hence we deduce the following

Corollary 1.2. *The n-th Morava K-theory at the prime 3 of the S-arithmetic group $GL_2(\mathbb{Z}[\sqrt[3]{1}, 1/3])$ is given by*

$$K(n)^*BGL_2(\mathbb{Z}[\sqrt[3]{1}, 1/3]) \approx \frac{\mathbb{F}_3[v_n, v_n^{-1}][[a, c_2]]}{(a^{7^n}, c_2^{(7^n+1)/2} \mod a)}\langle\sigma_1, \sigma_2\rangle$$

where the degrees of the generators are $|v_n| = 2(3^n - 1)$, $|a| = 2$, $|c_2| = 4$, $|\sigma_1| = -1$, $|\sigma_2| = 1$, and the second generator of the ideal is up to an indeterminacy $\mod a$. Moreover, a similar formula holds for the 3-adic version $\hat{K}(n)$.

Notation 1.3. In what follows p is an odd regular prime when not otherwise stated and $A = \mathbb{Z}[\zeta_p, 1/p]$ where $\zeta_p = \exp(2\pi i/p)$ is a prescribed p-th root of unity in the field $\mathbb{C}$ of complex numbers.

2. Étale models for classifying spaces

2.1. The original definition

Let $R = \mathbb{Z}[1/p]$, G a group scheme over $Spec(R)$, and BG the classifying simplicial scheme obtained by a bar construction as in [8, 1.2]. Then the classifying space $BG(D)$ of the group $G(D)$ of the D-points of G where D is any finitely generated R-algebra can be thought of as the connected component of a simplicial function complex [6, 1.4]

$$BG(D) = Map^0(Spec(D), BG)_{Spec(R)} \tag{2.1}$$

containing the natural base point induced by the unit map $Spec(R) \to G$ of G over $Spec(R)$. We recall that $Map(X, Y)_Z$ is a simplicial set given in dimension i by the set of simplicial scheme maps $X \otimes \Delta[i] \to Y$ over Z where X and Y are simplicial schemes over Z (a scheme is regarded as a constant simplicial scheme) and $\Delta[i]$ is the standard simplicial i-simplex. The tensor product between a simplicial scheme and a simplicial set is defined in [8, 1.1].

Also, we recall that the étale topological type $X_{\text{ét}}$ in the sense of Friedlander [8, 4.4] is a pro-space (i.e., inverse system of simplicial sets) which is naturally associated to a noetherian simplicial scheme X and reflects the étale cohomology of X. For any finitely generated R-algebra D, let $D_{\text{ét}}$ denote the étale topological type $Spec(D)_{\text{ét}}$. By replacing $Spec(D)$, BG, and $Spec(R)$ in (2.1) by their étale topological types $D_{\text{ét}}$, $(BG)_{\text{ét}}$, and $R_{\text{ét}}$, the space $BG(D_{\text{ét}})$ is defined in [6, 1.2] as the connected component of the simplicial complex of p-adic functions over $R_{\text{ét}}$

$$BG(D_{\text{ét}}) = Hom_p^0(D_{\text{ét}}, (BG)_{\text{ét}})_{R_{\text{ét}}} \tag{2.2}$$

containing the corresponding natural base point. This construction is similar to (2.1) and we can associate with each i-simplex of $BG(D)$ an i-simplex of $BG(D_{\text{ét}})$ regarded by definition as a map of pro-spaces over $R_{\text{ét}}$ from $D_{\text{ét}} \times \Delta[i]$ to the fibrewise p-adic completion of $(BG)_{\text{ét}}$ over $R_{\text{ét}}$ denoted by $(\mathbb{Z}/p)^{\bullet}(BG)_{\text{ét}}$ [4, 2.4]. This assignment is natural in both G and D and gives a map [4, 2.5]

$$f_D^G : BG(D) \to BG(D_{\text{ét}})$$

from the classifying space of the group $G(D)$ to its étale model $BG(D_{\text{ét}})$ at p. In the case when $G = GL_m$ is the group scheme over $SpecR$ corresponding to the general linear group $G(R) = GL_m(R)$ and $D = A$, we obtain the map (1.2). These definitions actually hold for *any* prime p.

2.2. A model structure definition

For convenience we will give an alternative way of thinking of (2.2) pointed out by Isaksen and based on his model structure. Namely, if pro-$\mathcal{SS}$ is the category of pro-spaces then there is a proper simplicial model structure on pro-$\mathcal{SS}$ introduced in [9]. This means that there are three classes of morphisms in pro-$\mathcal{SS}$ called *weak equivalences*, *cofibrations*, and *fibrations* subject to various axioms. Also there is a notion of *simplicial function complex* i.e., a natural assignment to each two pro-spaces X and Y of a simplicial set $Map(X, Y)$ interacting appropriately with the model structure [9, 16.2].

For the purpose of this paper we will use the induced proper simplicial model structure on the over-category pro-$\mathcal{SS}_V$ of pro-spaces over a fixed pro-space V. With respect to this model structure there is a *relative* simplicial function complex $Map(X, Y)_V$ naturally associated with every pair of objects X, Y in pro-$\mathcal{SS}_V$. Keeping the same notations as in the previous subsection we have the following

Proposition 2.1. *For any finitely generated R-algebra D, the space $BG(D_{\text{ét}})$ is weakly equivalent to the connected component of the natural base point of the simplicial function complex $Map(D_{\text{ét}}, T_p(BG)_{\text{ét}})_{R_{\text{ét}}}$ in the over-category of pro-spaces over $R_{\text{ét}}$,*

$$BG(D_{\text{ét}}) \simeq Map^0(D_{\text{ét}}, T_p(BG)_{\text{ét}})_{R_{\text{ét}}}$$

Here $T_p(BG)_{\acute{e}t}$ is a fibrant replacement of $(\mathbb{Z}/p)^\bullet(BG)_{\acute{e}t}$ over $R_{\acute{e}t}$ in the sense of the simplicial model structure of [9].

Proof. Let $X = D_{\acute{e}t} = \{X_\alpha\}$, $Y = (BG)_{\acute{e}t}$, and $V = R_{\acute{e}t}$. Then $Y \to V$ is a (strict) map of pro-spaces and let $T'_p(Y)$ be the level-space Moore-Postnikov tower naturally associated to the fibrewise p-adic completion of Y over V. Then we can think of $T'_p(Y) = \{T'_p(Y)_\delta\}$ as a pro-space over $V = \{V_\delta\}$ and by definition [4, 2.3]

$$Hom_p(X,Y)_V = holim_\delta colim_\alpha Map(X_\alpha, T'_p(Y)_\delta)_{V_\delta} \tag{2.3}$$

where Map is the usual relative simplicial function complex of simplicial sets and $holim$ denotes the homotopy inverse functor from pro-spaces to spaces [2, §6]. By [9, 10.6], the pro-space $T_p(Y)$ is the fibrant replacement of $T'_p(Y)$ in the model structure of Edwards-Hastings. By standard arguments, the space (2.3) is weakly equivalent to

$$Map(X, T_p(Y))_V = lim_\delta colim_\alpha Map(X_\alpha, T_p(Y)_\delta)_{V_\delta}$$

and the conclusion follows from (2.2). □

3. A homotopy fibre square

3.1. Preliminaries

We collect here a couple of known facts which will be used in the construction of a computable model for $BGL_m(A_{\acute{e}t})$ given in the next subsection. This model is naturally associated to the action of $\pi_1(R_{\acute{e}t})$ on the p-primary roots of unity.

Let D be a finitely generated normal R-algebra and $pt : Spec(k) \to Spec(D)$ a geometric point corresponding to a homomorphism from D to a separable closed field k. Then pt determines a base point of $D_{\acute{e}t}$ and we recall that $\pi_1(D_{\acute{e}t}, pt)$ is the pro-finite Grothendieck fundamental group of D pointed by pt [8, §5]. This group classifies finite étale covering spaces of $Spec(D)$.

Let μ_{p^ν} be the set of all complex numbers z such that $z^{p^\nu} = 1$ and μ_{p^∞} the union of all μ_{p^ν} for $\nu \geq 0$. Let R_∞ denote the ring obtained from R by adjoining the set μ_{p^∞} of all p-primary roots of unity,

$$R_\infty = R[\sqrt[p^\infty]{1}] = \mathbb{Z}[1/p, \mu_{p^\infty}],$$

and Γ the Galois (pro-)group

$$\Gamma = Gal(R_\infty, R) = \{Aut(\mu_{p^\nu}), \nu \geq 1\} \approx \{(\mathbb{Z}/p^\nu)^*, \nu \geq 1\}$$

In this context, observe that $\pi_1(R_{\acute{e}t})$ is the Galois group of the maximal unramified extension of R and let

$$\theta : \pi_1(R_{\acute{e}t}) \to \Gamma \tag{3.1}$$

be the homomorphism given by the action of this Galois group on the p-primary roots of unity. In other words, $R_{\acute{e}t}$ is provided with the natural structure map

$R_{\text{ét}} \to K(\Gamma, 1)$ which "classifies" the finite étale extensions $R \to R[\mu_{p^\nu}]$. Also, $A_{\text{ét}}$ is provided with a natural structure map

$$A_{\text{ét}} \to R_{\text{ét}} \to K(\Gamma, 1)$$

If k is a field, then $k_{\text{ét}}$ is a pro-space of type $K(\pi, 1)$, where π is the Galois group over k of the separable algebraic closure of k. In particular, $\mathbb{C}_{\text{ét}}$ is contractible and $(\mathbb{F}_q)_{\text{ét}}$ is equivalent to the pro-finite completion of a circle. If $R \to \mathbb{F}_q$ is a residue field map, then $(\mathbb{F}_q)_{\text{ét}}$ is provided with a natural structure map

$$(\mathbb{F}_q)_{\text{ét}} \to R_{\text{ét}} \to K(\Gamma, 1)$$

as well. This structure map sends the Frobenius element of the Galois group of $\bar{\mathbb{F}}_q$ over $\mathbb{F}_q$ identified with $\pi_1((\mathbb{F}_q)_{\text{ét}})$ to $q \in Aut(\mu_{p^\nu}) \cong (\mathbb{Z}/p^\nu)^*$ in Γ [6, 3.2].

3.2. A homotopy fibre square

Let U_m be the Lie group of $m \times m$ unitary matrices and $\hat{B}U_m$ the p-completion of its classifying space. The following proposition is the unstable analogue of [5, 4.5] and its proof is almost the same. For convenience, we review here the main arguments.

Proposition 3.1. *Let p be an odd regular prime, $A = \mathbb{Z}[\zeta_p, 1/p]$, and q a rational prime $\equiv 1 \mod p$ but $\not\equiv 1 \mod p^2$. Then there is a homotopy fibre square*

$$\begin{array}{ccc} BGL_m(A_{\text{ét}}) & \longrightarrow & \hat{B}U_m^W \\ \downarrow & & \downarrow \\ \hat{B}GL_m(\mathbb{F}_q) & \longrightarrow & \hat{B}U_m \end{array}$$

where W is the wedge of $(p-1)/2$ circles, $\hat{B}U_m^W$ denotes the simplicial function complex of unpointed maps from W to $\hat{B}U_m$, and the right-hand vertical map is the evaluation at the base-point.

Proof. As in [5, p. 145] we construct a map

$$(\mathbb{F}_q)_{\text{ét}} \vee W \to K(\Gamma, 1)$$

by sending the first summand via the natural structure map and mapping the other summand trivially. By a class-field argument (assuming the properties of q from hypothesis), there exists a map

$$g : (\mathbb{F}_q)_{\text{ét}} \vee W \to A_{\text{ét}}$$

over $K(\Gamma, 1)$ which is a mod p cohomology equivalence [5, p. 145]. In other words g is a "good mod p model" for A in the sense of [6, 1.9]. This means that by a spectral sequence argument [4, 2.11] and using 2.1 for $G = GL_m$ the map g induces a homotopy equivalence

$$Map^0(A_{\text{ét}}, T_p(BGL_m)_{\text{ét}})_{R_{\text{ét}}} \simeq Map^0((\mathbb{F}_q)_{\text{ét}} \vee W, T_p(BGL_m)_{\text{ét}})_{R_{\text{ét}}}$$

which can be reformulated by saying that we get a homotopy fibre square

$$\begin{array}{ccc} BGL_m(A_{\text{ét}}) & \longrightarrow & Map^0(W, T_p(BGL_m)_{\text{ét}})_{R_{\text{ét}}} \\ \downarrow & & \downarrow \\ BGL_m((\mathbb{F}_q)_{\text{ét}}) & \longrightarrow & Map^0(pt, T_p(BGL_m)_{\text{ét}})_{R_{\text{ét}}} \end{array}$$

where the right-hand vertical map is the evaluation at the base-point (the map $pt \to R_{\text{ét}}$ is induced from $R \subset \mathbb{C}$ recalling that $\mathbb{C}_{\text{ét}}$ is contractible). To finish the proof, we need only to identify the appropriate corners of this square.

For the two right-hand corners, we start with the fibration sequence [3, 2.3]

$$\{(\mathbb{Z}/p)_s(BGL_{m,\bar{\mathbb{F}}_q})_{\text{ét}}\}_s \to (\mathbb{Z}/p)^\bullet(BGL_m)_{\text{ét}} \to R_{\text{ét}} \tag{3.2}$$

where $\{(\mathbb{Z}/p)_s(-)\}_s$ denotes the Bousfield-Kan p-completion tower and $BGL_{m,\bar{\mathbb{F}}_q}$ is the classifying object of GL_m over $\bar{\mathbb{F}}_q$. Hence we get that

$$Map^0(pt, T_p(BG)_{\text{ét}})_{R_{\text{ét}}} \simeq holim\{(\mathbb{Z}/p)_s BGL_{m,\bar{\mathbb{F}}_q})_{\text{ét}}\}_s \simeq \hat{B}U_m$$

where the last equivalence is proved in [8, 8.8]. Because the composite map

$$\pi_1(W) \to \pi_1(A_{\text{ét}}) \to \pi_1(R_{\text{ét}}) \xrightarrow{\theta} \Gamma$$

is trivial by construction, as in [4, p. 146] we get a homotopy equivalence

$$Map^0(W, T_p(BGL_m)_{\text{ét}})_{R_{\text{ét}}} \simeq \hat{B}U_m^W$$

where $\hat{B}U_m^W$ denotes the function complex of unpointed maps from W to $\hat{B}U_m$ (basically $\pi_1(R_{\text{ét}})$ acts on the fibre of (3.2) via θ).

Finally, for the lower left-hand corner, there is a homotopy equivalence

$$\hat{B}GL_m(\mathbb{F}_q) \simeq BGL_m((\mathbb{F}_q)_{\text{ét}})$$

given in [3, 2.11] by exploiting the action of the Frobenius element on the fibre of (3.2) via the composite

$$\pi_1((\mathbb{F}_q)_{\text{ét}}) \to \pi_1(R_{\text{ét}}) \xrightarrow{\theta} \Gamma$$

and Quillen's homotopy fix point description of $\hat{B}GL_m(\mathbb{F}_q)$ [10]. □

4. The proof of the main theorem and its corollary

4.1. Proof of 1.1

The proof of the main theorem is based on Strickland's analysis of unitary bundles in [11] applied to the homotopy fibre square 3.1.

Let V be a complex vector bundle over a space X and write PV for the associated bundle of projective spaces and $U(V)$ for the associated bundle of unitary groups

$$U(V) = \{(x, g) | x \in X \text{ and } g \in U(V_x)\}$$

Let $EU(V)$ denote the geometric realization of the simplicial space $\{U(V)^{n+1}\}_{n\geq 0}$ and put $BU(V) = EU(V)/U(V)$ the usual simplicial model for the classifying space of $U(V)$.

Let E^* be an even periodic cohomology theory with complex orientation $x \in \tilde{E}^0\mathbb{C}P^\infty$. We are interested in describing $E^*U(V)$ as a Hopf algebra over E^*X (using the group structure on $U(V)$). The main result involves the exterior algebra over the ring E^*X generated by the module E^*PV which we denote by $\lambda^*_{E^*X}E^{*-1}PV$ and which is a Hopf algebra over E^*X by declaring E^*PV to be primitive.

Proposition 4.1 ([11, 4.4]). *There is a natural isomorphism of Hopf algebras over E^*X*

$$\lambda^*_{E^*X}E^{*-1}PV \approx E^*U(V)$$

We apply this proposition to the tautological bundle

$$V = \gamma_m = EU_m \times_{U_m} \mathbb{C}^m$$

over $X = BU_m$. In this case, we have

$$E^*P\gamma_m \approx \frac{E^*BU_m[x]}{(x^m + c_1x^{m-1} + \cdots + c_m)} \approx E^*BU_m\{1, \ldots, x^{m-1}\}$$

where c_i is the i-th Chern class of γ_m and the last isomorphism indicates that $E^*P\gamma_m$ is a free module over E^*BU_m with basis $1, x, \ldots, x^{m-1}$. In particular,

$$(4.1) \qquad E^*U(\gamma_m) \approx \lambda^*_{E^*BU_m}E^{*-1}P\gamma_m \approx E^*BU_m\langle\sigma(1), \ldots, \sigma(x^{m-1})\rangle$$

where σ lowers the degree by 1.

Going back to the homotopy fibre square 3.1, we observe that $\hat{B}U_m^W$ is the $(p-1)/2$-fold fibre product of $\hat{B}U_m^{S^1} = \hat{U}(\gamma_m)$ over $\hat{B}U_m$ and for $E^* = K(n)$ or $\hat{K}(n)$ and any space X we have $E^*\hat{X} = E^*X$. In this case, if we apply E^* to $\hat{B}U_m^W$ and use (4.1) we obtain

$$\begin{aligned} E^*(\hat{B}U_m^W) &\approx (\lambda^*_{E^*BU_m}E^{*-1}P\gamma_m)^{\otimes(p-1)/2} \\ &\approx E^*BU_m\langle\sigma(1), \ldots, \sigma(x^{m-1})\rangle^{\otimes(p-1)/2} \end{aligned}$$

where the tensor product is over E^*BU_m. In particular, $E^*(\hat{B}U_m^W)$ is a free module over $E^*\hat{B}U_m = E^*BU_m$ and therefore 1.1 follows from the above formula by a base-change induced from 3.1:

$$E^*BGL_m(A_{\text{ét}}) \approx E^*\hat{B}GL_m(\mathbb{F}_q) \otimes_{E^*\hat{B}U_m} E^*(\hat{B}U_m^W)$$

where q can be always chosen with the prescribed properties (by Dirichlet's density theorem for instance).

4.2. Proof of 1.2

This has been already explained in the introduction, except for the analysis of the formula (1.1) in the case $p = 3$, $m = 2$, and $q = 7$. The goal of this subsection is to complete this analysis.

Proposition 4.2. *Let p be an odd rational prime and $\mathbb{F}_q$ a finite field with q elements such that $q \equiv 1 \mod p^r$ but $\not\equiv 1 \mod p^{r+1}$ for some integer $r > 0$. Then*

$$K(n)^* BGL_2(\mathbb{F}_q) \approx \frac{K(n)^*(pt)[[a, c_2]]}{(a^{p^{nr}}, c_2^{(p^{nr}+1)/2} \mod a)}$$

Proof. According to Tanabe's formula

$$K(n)^* BGL_2(\mathbb{F}_q) \approx (K(n)^* BU_2)_\psi \tag{4.2}$$

where the co-invariants are calculated with respect to the q-th Adams operation ψ [12]. Recall that

$$K(n)^* BU_2 \approx K(n)^*(pt)[[c_1, c_2]] \tag{4.3}$$

where $c_1 = x + y$ and $c_2 = xy$ are expressed in terms of the generators of the ring

$$K(n)^*(\mathbb{C}P^\infty \times \mathbb{C}P^\infty) \approx K(n)^*(pt)[[x, y]]$$

which are induced by a complex orientation on $K(n)^*(\mathbb{C}P^\infty)$ [12, 2.12]. It is easy to see that we can replace c_1 in (4.3) by the formal group sum $a = x +^{K(n)} y$ of x and y (induced from the tensor product of complex line bundles). Then the proposition follows from (4.2) and the following lemmas

Lemma 4.3. $a - \psi(a) = (unit) \times a^{p^{nr}}$

Lemma 4.4. $c_2 - \psi(c_2) \equiv (unit) \times c_2^{(p^{nr}+1)/2} \mod a$

where "unit" means invertible element in (4.3). □

Proof of 4.3. Let us expand q in the ring $\mathbb{Z}_p$ of p-adic integers as

$$q = \sum_{k=0}^{\infty} \alpha_k p^k$$

where the coefficients $\alpha_k \in \mathbb{Z}$ are subject to $0 \leq \alpha_k < p$, $\alpha_0 = 1$, $\alpha_r \neq 0$, and $\alpha_k = 0$ for $0 < k < r$. Then for $t = x$ or y we have

$$\psi(t) = [q](t) = \sum\nolimits^{K(n)} [\alpha_k](t^{p^{nk}}) = t + \alpha_r t^{p^{nr}} + \cdots$$

where $[q](t)$ means the formal group q-multiple of t. Hence,

$$\psi(a) = \psi(x) +^{K(n)} \psi(y) = \sum\nolimits^{K(n)} [\alpha_k](a^{p^{nk}}) = a + \alpha_r a^{p^{nr}} + \cdots$$

and the conclusion follows. □

Proof of 4.4. With the same notations as in the previous proof, we have

$$x \equiv [-1](y) \equiv -y \mod a$$

and hence

$$c_2 - \psi(c_2) \equiv -x^2 + \psi(x)\psi(x) \equiv (unit) \times x^{p^{nr}+1} \equiv (unit) \times c_2^{(p^{nr}+1)/2} \mod a.$$

□

References

[1] Anton, M.F.: *On a conjecture of Quillen at the prime* 3, J. Pure Appl. Algebra **144** (1999), 1–20

[2] Bousfield, A.K.; Kan, D.M.: *Homotopy Limits, Completions and Localizations*, Lecture Notes in Math. 304, Springer, Berlin, 1972

[3] Dwyer, W.G.: *Exotic cohomology for* $GL_n(\mathbb{Z}[\frac{1}{2}])$, Proc. Amer Math. Soc. **272** (1998), 2159–2167

[4] Dwyer, W.G.; Friedlander, E.M.: *Algebraic and etale K-theory*, Trans. Amer. Math. Soc. **272** (1985), 247–280

[5] Dwyer, W.G.; Friedlander, E.M.: *Conjectural calculations of general linear group homology* in Applications of Algebraic K-Theory to Algebraic Geometry and Number Theory, Contemporary Mathematics 55, Part I, Amer. Math. Soc., Providence, (1986), pp. 135–147

[6] Dwyer, W.G.; Friedlander, E.M.: *Topological models for arithmetic*, Topology **33** (1994), 1–24

[7] Dwyer, W.G.; Mitchell, S.A.: *On the K-theory spectrum of a ring of algebraic integers*, K-Theory **14** (1998), no. 3, 201–263

[8] Friedlander, E.M.: *Etale Homotopy of Simplicial Schemes*, Princeton Univ. Press, Princeton, 1982

[9] Isaksen, D.C.: *A model structure on the category of pro-simplicial sets*, Trans. Amer. Math. Soc. **353** (2001), 2805–2841

[10] Quillen, D.: *On the cohomolgy and K-theory of the general linear groups over a finite field*, Annals of Math. **96** (1972), 552–586

[11] Strickland, N.P.: *Common subbundles and intersections of divisors*, preprint (University of Sheffield), (2000)

[12] Tanabe, M.: *On Morava K-theories of Chevalley groups*, Amer. J. Math. **117** (1995), 263–278

Marian F. Anton
Department of Pure Mathematics
University of Sheffield
Hicks Building
Sheffield S3 7RH, UK
and
IMAR
P.O.Box 1-764
Bucharest, RO 70700
e-mail: Marian.Anton@imar.ro

Progress in Mathematics, Vol. 215
Algebraic Topology: Categorical Decomposition Techniques, 31–45

Ext Groups for the Composition of Functors

Stanislaw Betley

0. Introduction

In recent years we observe the growing interest in homological algebra in various categories of functors from small categories to vector spaces. Let Γ and $\mathcal{V}_p$ be the category of finite pointed sets and the finite-dimensional vector spaces over the prime field F_p respectively. The categories of functors from Γ or $\mathcal{V}_p$ to vector spaces over F_p (denoted $\tilde{\Gamma}$ and $\mathcal{F}$ respectively) are of the special interest because of their relations to Steenrod algebra, stable derived functors and many other questions from algebraic topology, see for example [K], [BS], [B1], [P1] etc. Moreover the homological algebra in these categories turned out to be fairly well computable, see for example [FLS], [FFSS], [P1], [B1], [BS]. But the most general calculations of $Ext_{\mathcal{F}}$-groups obtained in [FFSS] are still not satisfactory – for the purpose of studying filtrations on Eilenberg-MacLane spaces one has to study Ext and Tor groups in categories of functors when at least one variable is given as a composition of a functor with the symmetric power.

Let S^n denote the nth symmetric power functor as an object of $\mathcal{F}$. The problem of calculating

$$Ext^*_{\mathcal{F}}(Id, S^n \circ S^m)$$

was successfully solved by Alain Troesch in [T]. He used the approach which turned out to be fruitful before (see [FLS] and [FFSS]). Roughly speaking one replaces $\mathcal{F}$ by the category $\mathcal{P}$ of polynomial functors and performs calculations of $Ext^*_{\mathcal{P}}(.,.)$. This problem is easier than the original one because the category $\mathcal{P}$ splits into homogeneous pieces. Having $Ext_{\mathcal{P}}$-groups one can use old theorems from algebraic geometry which compare $Ext_{\mathcal{F}}$ and $Ext_{\mathcal{P}}$. On the other hand homological algebra in $\tilde{\Gamma}$ is also easily accessible, see for example [B1]. In the following note we show how to use $\tilde{\Gamma}$ instead of $\mathcal{P}$ to handle calculations of $Ext^*_{\mathcal{F}}$-groups when one of the variables is a composition of a functor with S^n. But we should warn the reader here: the category $\tilde{\Gamma}$ will be used directly only in Chapter I, because we compare $Ext^*_{\tilde{\Gamma}}$-groups and $Ext^*_{\mathcal{F}}$-groups there. Later we will use only known calculations of $Ext^*_{\tilde{\Gamma}}$ groups and the crucial role of the category $\tilde{\Gamma}$ will be deeply hidden.

In order to get some new groups we will deal with the dual situation comparing to [T]. It means we will work with the groups $Ext^*_{\mathcal{F}}(S^n \circ S^m, Id)$. We hope

The author was partially supported by the Polish Scientific Grant (KBN) 2 P03A 002 18.

that techniques developed in this note will be fruitful in other calculations as well. The really new tools used here are comparison of $Ext_{\mathcal{F}}$ and $Ext_{\tilde{\Gamma}}$ groups, the spectral sequence coming from this comparison and good understanding of the effect of the Frobenius map $S^n \to S^{pn}$ on Ext-groups. It is important to underline that the purpose of this note is to show new methods rather than getting the most general calculations. The project of calculating $Ext_{\tilde{\Gamma}}$-groups between classical functors is still not finished but we hope that it is easier than working over $\mathcal{F}$. Then the methods described below should give us for example a new way of attacking $Ext_{\mathcal{F}}$-groups from more "injective" to more "projective" functors (compare [FFSS, beginning of Section 5]).

Let us finish the introduction by saying what stimulated us to perform calculations presented below. Our interest in doing them came from the effort to fulfill our promise from [B1, 5.7]. We promised there to describe the multiplicative structure of the Steenrod algebra in the framework of functors. We expected to get Steenrod algebra multiplication as a kind of Yoneda product. It turned out to be more complicated that we expected while writing [B1] because we could not find any good description of the generators. In the present paper we showed how to see in the natural way the Steenrod algebra generators in $Ext_{\mathcal{F}}$-groups. Our Theorem 2.1 below and other calculations can be interpreted in the following way. The even-dimensional Steenrod algebra generators from dimensions $2p^k - 2 < j < 2p^{k+1} - 2$ can be found in $Ext^*_{\mathcal{F}}(S^p \circ L^{p^k}, Id)$ while the odd-dimensional generators from the same range of dimensions appears naturally in the groups $Ext^*_{\mathcal{F}}(S^p \circ L^{p^{k+1}}, Id)$. The symbol L^n denotes here the nth reduced symmetric power functor (see paragraphs before Theorem 1.3 for the definition). But this Steenrod algebra point of view will not be discussed further in the paper because so far we do not have any good application of it.

I would like to express my thanks to the referee for the careful reading of the manuscript and many useful suggestions improving exposition.

1. Preliminaries

Let us start from recalling the basic notation and definitions. For a given prime number p, $\mathcal{V}_p$ denotes the category of finite-dimensional vector spaces over the prime field F_p. By $\mathcal{F}$ we denote the abelian category of functors from $\mathcal{V}_p$ to vector spaces over F_p. Γ denotes the category of finite based sets and $\tilde{\Gamma}$ stands for the category of functors from Γ to vector spaces over F_p. For the whole paper p denotes the odd prime number. Our methods work also for the prime 2 but the formulas are slightly different due to the fact that the description of the Steenrod algebra changes when one moves from 2 to odd primes.

Let T be a functor from the category $\mathcal{F}$. Let $L : \Gamma \to \mathcal{V}_p$ be a functor which sends a pointed finite set X to $F_p[X]/F_p[0]$ where 0 is a base point in X. By L^* we shall denote the dual of L in the ordinary linear algebra sense. We will treat T as an object of $\tilde{\Gamma}$ just by composing it with L. Let $Z : \mathcal{V}_p \to \Gamma^{op}$ be a functor which has value V^* on any vector space V. The following simple proposition is crucial for the whole project because it compares homological algebras in $\mathcal{F}$ and $\tilde{\Gamma}^{op}$.

1.1. Proposition. For any natural number i, $F \in \mathcal{F}$ and any $G \in \tilde{\Gamma}^{op}$ we have

$$Ext^i_{\mathcal{F}}(F, G \circ Z) = Ext^i_{\tilde{\Gamma}^{op}}(F \circ L^*, G).$$

Proof. One directly checks that the functors L^* and Z are adjoint. Moreover observe that precomposition with L^* is exact and takes the basic projective object in $\mathcal{F}$ defined by $V \in \mathcal{V}_p$ to the basic projective object in $\tilde{\Gamma}^{op}$ defined by V^* treated as a finite set:

$$\begin{aligned} F_p[Hom_{F_p}(V, .)] \circ L^*(X) &= F_p[Hom_{F_p}(V, L^*(X))] \\ &= F_p[Hom_{F_p}(L(X), V^*)] = F_p[Hom_{\Gamma}(X, V^*)] \\ &= F_p[Hom_{\Gamma}(., V^*)](X). \end{aligned}$$

Let now $\cdots \to P_1 \to P_0 \to F$ be the projective resolution of F in $\mathcal{F}$ consisting of the sums of basic projectives. Then $\cdots \to P_1 \circ L^* \to P_0 \circ L^* \to F \circ L^*$ is a projective resolution of $F \circ L^*$ in $\tilde{\Gamma}^{op}$ consisting of sums of basic projectives. Applying $Hom_{\mathcal{F}}(., G \circ Z)$ to the first and $Hom_{\tilde{\Gamma}^{op}}(., G)$ to the second resolution we get two complexes of vector spaces which are canonically isomorphic. Hence they have the same homology and our proposition is proved. Observe that taking $i = 0$ we have shown that precomposition with Z and L^* is a pair of adjoint functors between $\mathcal{F}$ and $\tilde{\Gamma}^{op}$.

The proposition above is also obviously true without dualizing L. The only reason for writing it in a such contravariant fashion is the fact that for applications we are going to use very contravariant Remark 4.13 from [B1]. As an immediate corollary of 1.1 we get the following "calculation" of Ext_{F}-groups of some composition of functors.

1.2. Corollary.

$$Ext^*_{\mathcal{F}}(Id, S^{n*} \circ L \circ Z) = Ext^*_{\tilde{\Gamma}^{op}}(L^*, (S^n \circ L)^*).$$

We called this corollary "calculation" because the right-hand side of the equality above is known from [B1,Part 4]. These groups are nontrivial only for n being a power of p and for $n = p^i$ they are the part of the Steenrod algebra. Observe that by the Kuhn's duality defined in [K, Part 3.4] the left-hand side is isomorphic to

$$Ext^*_{\mathcal{F}}(S^n \circ L, Id)$$

where we treat here L as a functor $\mathcal{V}_p \to \mathcal{V}_p$ taking V to $F_p[V]/F_p[0]$.

We can treat any $V \in \mathcal{V}_p$ as an abelian group. Having this point of view in mind we get a new object I of $\mathcal{F}$ given by the kernel of the augmentation map

$$F_p[V] \to F_p.$$

It is easy to observe that the composition of the natural embedding and quotient $I(V) \to F_p[V] \to L(V)$ is an isomorphism of functors. As usual we can filter the augmentation ideal by its powers and hence we have the corresponding filtration on the functor L. By [Q] we know that the subquotients of this filtration are isomorphic as functors to the reduced symmetric powers L^n. Let us recall briefly what are L^n's. We consider an ideal $A(V)$ in the full symmetric algebra $S^*(V) = \oplus_{n=0}^{\infty} S^n(V)$ generated by the image of the Frobenius map $S^1(V) \to S^p(V)$. Then $S^*(V)/A(V)$ is a graded vector space and its nth graded part defines $L^n(V)$.

Let us chose a natural number n. Any filtration on the functor L determines a filtration on $S^n \circ L$. Of course it is not true that the quotients of the filtration on $S^n \circ L$ coming from the filtration by the powers of the augmentation ideal are of the form $S^n \circ L^k$. But homologically they are – we have only to use the beautiful observation from [T] (Theorem 4).

1.3. Theorem. Let $0 \to F_1 \to F_2 \to F_3 \to 0$ be an exact sequence of objects in $\mathcal{F}$ taking values in finite-dimensional vector spaces. Let P be an object of $\mathcal{F}$ of finite degree. Call H the homology of $0 \to P \circ F_1 \to P \circ F_2 \to P \circ F_3 \to 0$ at $P \circ F_2$. Then

$$Ext^*_{\mathcal{F}}(Id, H) = 0.$$

The same statement as in 1.3 is true for $Ext^*_{\mathcal{F}}(H, Id)$ and the proof is the same. Actually, the case of 1.3 with $P = S^n$ is much easier than the general case. We strongly suggest the reader to prove it yourself using the observation of Pirashvili from the appendix to [BP]. We formulate a simplified version of it here because it will be useful further on.

1.4. Proposition. Let $N : (\mathcal{V}_p)^n \to F_p-vect$ be a functor such that $N(V_1, \dots, V_n) = 0$ whenever one of V_i's is 0. Define $G \in \mathcal{F}$ to be N precomposed with the diagonal map $\mathcal{V}_p \to (\mathcal{V}_p)^n$. Let $F \in \mathcal{F}$ be a functor of degree less than n. Then

$$Ext_{\mathcal{F}}(G, F) = Ext_{\mathcal{F}}(F, G) = 0.$$

As an immediate corollary of 1.3 we get

1.5. Lemma. Let $0 \to F_1 \to F_2 \to F_3 \to 0$ be an exact sequence of objects in $\mathcal{F}$ taking values in finite-dimensional vector spaces. Let $P \in \mathcal{F}$ be a functor of finite degree. Let $Q = P \circ F_2 / P \circ F_1$. Then the natural map $Q \to P \circ F_3$ induces an isomorphism $Ext^*_{\mathcal{F}}(P \circ F_3, Id) \to Ext^*_{\mathcal{F}}(Q, Id)$

Proof. Let K be the kernel of $P \circ F_2 \to P \circ F_3$. Then $P \circ F_1$ embeds naturally in K and by 1.3 $Ext^*_{\mathcal{F}}(K/P \circ F_1, Id) = 0$. Hence the embedding $P \circ F_1 \to K$ induces an isomorphism $Ext^*_{\mathcal{F}}(K, Id) \to Ext^*_{\mathcal{F}}(P \circ F_1, Id)$. This proves our lemma by a long exact sequence argument and 5-lemma.

Theorem 1.3 and Lemma 1.5 are formulated for short exact sequences of functors. We will use them usually for long exact sequences so the corollary below is formulated here for the readers convenience.

1.6. Corollary. Let $0 \to F \to A_1 \to A_2 \to \cdots \to A_k \to G \to 0$ be an exact sequence of objects in $\mathcal{F}$. Let P be an object of $\mathcal{F}$ of finite degree. Assume that for any $1 \leq i \leq k$ we have $Ext^*_{\mathcal{F}}(Id, A_i) = 0$ and $Ext^*_{\mathcal{F}}(Id, P \circ A_i) = 0$. Then for any j

$$Ext^j_{\mathcal{F}}(Id, P \circ F) = Ext^{j-k}_{\mathcal{F}}(Id, P \circ G).$$

Proof. Let $K = ker(A_k \to G)$. Then we have the short exact sequence $0 \to K \to A_k \to G \to 0$. Let $0 \to P \circ K \to P \circ A_k \to Q \to 0$ be exact. By 1.5 we know that $Ext^a_{\mathcal{F}}(Id, Q) = Ext^a_{\mathcal{F}}(Id, P \circ G)$. By the general homological algebra and the hypothesis of our corollary we have

$$Ext^{a-1}_{\mathcal{F}}(Id, Q) = Ext^a_{\mathcal{F}}(Id, P \circ K).$$

Now we can finish the proof by applying the same procedure to the short exact sequences $0 \to K_1 \to A_{k-1} \to K \to 0$, $0 \to K_2 \to A_{k-2} \to K_1 \to 0$ etc.

Of course Corollary 1.6 remains true also in the $Ext^*_{\mathcal{F}}(., Id)$-situation. Now we can have a closer look at $S^n \circ L$. Using the filtration coming from the augmentation ideal we get a filtration of $S^n \circ L$. We will say that two functors F and G are homologically the same if there is a map between them inducing an isomorphism on $Ext^*_{\mathcal{F}}(., Id)$. By 1.5 we know that our filtration on $S^n \circ L$ has quotients which are homologically the same as $S^n \circ L^m$. By basic homological algebra for the filtered object we get a spectral sequence with the second table given by

1.6.1. $E_2^{s,t} = Ext^{t+s}_{\mathcal{F}}(S^n \circ L^{s+1}, Id) \qquad 0 \leq s.$

Observe that all columns of this spectral sequences are modules over the well-known ring $Ext^*_{\mathcal{F}}(Id, Id)$ (see [FLS, Theorem 7.3]) and the differentials preserve this module structure.

1.7. Lemma. Assume that at least one of the numbers n and m is not a power of p. Then $Ext^*_{\mathcal{F}}(S^n \circ L^m, Id) = 0$.

Proof. If n is not a power of p then S^n is a direct summand in a tensor product of functors with value 0 at 0 (see for example [FLS, Proposition 6.1]). The same is true for the reduced symmetric powers. Hence if either n or m is not a power of p then $S^n \circ L^m$ satisfies the hypothesis of our vanishing criterion 1.4.

Now we know that our spectral sequence is nontrivial only for $n = p^i$ and an s-column is (possibly) nontrivial only for $s+1$ being a power of p. This means that we have $E_2^{s,t} = E_3^{s,t} = \cdots = E_p^{s,t}$. Let me underline here that we will use these identifications very often in our future calculations.

In order to be able to have any use from the spectral sequence described above we must check that it converges to $Ext^*_{\mathcal{F}}(S^n \circ L, Id)$. This is not obvious because our spectral sequence is not concentrated in the first quadrant but only above the line $s + t = 0$, $0 \leq s$. On the other hand we will know that it converges as we wish if we show that there are only finitely many nontrivial groups on any line described by the equation $s + t = c$. To see this we need two lemmas which will be useful also in further calculations.

1.8. Lemma. $Ext^k_{\mathcal{F}}(S^{p^i} \circ S^{p^j}, Id) = 0$ for $k \leq 2(p^i + p^j) - 5$.

Proof. Let D^n denote the Kuhn's dual of S^n. This means $D^n(V) = (S^n(V^*))^*$. By the general abstract nonsense we know ([K]) that $Ext^*_{\mathcal{F}}(F,G) = Ext^*_{\mathcal{F}}(D(G),D(F))$ where $D(F)$ is the dual functor of F. Observe that $D(S^n \circ S^m) = D^n \circ D^m$ and $D(Id) = Id$ hence we can study the groups $Ext^*_{\mathcal{F}}(Id, D^n \circ D^m)$ in order to prove our lemma.

We have a Koszul exact sequence connecting S^n and exterior power functor Λ^n:

$$0 \to \Lambda^n \to \Lambda^{n-1} \otimes S^1 \to \cdots \to \Lambda^1 \otimes S^{n-1} \to S^n \to 0.$$

The functor Λ^n is isomorphic to its dual so the dual exact sequence has the form:

$$0 \to D^n \to D^{n-1} \otimes \Lambda^1 \to \cdots \to \Lambda^n \to 0$$

which is also exact.

Connecting both these sequences into one at Λ^n we get a sequence connecting D^n and S^n with all middle terms given by tensor products. If we precompose it with D^m we obtain an exact sequence connecting $D^n \circ D^m$ with $S^n \circ D^m$ with all middle terms M_s given by some tensor product of functors with value 0 at 0. Hence by 1.4 for any s

$$Ext^*_{\mathcal{F}}(Id, M_s) = 0$$

and we get

$$Ext^k_{\mathcal{F}}(Id, D^n \circ D^m) = Ext^{k-2n+2}_{\mathcal{F}}(Id, S^n \circ D^m).$$

Observe that we can do the same trick starting now from the sequence connecting D^m and S^m and composing it with S^n. Composition is of course not exact but by 1.6 has the same effect on $Ext^*_{\mathcal{F}}(Id, .)$-groups as if it was. Hence

$$Ext^l_{\mathcal{F}}(Id, S^n \circ D^m) = Ext^{l-2m+2}_{\mathcal{F}}(Id, S^n \circ S^m).$$

So to get some nontrivial groups we must have $0 \leq l - 2m + 2$ and hence $2m + 2n - 4 \leq k$.

1.9. Lemma. $Ext^k_{\mathcal{F}}(S^{p^i} \circ L^{p^j}, Id) = 0$ for $k \leq 2(p^i + p^{j-1}) - 4$.

Proof. Let K be the kernel of the quotient map $S^{p^j} \to L^{p^j}$. The Frobenius morphism $S^{p^{j-1}} \to S^{p^j}$ factors through K. It was shown in [B1, Lemma 2.5] that K has a filtration with quotients given by tensor products of functors with value 0 at 0. Hence the map $S^{p^{j-1}} \to K$ induces an isomorphism on $Ext^*_{\mathcal{F}}(., Id)$-groups and by the same argument which was used in 1.8 (in general by 1.3, 1.4 and 1.6) we get that the map

$$Ext^*_{\mathcal{F}_y}(S^{p^i} \circ K, Id) \to Ext^*_{\mathcal{F}}(S^{p^i} \circ S^{p^{j-1}}, Id)$$

is an isomorphism (compare 4.2). On the other hand by 1.3 the non-exact sequence

$$S^{p^i} \circ K \to S^{p^i} \circ S^{p^j} \to S^{p^i} \circ L^{p^j}$$

gives us a long exact sequence of $Ext^*_{\mathcal{F}}(., Id)$-groups. Then by the previous lemma we get our statement.

The more general version of the theorem below was proved in [T, Theorem 3]. In reality it was proved there for the groups $Ext^*_{\mathcal{F}}(Id, F)$ but the proof in our case is completely the same by the Kuhn's duality. This theorem is not necessary for our applications but simplifies a lot the description of the final result.

1.10. Theorem. Let F be a homogeneous functor of degree p^i. There exists a graded vector space $A(F)$ such that

$$Ext^*_{\mathcal{F}}(F, Id) = A(F) \otimes Ext^*_{\mathcal{F}}(D^{p^i}, Id).$$

Moreover $A(F)$ is trivial in dimensions bigger than $2p^i - 2$.

Troesch calculated $A(F)$ for F being a composition of symmetric powers in terms of $Ext_{\mathcal{P}}$-groups. We are going to use $Ext_{\tilde{\Gamma}}$ instead – this is the point where these two approaches take different routes.

2. Calculations

The goal of this section is to present methods of calculating $Ext^*_{\mathcal{F}}(S^p \circ L^{p^i}, Id)$ and to calculate fully $Ext^*_{\mathcal{F}}(S^p \circ L^p, Id)$. But first let us emphasize what will be the main tool standing behind all our calculations. We proved in the previous section, starting after Corollary 1.6, that there exists a spectral sequence with the second table given by

$$E_2^{s,t} = Ext^{t+s}_{\mathcal{F}}(S^n \circ L^{s+1}, Id) \qquad 0 \leq s$$

which has nontrivial elements only when n is a power of p and we look at the column s with $s+1$ being a power of p. The second differential in our spectral sequence maps $E_2^{s,t} \to E_2^{s-1,t+2}$. We know by 1.2 that our spectral sequence converges to

$$Ext^*_{\tilde{\Gamma}^{op}}(L^*, (S^n \circ L)^*),$$

all columns in it are modules over $Ext^*_{\mathcal{F}}(Id, Id)$ and differentials respect these module structures. Moreover we know from [B1, Section 3] that $Ext^k_{\tilde{\Gamma}^{op}}(L^*, (S^p \circ L)^*) = F_p$ for $k = 2l(p-1),\ 2l(p-1)+1,\ l = 1, 2, \ldots$ and is 0 otherwise. Hence we have a spectral sequence with interesting entries and we know the groups to which our spectral sequence converges. Our goal is to convince the reader that our information about this spectral sequence is strong enough to calculate its second table.

By [FLS] we know the ring structure of $Ext^*_{\mathcal{F}}(Id, Id)$ coming from the Yoneda multiplication. Let us recall it:

$$Ext^*_{\mathcal{F}}(Id, Id) = F_p[e_0, e_1, \ldots, e_h, \ldots] / \langle (e_h)^p; h \in N \rangle .$$

Moreover we know (also by [FLS]) the first column of our spectral sequence:

$$E_2^{0,t} = Ext^t_{\mathcal{F}}(S^p, Id) = F_p \qquad \text{for} \qquad t = 2ip-2, \qquad i = 1, 2, \ldots$$

and the $Ext^*_{\mathcal{F}}(Id, Id)$-module structure is also clear from [FLS]. Perhaps a short explanation is needed here because to be honest in [FLS] the dual situation was considered: they were working with $Ext^*_{\mathcal{F}}(Id, S^p)$. But observe that $Ext^*_{\mathcal{F}}(S^p, Id) = Ext^*_{\mathcal{F}}(Id, D(S^p))$. Then one can connect the Koszul exact sequence with its dual and get an exact sequence connecting $D(S^p)$ with S^p giving us the formula

$$Ext^*_{\mathcal{F}}(S^p, Id) = Ext^{*+2p-2}_{\mathcal{F}}(Id, S^p)$$

which is equivariant with respect to $Ext^*_{\mathcal{F}}(Id, Id)$-module structures. Moreover using the same argument as in [FLS, Section 6] but working with the dual of the de Rham sequence one gets that the Frobenius $Id \to S^p$ induces an isomorphism

$$Ext^*_{\mathcal{F}}(S^p, Id) \to Ext^*_{\mathcal{F}}(Id, Id)$$

in all dimensions when the first group is nontrivial.

Our goal now is to calculate the $p-1$-column of our spectral sequence consisting of groups $Ext^*_{\mathcal{F}}(S^p \circ L^p, Id)$ starting from the spectral sequence's indices $\{p-1, -p+1\}$. Observe that by 1.10 we have to calculate our groups up to degree $2p^2 - 2$. By 1.9 our fourth nontrivial column ($p^3 - 1$-column) has 0 up to total degree $2(p+p^2) - 4$. So we have to understand only beginning parts of columns $p-1$ and $p^2 - 1$.

2.1. Theorem. Up to dimension $2p^2 - 2$ we have $Ext^i_{\mathcal{F}}(S^p \circ L^p, Id) = F_p$ for $i = 2p-1, 2k(p-1), 2kp-3, 2kp-1$ where $k = 2, 3, 4, \ldots, p-1$ and is 0 otherwise.

Proof. The beginning is simple: By 1.9 we can easily say that up to total degree $4p-4$ all columns but 0 and $p-1$ are trivial. The latter one is trivial in all dimensions smaller than $2p-1$. The 0 column has F_p in degree $2p-2$ and is 0 otherwise up to $4p-2$. We know the abutment of our spectral sequence: there are F_p's there in degrees $2p-2$, $2p-1$ and $4p-4$. This forces the following calculation:

$$Ext^{2p-1}_{\mathcal{F}}(S^p \circ L^p, Id) = F_p$$

$$Ext^{4p-4}_{\mathcal{F}}(S^p \circ L^p, Id) = F_p$$

$$Ext^i_{\mathcal{F}}(S^p \circ L^p, Id) = 0 \qquad \text{for} \qquad i = 0, 1, \ldots, 4p-5, \qquad i \neq 2p-1.$$

In order to go further with our calculations we need several lemmas and claims.

2.2. Lemma. Up to dimension $6p-8$ the only nontrivial groups in the $p-1$ and p^2-1 columns are given by

$$Ext^{4p-3}_{\mathcal{F}}(S^p \circ L^p, Id) = Ext^{4p-1}_{\mathcal{F}}(S^p \circ L^p, Id) = F_p$$
$$Ext^{4p-3}_{\mathcal{F}}(S^p \circ L^{p^2}, Id) = Ext^{4p-2}_{\mathcal{F}}(S^p \circ L^{p^2}, Id) = F_p.$$

Proof. As a main tool here we will use Koszul and de Rham sequences connecting symmetric and exterior powers. Let us recall that for any n we have an exact sequence of Koszul

$$0 \to \Lambda^n \to \Lambda^{n-1} \otimes S^1 \to \cdots \to \Lambda^1 \otimes S^{n-1} \to S^n \to 0$$

which gives us the formula

$$Ext^i_{\mathcal{F}}(S^n, Id) = Ext^{i-n+1}_{\mathcal{F}}(\Lambda^n, Id)$$

and hence also for any m

$$Ext^i_{\mathcal{F}}(S^n \circ L^m, Id) = Ext^{i-n+1}_{\mathcal{F}}(\Lambda^n \circ L^m, Id).$$

Also for any n we have de Rham sequence R^n:

$$0 \to S^n \to S^{n-1} \otimes \Lambda^1 \to \cdots \to S^1 \otimes \Lambda^{n-1} \to \Lambda^n \to 0$$

which is exact for n different than a power of p and satisfies

$$H_*(R^{p^n}) = R^{p^{n-1}}$$

as graded vector spaces, with homology starting from the place corresponding to S^{p^n}. In other words we give R^n the homological grading with Λ^n in dimension 0 and S^n in dimension n. Then $H(R^{p^n})$ is nontrivial in dimensions $p^n - p^{n-1}, \dots, p^n$.

If we denote as $R^n \circ L^m$ the sequence

$$0 \to S^n \circ L^m \to (S^{n-1} \otimes \Lambda^1) \circ L^m \to \cdots \to \Lambda^n \circ L^m \to 0$$

then we have the corresponding formula for homology

$$H_*(R^{p^n} \circ L^m) = R^{p^{n-1}} \circ L^m.$$

We can associate two hyper-cohomology spectral sequences with R^{p^n} and $R^{p^n} \circ L^{p^j}$ the same way as it was done in all papers with computations of $Ext_{\mathcal{F}}$-groups. Both of them converge to the same groups and are given by the following formulas expressing the first of the one and the second table of the other:

$$\begin{aligned} {}^1E_1^{s,t} &= Ext^t_{\mathcal{F}}((R^{p^n} \circ L^{p^j})_s, Id) \\ {}^2E_2^{s,t} &= Ext^t_{\mathcal{F}}(H_s(R^{p^n} \circ L^{p^j}), Id). \end{aligned}$$

It is important to underline here that these spectral sequences are coming from two standard different filtration on the same bicomplex so to get the formulas above we had to renumerate the second spectral sequence (comparing to conventions used in [FLS, Sections 5,6]). The result is that the differential in the first spectral sequence maps ${}^1E_a^{s,t}$ to ${}^1E_a^{s+a,t-a+1}$ while in the second we have ${}^2E_a^{s,t} \to {}^2E_a^{s-a+1,t+a}$. We can fully calculate the second spectral sequence in the case $n = 1 = j$.

2.2.1. Claim. There are only two nontrivial columns in ${}^2E_a^{s,t}$ namely the pth and $p-1$. Both of them consist of groups $Ext^k_{\mathcal{F}}(L^p, Id)$.

Proof of the claim. This follows immediately from 1.4 and the formula for the homology of the de Rham sequence.

2.2.2. Claim. $Ext^k_{\mathcal{F}}(L^p, Id) = F_p$ for k odd and $k \neq 2lp - 1$.

Proof of the claim. Consider the exact sequence of functors $0 \to Id \to S^p \to L^p \to 0$. The first map is given by Frobenius and we know from [FLS] that it induces the isomorphism on $Ext^k_{\mathcal{F}}(., Id)$-groups in all dimensions k in which $Ext^k_{\mathcal{F}}(S^p, Id)$

is nontrivial. Hence the rest follows from the long exact sequence of Ext-groups associated with the given short exact sequence of functors.

2.2.3. Claim. ${}^2E_a^{s,t}$ converges to the groups F_p in dimensions $(2l+1)p$ and $(2k+1)p-3$ for $l=0,1,\dots$ and $k=1,2,\dots$.

Proof of the claim. By our previous calculations we know that $Ext^k_{\mathcal{F}}(S^p \circ L^p, Id)$ is trivial up to dimension $4p-4$ but the dimension $2p-1$ where it is equal to F_p. In the first spectral sequence we have also only two nontrivial columns:

$$\begin{aligned} {}^1E_1^{0,t} &= Ext^t_{\mathcal{F}}(\Lambda^p \circ L^p, Id) \\ {}^1E_1^{p,t} &= Ext^t_{\mathcal{F}}(S^p \circ L^p, Id). \end{aligned}$$

By the first observation and the Koszul exact sequence we get ${}^1E_1^{0,p} = F_p$ and ${}^1E_1^{0,p+1} = 0$. Hence the second differential

$${}^2E_2^{p,1} = F_p \to F_p = {}^2E_2^{p-1,3}$$

is nontrivial. The rest is forced by the formula for $Ext^*_{\mathcal{F}}(L^p, Id)$ and the fact that the differential is a module map over $Ext^*_{\mathcal{F}}(Id, Id)$.

Now we can come back to the proof of 2.2. We will use the first spectral sequence ${}^1E_a^{s,t}$. We know the groups to which it converges by 2.2.3, because ${}^1E_a^{s,t}$ and ${}^2E_a^{s,t}$ converge to the same groups. By our previous calculations and the proof of 2.2.3 we know that ${}^1E_1^{s,t}$ has in pth column zeros up to dimension $4p-4$ with the one exception. There is F_p in dimension $2p-1$. From this we get that the first column of it has F_p in dimensions p, $3p-3$, $3p-2$ and $3p$ and all the rest of the groups are trivial up to dimension $5p-5$. Hence again by Koszul exact sequence we get

$$Ext^{4p-3}_{\mathcal{F}}(S^p \circ L^p, Id) = Ext^{4p-1}_{\mathcal{F}}(S^p \circ L^p, Id) = F_p.$$

To get the calculation for $S^p \circ L^{p^2}$ in the given range we have to proceed similarly. We have again two spectral sequences

$$\begin{aligned} {}^1E_1^{s,t} &= Ext^t_{\mathcal{F}}((R^p \circ L^{p^2})_s, Id) \\ {}^2E_2^{s,t} &= Ext^t_{\mathcal{F}}(H_s(R^p \circ L^{p^2}), Id). \end{aligned}$$

The second one has only trivial differentials by dimensional reasons. It has only two nontrivial columns ($(p-1)$st and pth) filled with $Ext^k_{\mathcal{F}}(L^{p^2}, Id) = F_p$ for $k = 2ip-1$ and p^2 does not divide $2ip$. Hence using the first spectral sequence we get

$$Ext^{3p-2}_{\mathcal{F}}(\Lambda^p \circ L^{p^2}, Id) = Ext^{3p-1}_{\mathcal{F}}(\Lambda^p \circ L^{p^2}, Id) = F_p$$

forcing

$$Ext^{4p-3}_{\mathcal{F}}(S^p \circ L^{p^2}, Id) = Ext^{4p-2}_{\mathcal{F}}(S^p \circ L^{p^2}, Id) = F_p.$$

Now the easy part of our calculations is over. But the rest of the proof of Theorem 2.1 is still easy if we know

2.3. Theorem. The differential

$$d_p : F_p = Ext_{\mathcal{F}}^{4p-3}(S^p \circ L^p, Id) = E_p^{p-1,3p-2} \to E_p^{0,4p-2} = Ext_{\mathcal{F}}^{4p-2}(S^p, Id) = F_p$$

is an isomorphism.

Section 3 is devoted to the proof of Theorem 2.3. Assuming it we can proceed as follows (remember we are in the range up to dimension $2p^2 - 2$):

2.4. Lemma. The groups $Ext^i_{\mathcal{F}}(S^p \circ L^p, Id) = F_p$ for $i = 2p-1$, $2k(p-1)$, $2kp-3$, $2kp-1$ with $k = 2, 3, \ldots$.

Proof. In other words we are proving now that our groups are as they were described in 2.1 if they are nontrivial. This is obvious using spectral sequences coming from the de Rham sequence $R^p \circ L^p$ and the equivalence obtained from Koszul exact sequence $Ext^i_{\mathcal{F}}(S^p \circ L^p, Id) = Ext_{\mathcal{F}}^{i-p+1}(\Lambda^p \circ L^p, Id)$.

2.5. Lemma. The groups $Ext^i_{\mathcal{F}}(S^p \circ L^{p^2}, Id) = F_p$ for $i = 2kp-2, 2k(p-1)+1$ with $k = 2, 3, \ldots$.

As before it comes immediately as in the previous lemma from the spectral sequences related to $R^p \circ L^{p^2}$ by using the isomorphism $Ext^i_{\mathcal{F}}(S^p \circ L^{p^2}, Id) = Ext_{\mathcal{F}}^{i-p+1}(\Lambda^p \circ L^{p^2}, Id)$ coming from the Koszul exact sequence.

So the main point now is to show that the groups from the $(p-1)$st and (p^2-1)st columns of our spectral sequence which are not listed in 2.4 and 2.5 are trivial. Observe that from de Rham sequences we get immediately triviality of some groups:

on column $p-1$:

$$Ext_{\mathcal{F}}^{2kp-2}(S^p \circ L^p, Id) = 0$$

$$Ext^i_{\mathcal{F}}(S^p \circ L^p, Id) = 0 \qquad \text{for} \qquad 2kp-1 < i < 2(k+1)(p-1)$$

on column p^2-1:

$$Ext^i_{\mathcal{F}}(S^p \circ L^{p^2}, Id) = 0 \qquad \text{for} \qquad 2kp-2 < i < 2(k+1)(p-1)+1.$$

Moreover we know that possibly nontrivial groups are coming in pairs (this is normal for spectral sequence arguments with only two nontrivial columns). For example either $Ext_{\mathcal{F}}^{6p-4}(S^p \circ L^p, Id) = Ext_{\mathcal{F}}^{6p-5}(S^p \circ L^p, Id) = 0$ or both of this groups are F_p. The same is true for the groups in dimensions $6p-3$ and $6p-4$ on the next nontrivial column.

Now because differentials are module maps over $Ext^*_{\mathcal{F}}(Id, Id)$ and we know the module structure on the first column we get immediately that the differential

2.5.1.

$$F_p = E_p^{p-1,5p-2} \to E_p^{0,6p-2} = F_p$$

is nontrivial. But then $E_2^{p^2-1,6p-p^2-2} = Ext_{\mathcal{F}}^{6p-3}(S^p \circ L^{p^2}, Id) = 0$ because there is no nontrivial differential which comes in or goes out this group and the group is trivial in the limit. Hence $Ext_{\mathcal{F}}^{6p-4}(S^p \circ L^{p^2}, Id) = 0$ and this forces $Ext_{\mathcal{F}}^{6p-4}(S^p \circ L^p, Id) = 0$ by the limit argument so also $Ext_{\mathcal{F}}^{6p-5}(S^p \circ L^p, Id) = 0$. The arguments for higher groups work precisely the same way by observing that 2.5.1 works in any total dimension of the form $2kp-3$. This finishes the proof of 2.1. Observe also that as a side effect of our calculations we got:

2.6. Corollary. Up to dimension $2p^2-2$ the groups $Ext^i_{\mathcal{F}}(S^p \circ L^{p^2}, Id)$ are nontrivial only in dimensions $2kp-2$ and $2k(p-1)+1$ for $k=2,3,\dots$.

3. Proof of Theorem 2.3

Let us come back for a while to Corollary 1.2. It is obviously true also if we replace S^n by any other functor F from $\mathcal{F}$. Hence we get a spectral sequence as in 1.5.1 for any F. We will label such spectral sequences with (F). Assume now that $F = Id$. Then we have $E^{s,t}_*(Id)$ and we know everything about it. The nontrivial columns in it are given by $Ext^*_{\mathcal{F}}(L^{p^i}, Id)$ which we know and the differentials are completely determined by the fact that $E^{s,t}_*(Id)$ converges by 1.1 to $Ext^*_{\tilde{\Gamma}^{op}}(L^*, L^*)$ which is equal to F_p in dimension 0 and is trivial in other dimensions. Of course we easily calculate that the differential $F_p = E^{p-1,3p-2}_p(Id) \to E^{0,4p-2}_p(Id) = F_p$ is an isomorphism.

Now observe that the Frobenius map $\phi : Id \to S^p$ induces a map of spectral sequences $E^{s,t}_*(S^p) \to E^{s,t}_*(Id)$. Hence in order to get 2.3 it is enough to show that

$$\phi_* : Ext^{4p-3}_{\mathcal{F}}(S^p \circ L^p, Id) \to Ext^{4p-3}_{\mathcal{F}}(L^p, Id)$$

is an isomorphism. But if we consider the exact sequence $0 \to L^p \to S^p \circ L^p \to L^p \circ L^p \to 0$ (where the first map is induced by Frobenius) and the long exact sequence of Ext-groups associated with it we get that our statement is equivalent to showing that

$$Ext^{4p-3}_{\mathcal{F}}(L^p \circ L^p, Id) = 0.$$

To get this we have to look a little closer at $E^{s,t}_*(L^p)$. This spectral sequence converges to $Ext^*_{\tilde{\Gamma}^{op}}(L^*, (L^p \circ L^p)^*)$ which is F_p in dimensions 1, $2i(p-1)$ and $2i(p-1)+1$ for $i = 1, 2, \dots$ and is 0 otherwise (see [B1, Section 3]). Observe that in the 0 column we have a class in dimension $4p-3$ and 0 in dimension $4p-2$. In the column $p-1$ we have the groups $Ext^*_{\mathcal{F}}(L^p \circ L^p, Id)$ with the one of the greatest importance to us. The column p^3-1 has trivial groups up to dimension $2p^2$ so we do not have to care about it. We know that in the limit our spectral sequence has F_p in dimension $4p-3$. Hence we can finish the proof by proving the following lemma:

3.1. Lemma. $Ext^{4p-3}_{\mathcal{F}}(L^p \circ L^{p^2}, Id) = F_p$ and $Ext^{4p-4}_{\mathcal{F}}(L^p \circ L^{p^2}, Id) = 0$.

Proof. We will proceed like in our preliminary calculations for $Ext^{4p-3}_{\mathcal{F}}(S^p \circ L^{p^2}, Id)$. Observe that from the de Rham sequence R^p we can get a sequence

$$0 \to L^p \to S^p \otimes S^1 \to \cdots \to \Lambda^p \to 0$$

just by dividing the first place by the image of the Frobenius map. Then this new sequence has only one nontrivial homology group at the place corresponding to

$S^p \otimes S^1$. It is equal to Id. We can precompose this new sequence with L^{p^2} and get a sequence with one nontrivial homology equal to L^{p^2}. Call this sequence $\tilde{R}^p \circ L^{p^2}$.

As previously we have two spectral sequences connected with $\tilde{R}^p \circ L^{p^2}$. The second one has only one nontrivial column. This is the column $p-1$ with $Ext^*_{\mathcal{F}}(L^{p^2}, Id)$ staying in it. These groups we know, in the interesting range they are nontrivial (equal to F_p) in dimensions $2p-1$ and $4p-1$. Hence both spectral sequences converge in the range up to $5p$ to the groups F_p in dimensions $3p-2$ and $5p-2$.

In the first spectral sequence we have two nontrivial columns, namely 0 and p. In the first one we have the groups $Ext^*_{\mathcal{F}}(\Lambda^p \circ L^{p^2}, Id)$ which are nontrivial by our previous calculations (and relation via Koszul exact sequence to $Ext^*_{\mathcal{F}}(S^p \circ L^{p^2}, Id)$) in dimensions $3p-2$, $3p-1$, $5p-4$, $5p-1$ and perhaps $5p-2$ and $5p-3$. The column p consists of the groups $Ext^*_{\mathcal{F}}(L^p \circ L^{p^2}, Id)$. By the formulas for the 0 column and the abutment we get immediately that $Ext^*_{\mathcal{F}}(L^p \circ L^{p^2}, I)$ are certainly nontrivial in dimensions $2p$ and $4p-3$ and $Ext^{4p-4}_{\mathcal{F}}(L^p \circ L^{p^2}, Id) = 0$. This latter nontrivial group has to kill F_p staying in the first column in dimension $5p-4$.

3.2. Remark. From all the calculations we have already done one could get an impression that the *Ext*-groups we are considering are either F_p or 0. But that is not the case – if one looks closer at $E^{s,t}_*(L^p)$ one immediately realizes that $Ext^{4p-4}_{\mathcal{F}}(L^p \circ L^p, Id) = F_p \oplus F_p$.

3.3. Remark. The proof of 3.1 shows how did we come to our approach and why it worked. After 1.1, the starting point for calculations was the observation that we knew everything about the spectral sequence $E^{s,t}_*(Id)$. The second ingredient was to realize that the effect of the Frobenius map $Id \to S^p$ on $Ext_{\mathcal{F}}$-groups was also well understood. Both of them suggested that comparing three spectral sequences $E^{s,t}_*(Id)$, $E^{s,t}_*(S^{p^i})$ and $E^{s,t}_*(L^{p^i})$ by using maps from the exact sequence $Id \to S^p \to L^p$ should be enough for our calculations.

4. Final remarks

We start this section from moving from reduced symmetric powers to S^n.

4.1. Lemma. Up to dimension $2p^2-2$ the groups $Ext^i_{\mathcal{F}}(S^p \circ S^p, Id)$ are nontrivial only in dimensions $2k(p-1)$ and $2kp-3$ for $k>1$.

Proof. We shall write the proof in a very short way because the techniques are the same as in the previous two sections. Composing the Frobenius exact sequence

$$0 \to Id \to S^p \to L^p \to 0$$

with S^p and using 1.3 we get a long exact sequence

$$\cdots \to Ext^i_{\mathcal{F}}(S^p \circ L^p, Id) \to Ext^i_{\mathcal{F}}(S^p \circ S^p, Id) \\ \to Ext^i_{\mathcal{F}}(S^p, Id) \to Ext^{i+1}_{\mathcal{F}}(S^p \circ L^p, Id) \to \cdots$$

From it and 2.1 we get immediately our result if we prove that the boundary map

$$Ext_{\mathcal{F}}^{2kp-2}(S^p, Id) \to Ext_{\mathcal{F}}^{2kp-1}(S^p \circ L^p, Id)$$

is an isomorphism. Both of these groups are F_p hence it is enough to show that $Ext_{\mathcal{F}}^{2kp-1}(S^p \circ S^p, Id) = 0$. But this follows directly from the spectral sequences associated with the de Rham-type sequence connecting $S^p \circ S^p$ with $\Lambda^p \circ S^p$.

So we have calculated the groups $Ext_{\mathcal{F}}^*(S^p \circ S^p, Id)$ but at the beginning we promise to develop techniques good enough for computing $Ext_{\mathcal{F}}^*(S^{p^i} \circ L^{p^j}, Id)$ and $Ext_{\mathcal{F}}^*(S^{p^i} \circ S^{p^j}, Id)$. Techniques coming from Koszul and de Rham sequences exist obviously in this generality so it is easy to believe that most of our arguments work well in this more general situation. The only thing which remains to observe is the lemma below (which was used in some version in 1.9 and was not needed so far for other calculations).

4.2. Lemma. Let F be an object of $\mathcal{F}$ of finite degree. The Frobenius morphism $S^{p^k} \to S^{p^{k+1}}$ induces two long exact sequences

$$\cdots \to Ext_{\mathcal{F}}^i(F \circ L^{p^{k+1}}, Id) \to Ext_{\mathcal{F}}^i(F \circ S^{p^{k+1}}, Id) \\ \to Ext_{\mathcal{F}}^i(F \circ S^{p^k}, Id) \to Ext_{\mathcal{F}}^{i+1}(F \circ L^{p^{k+1}}, Id) \to \cdots$$

and

$$\cdots \to Ext_{\mathcal{F}}^i(L^{p^{k+1}} \circ F, Id) \to Ext_{\mathcal{F}}^i(S^{p^{k+1}} \circ F, Id) \\ \to Ext_{\mathcal{F}}^i(S^{p^k} \circ F, Id) \to Ext_{\mathcal{F}}^{i+1}(L^{p^{k+1}} \circ F, Id) \to \cdots .$$

Proof. The argument for both sequences is the same so we will talk only about the first one. If the sequence $0 \to S^{p^k} \to S^{p^{k+1}} \to L^{p^{k+1}} \to 0$ was exact (what is the case for $k = 0$) then our lemma would follow immediately from 1.3. Let K be the kernel of the quotient map $S^{p^{k+1}} \to L^{p^{k+1}}$. Then obviously Frobenius $S^{p^k} \to S^{p^{k+1}}$ factors uniquely through K. Let $\phi : S^{p^k} \to K$ be the corresponding map. It was shown in [B1, Lemma 2.5] that K has an increasing filtration starting from S^{p^k} with subquotients being tensor products of functors which are trivial at 0. Hence 1.3 and 1.4 tell us that ϕ induces an isomorphism $Ext_{\mathcal{F}}^*(K, Id) \to Ext_{\mathcal{F}}^*(S^{p^k}, Id)$. So we get our long exact sequence from the short exact sequence $0 \to K \to S^{p^{k+1}} \to L^{p^{k+1}} \to 0$ after composing with F.

References

[BP] S. Betley, T. Pirashvili. Stable K-theory as a derived functor. *J. Pure and Applied Algebra* 96 (1994) 245–258.

[BS] S. Betley, J. Słomińska. New approach to the groups $H_*(\Sigma_n, Lie_n)$ by the homology theory of the category of functors. *J. Pure and Applied Algebra* 161 (2001) 31–43.

[B1] S. Betley. Stable derived functors, the Steenrod algebra and homological algebra in the category of functors. *Fundamenta Math.* 168 (2001) 279–293.

[FLS] V. Franjou, J. Lannes, L. Schwartz. Autour de la cohomologie de MacLane des corps finis. *Invent. Math.* 115 (1994) 513–538.

[FFSS] V. Franjou, E. Friedlander, A. Scorichenko, A. Suslin. General linear and functor cohomology over finite fields. *Ann. of Math.* 150 (1999) 663–728.

[K] N. Kuhn. Generic representation theory of the finite general linear groups and the Steenrod algebra: I. Amer. J. Math. 116 (1994) 327–360.

[P1] T. Pirashvili. Hodge decomposition for higher order Hochschild homology. *Ann. Scient. Ec. Norm. Sup.* 33 (2000) 151–179.

[Q] D.G. Quillen. On the associated graded ring of a group ring. *Journal of Algebra* 10 (1968) 411–418.

[T] A. Troesch. Quelques calculus de cohomologie de compositions de puissances symetriques. *Comm. in Algebra* 30 (2002) 3351–3382.

Stanislaw Betley
Instytut Matematyki
University of Warsaw
ul. Banacha 2
02-097 Warsaw, Poland
e-mail: `betley@mimuw.edu.pl`

Progress in Mathematics, Vol. 215
Algebraic Topology: Categorical Decomposition Techniques, 47–75

Homotopy Operations and Rational Homotopy Type

David Blanc

1. Introduction

In [HS] and [F1] Halperin, Stasheff, and Félix showed how an inductively-defined sequence of elements in the cohomology of a graded commutative algebra over the rationals can be used to distinguish among the homotopy types of all possible realizations, thus providing a collection of algebraic invariants for distinguishing among rational homotopy types of spaces. There is also a dual version, in the setting of graded Lie algebras (see [O]).

However, these authors provided no homotopy-theoretic interpretation of these invariants, which are defined in terms of differential graded algebras (resp. Lie algebras) and their possible perturbations.

The goal of this paper is to provide such an interpretation, in terms of higher rational homotopy operations, and thus to make sense of the following

Theorem A. *For any simply-connected space* $\mathbf{X}$*, there is a sequence of higher homotopy operations taking value in* $\pi_*\mathbf{X}$*, which, together with the rational homotopy Lie algebra* $\pi_{*-1}\mathbf{X} \otimes_{\mathbb{Z}} \mathbb{Q}$ *itself, determine the rational homotopy type of* $\mathbf{X}$.

(See Theorem 7.14 below). At the same time, we provide a more concrete (rational) version of the general theory of higher homotopy operations provided in [BM].

It should be noted that an integral version of the Lie algebra case is contained in [Bl5] (see also [BG, BDG]), and the mod p homology analogue of the Halperin-Stasheff-Félix theory appears in [Bl7]. Moreover, in [Bl2] we showed that the (integral) homotopy type of a space $\mathbf{X}$ is in fact determined by its homotopy groups $\pi_*\mathbf{X}$, together with the action of all primary homotopy operations on it, and of certain higher homotopy operations (see [Bl3, Bl6] for subsequent modifications and improvements).

However, if we are interested only in the *rational* homotopy type of a simply-connected space $\mathbf{X}$, Whitehead products are the only non-trivial primary homotopy operations on the rational homotopy groups $\pi_*\mathbf{X}_{\mathbb{Q}} = \pi_*\mathbf{X} \otimes \mathbb{Q}$, which, after

2000 *Mathematics Subject Classification.* Primary 55P62; Secondary 55Q35, 55P15, 18G50.
Key words and phrases. rational homotopy type, higher homotopy operations, homology of DGLs, non-associative algebras, obstruction theory.

re-indexing, constitute a graded Lie algebra over $\mathbb{Q}$. The relevant higher order operations are also simpler than in the integral case. Thus we hope that the rational version of this theory will be both easier to understand, and more accessible to computation.

Moreover, the higher operations we define are certain subsets of $\pi_*\mathbf{X}$, indexed by elements in homology groups of a certain inductively defined collection of differential graded Lie algebras (DGLs) defined below, so we provide an explicit correspondence between our higher operations and the corresponding elements in cohomology groups with coefficients in $\pi_*\mathbf{X}$ (provided by the Halperin-Stasheff-Félix theory) – a correspondence which was lacking in the integral case.

Finally, while a notion of higher homotopy operations for a differential graded Lie algebra L has been defined in the special case of higher Whitehead products (also known as "Lie-Massey products" – see [A1, A2, AA, R1, R2, T]), in general it is not clear how to represent such rational operations as *integral* higher order operations in $\pi_*\mathbf{X}$, if L represents the rational homotopy type of a topological space $\mathbf{X}$. In order to address this problem, we must consider a somewhat "flabbier" model of rational homotopy than that provided by differential graded Lie algebras, namely a certain class of differential graded non-associative algebras (see Section 7 below).

Thus we also provide a (somewhat incomplete) answer to the following question: what additional structure on the ordinary homotopy groups $\pi_*\mathbf{X}$ of a simply-connected space $\mathbf{X}$, beyond the Whitehead products, is needed to determine its homotopy type up to rational equivalence?

1.1. Notation and conventions

The ground field for all vector spaces, algebras, and tensor products will be $\mathbb{Q}$ (the rationals), unless otherwise stated.

$\mathcal{T}_*$ denotes the category of pointed CW complexes with base-point preserving maps, and by a *space* we shall always mean an object in $\mathcal{T}_*$, which will be denoted by a boldface letter: $\mathbf{X}, \mathbf{S}^n, \ldots$. The subcategory of 1-connected spaces is denoted by $\mathcal{T}_1$, and the rationalization of a space $\mathbf{X} \in \mathcal{T}_1$ is $\mathbf{X}_{\mathbb{Q}}$. The category of rational 1-connected topological spaces is denoted by $\mathcal{T}_{\mathbb{Q}}$.

Let Δ denote the category of ordered sequences $\mathbf{n} = \langle 0, 1, \ldots, n \rangle$ ($n \in \mathbb{N}$), with order-preserving maps. For any category $\mathcal{C}$, we let $s\mathcal{C}$ denote the category of simplicial objects over $\mathcal{C}$ – i.e., functors $\mathbf{\Delta}^{op} \to \mathcal{C}$ (cf. [Ma, §2]); objects therein will be written $A_\bullet, \ldots$. If we omit the degeneracies, we have a Δ-*simplicial* object, which we denote by $A_\bullet^\Delta, \ldots$.

The category of non-negatively graded objects over a category $\mathcal{C}$ will be denoted by $\operatorname{gr}\mathcal{C}$, with objects written $T_*, \ldots$; we will write $|x| = p$ if $x \in T_p$. An upward shift by one in the indexing will be denoted by $\Sigma : \operatorname{gr}\mathcal{C} \to \operatorname{gr}\mathcal{C}$, so that $(\Sigma X_*)_{k+1} = X_k$, and $(\Sigma X_*)_0 = 0$. The category of graded vector spaces is denoted by $\mathcal{V}$.

The category of chain complexes (over $\mathbb{Q}$) will be denoted by $d\mathcal{V}$, and that of double chain complexes by $dd\mathcal{V}$. The differential of any differential graded object is written ∂ (to distinguish it from the face maps d_i of a simplicial object).

If $\mathcal{C}$ is a closed model category (cf. [Q1, I] or [Q3, II, §1]), we denote by $\operatorname{ho}\mathcal{C}$ the corresponding homotopy category. If $X \in \mathcal{C}$ is cofibrant and $Y \in \mathcal{C}$ is fibrant, we denote by $[X, Y]_{\mathcal{C}}$ the set of homotopy classes of maps between them.

Let $\mathcal{S}et$ denote the category of sets, $\mathcal{V}ec$ the category of vector spaces (over $\mathbb{Q}$), $\mathcal{L}ie$ the category of Lie algebras, and $\mathcal{A}lg$ the category of non-associative algebras. We write $\mathcal{S}$ rather than $s\mathcal{S}et$ for the category of simplicial sets, and $\mathcal{S}_*$ for the category of *pointed* simplicial sets.

1.2. Organization

In Section 2 we review some background material on the Quillen DGL model for rational homotopy theory, and describe a bigraded variant of it; and in Section 3 we give some more background on simplicial resolutions. These are applied to the rational context in Section 4, where we also define higher order homotopy operations for DGLs. These appear as the obstructions to realizing certain algebraic equivalences, and serve to determine the rational homotopy type of a simply-connected space. We give a first approximation to Theorem A in §4.15.

In Section 5 we explain how to translate the usual bigraded and filtered DGL models into simplicial DGLs, which allows us to construct appropriate minimal simplicial resolutions. In Section 6 we define the homology and cohomology of a DGL (after Quillen), and show that the obstructions we define above actually take value in the appropriate cohomology groups. Finally, in Section 7 we describe a non-associative differential graded algebra model for rational homotopy theory, which facilitates the translation of the higher homotopy operations described above into integral homotopy operations. We summarize our main results in Theorem 7.14.

1.3. *Acknowledgements.* I would like to thank Ron Livné for arousing my interest in rational higher homotopy operations, Martin Arkowitz and Jean-Michel Lemaire for providing me with copies of hard-to-get theses, and Steve Halperin for some useful comments. I would also like to thank the referee for his remarks and corrections.

2. Lie models

In this section we briefly recall some well-known definitions and facts of rational homotopy theory, and describe variants thereof.

2.1. Differential graded Lie algebras

Let $\mathcal{L}$ denote the category of graded Lie algebras, or *GL*'s. An object $L_* \in \mathcal{L}$ is thus a graded vector space: $L_* = \oplus_{n=0}^{\infty} L_n$ over $\mathbb{Q}$, equipped with a bilinear graded product $[\ ,\ ] : L_p \otimes L_q \to L_{p+q}$ for each $p, q, \geq 0$, such that $[x, y] = (-1)^{|x||y|+1}[y, x]$

and $(-1)^{|x||z|}[[x,y],z]+(-1)^{|y||x|}[[y,z],x]+(-1)^{|z||y|}[[z,x],y]=0$. We denote by $\mathcal{L}_0$ the full subcategory of all *connected* graded Lie algebras – that is, those for which $L_0=0$.

The free graded Lie algebra generated by a graded set X_* is denoted by $\mathbb{L}\langle X_*\rangle$. The functor $\mathbb{L}:\operatorname{gr}\mathcal{S}et\to\mathcal{L}$ is left adjoint to the forgetful "underlying graded set" functor $U:\mathcal{L}\to\operatorname{gr}\mathcal{S}et$, and it factors through $\mathcal{V}$: that is, $\mathbb{L}\langle X_*\rangle = L(\mathbb{V}\langle X_*\rangle)$, where $\mathbb{V}\langle X_*\rangle\in\mathcal{V}$ is the graded vector space with basis X_* and $L(V_*)$ is the free Lie algebra on the graded vector space V_* (defined as the appropriate quotient of the graded tensor algebra).

The category of *differential graded Lie algebras*, or *DGLs*, will be denoted by $d\mathcal{L}$, with $d\mathcal{L}_0$ the subcategory of *connected* Lie algebras (i.e., those with $L_0=0$). An object $L=(L_*,\partial_L)\in d\mathcal{L}$ is a graded Lie algebra $L_*\in\mathcal{L}$, together with a differential $\partial_L=\partial_L^n:L_n\to L_{n-1}$, for each $n>0$, such that $\partial_L^{n-1}\circ\partial_L^n=\{0\}$ and $\partial_L[x,y]=[\partial_L x,y]+(-1)^{|x|}[x,\partial_L y]$.

The homology of the underlying chain complex of a DGL $L=(L_*,\partial)$ will be denoted H'_*L, to distinguish it from the DGL homology defined in §6.5 below. Because the differential ∂ is a derivation, H'_*L inherits from L the structure of a graded Lie algebra.

A morphism of DGLs which induces an isomorphism in homology will be called a *quasi-isomorphism*, or *weak equivalence*, denoted by $f:L\xrightarrow{\simeq}L'$.

In [Q3, II, §4–5], Quillen defined closed model category structures for the categories $d\mathcal{L}_0$ and $s\mathcal{L}ie$, as well as for topological spaces (and thus for $\mathcal{T}_\mathbb{Q}$), and proved:

2.2. Proposition. *There are pairs of adjoint functors $\mathcal{T}_\mathbb{Q}\rightleftharpoons s\mathcal{L}ie$ and $s\mathcal{L}ie\rightleftharpoons d\mathcal{L}_0$, which induce equivalences between the corresponding homotopy categories:* $\operatorname{ho}\mathcal{T}_\mathbb{Q}\approx\operatorname{ho}(s\mathcal{L}ie)\approx\operatorname{ho}(d\mathcal{L}_0)$.

2.3. Notation. To every simply-connected space $\mathbf{X}\in\mathcal{T}_1$ one can thus associate a DGL $(L_*,\partial_L)\in d\mathcal{L}_0$, unique up to quasi-isomorphism, which determines its rational homotopy type. We denote any such DGL by L_X. In particular, $H'_*(L_X)\cong\pi_{*-1}\mathbf{X}\otimes_\mathbb{Z}\mathbb{Q}$, the rational homotopy algebra of $\mathbf{X}$, which we denote by $\Pi_*^X\in\mathcal{L}$.

2.4. Definition. The graded Lie algebra $H'_*(L_X)$ does not suffice to determine the rational homotopy type of $\mathbf{X}\in\mathcal{T}_1$: in fact, there may be infinitely many DGLs $\{L^{(n)}\}_{n=1}^\infty$ with $H'_*(L^{(n)})\cong H'_*(L_X)$, no two of which are quasi-isomorphic as DGLs; see, e.g., [LS]. We shall denote by $d\mathcal{L}_0(\mathbf{X})$ the full subcategory of $d\mathcal{L}_0$ whose objects A satisfy $H'_*A\cong H'_*(L_X)$, with the isomorphism in $\mathcal{L}$ (see [SS], [LS], or [F1] for treatments of the cohomology analogue of $d\mathcal{L}_0(\mathbf{X})$ in terms of algebraic varieties). The objects of ho $d\mathcal{L}_0(\mathbf{X})$ are thus all rational homotopy types which are indistinguishable from $\mathbf{X}_\mathbb{Q}$ on the primary homotopy operation level. Among these there is a distinguished simplest one: recall that a space $\mathbf{X}_\mathbb{Q}\in\mathcal{T}_\mathbb{Q}$ (or its corresponding DGL model $L_X\in d\mathcal{L}$) is called *coformal* (cf. [MN]) if L_X is weakly equivalent to the trivial DGL $(L_*,0)$ (where of course $L_*=H'_*(L_X)$).

2.5. Minimal models

Baues and Lemaire (in [BL, Cor. 2.4]; see also [N, Props. 5.6, 8.1 & 8.8]) showed that each connected DGL (L_*, ∂) has a *minimal model* $(\hat{L}_*, \hat{\partial})$, such that $\hat{L}_*$ is a free graded Lie algebra, $\hat{\partial} : \hat{L} \to \hat{L}$ factors through $[\hat{L}, \hat{L}]$, and there is a quasi-isomorphism of DGLs $\varphi : (\hat{L}_*, \hat{\partial}) \to (L_*, \partial)$ (unique up to chain homotopy). In particular, we can choose such a minimal model $\hat{L}_X$ for any space $\mathbf{X} \in \mathcal{T}_1$ (cf. §2.3).

As Neisendorfer observes in [N, §5], in general minimal models do not exist for non-connected DGLs (but see [Me] or [GHT] for ways around this).

2.6. Bigraded Lie algebras

A differential bigraded Lie algebra, or *DBGL*, is a bigraded vector space $L_{*,*} = \oplus_{p=0}^{\infty} \oplus_{s=0}^{\infty} L_{p,s}$, equipped with a differential $\partial_L = \partial_L^{p,s} : L_{p,s} \to L_{p-1,s}$ and a bilinear graded product $[\ ,\ ] : L_{p,s} \otimes L_{q,t} \to L_{p+q,s+t}$ for each $p, q, s, t \geq 0$ satisfying:

$$
\begin{aligned}
[x, y] &= (-1)^{(p+s)(q+t)+1}[y, x] \\
(-1)^{(p+s)(r+u)}&[[x, y], z] + (-1)^{(p+s)(q+t)}[[y, z], x] \\
&\quad + (-1)^{(q+t)(r+u)}[[z, x], y] = 0 \\
\partial_L \circ \partial_L &= 0 \\
\partial_L[x, y] &= [\partial_L x, y] + (-1)^{p+s}[x, \partial_L y]
\end{aligned} \tag{2.7}
$$

for $x \in L_{p,s}$, $y \in L_{q,t}$, and $z \in L_{r,u}$. The category of such DBGLs will be denoted by $\mathrm{db}\,\mathcal{L}$, with $\mathrm{db}\,\mathcal{L}_0$ the subcategory with $L_{p,0} = 0$ for all p.

2.8. Definition. For each DBGL $(L_{*,*}, \partial_L)$ there is an *associated* DGL (L_*, ∂_L), defined $L_n = \bigoplus_{p+q=n} L_{p,q}$ (same ∂_L); some authors re-index $L_{*,*}$ so that $\hat{L}_{p,s} = L_{p,p+s}$, and then L_* is obtained from $\hat{L}_{*,*}$ by disregarding the first (homological) grading.

As for ordinary graded Lie algebras, one can define closed model category structures on $s\mathcal{L}_0$ and $\mathrm{db}\,\mathcal{L}_0$ (see [BS, §2], and [Bl4, §4]), and we have the following analogue of [Q3, I, Props. 2.3 & 4.6, Thm. 4.4]:

2.9. Proposition. *There are adjoint functors* $s\mathcal{L}_0 \underset{N^*}{\overset{N}{\rightleftarrows}} \mathrm{db}\,\mathcal{L}_0$*, which induce equivalences of the corresponding homotopy categories* $\mathrm{ho}(s\mathcal{L}_0) \approx \mathrm{ho}(\mathrm{db}\,\mathcal{L}_0)$*.* N^* *takes free DBGLs to free simplicial graded Lie algebras.*

Proof. (We give the proof mainly to fix notation which will be needed later.) Given a simplicial graded Lie algebra $L_{\bullet,*} \in s\mathcal{L}_0$, we define the simplicial Lie bracket $[\![\ ,\ ]\!] : L_{p,s} \otimes L_{q,t} \to L_{p+q,s+t}$ on $L_{\bullet,*}$ by combining the Lie brackets with the simplicial structure on $L_{\bullet,*}$ via the Eilenberg-Zilber map:

$$
[\![x, y]\!] = \sum_{(\sigma,\tau) \in S_{p,q}} (-1)^{\varepsilon(\sigma)+pt}[s_{\tau_q} \dots s_{\tau_1} x, s_{\sigma_p} \dots s_{\sigma_1} y] \tag{2.10}
$$

where $S_{p,q}$ denotes the set of all (p,q)-shuffles – that is, partitions of $\{0,1,\ldots, p+q-1\}$ into disjoint sets $\sigma = \{\sigma_1, \sigma_2, \ldots, \sigma_p\}$ and $\tau = \{\tau_1, \tau_2, \ldots, \tau_q\}$ with $\sigma_1 < \sigma_2 < \cdots < \sigma_p$, $\tau_1 < \tau_2 < \cdots < \tau_q$ – and $\varepsilon(\sigma) = p + \sum_{i=1}^{p}(\sigma_i - i)$, so $(-1)^{\varepsilon(\sigma)}$ is the sign of the permutation corresponding to (σ, τ). (See [Mc1, VIII, §8]).

Now let $(C_{*,*}, \partial)$ be the Moore chain complex (cf. [Ma, §22]) of $L_{\bullet,*}$, defined by:

$$(2.11) \qquad C_{p,s} = \bigcap_{i=1}^{p} [\mathrm{Ker}(d_i^p)]_s \qquad \text{with } \partial_p = (-1)^s \, d_0^p|_{C_{p,s}}.$$

2.12. Lemma. *If $A_\bullet \in s\mathcal{L}$ is a simplicial graded Lie algebra, $x \in A_p$ with $d_i x = 0$ for $1 \leq i \leq p-1$, and $y \in A_q$ with $d_j y = 0$ for $1 \leq j \leq q-1$, then $d_k([\![x,y]\!]) = 0$ for $1 \leq k \leq p+q-1$.*

Proof. By definition (2.10) we have

$$(2.13) \qquad [\![x,y]\!] = \sum_{(\sigma,\tau)\in S_{p,q}} (-1)^{\varepsilon(\sigma)+p|y|} [s_{\tau_q} \ldots s_{\tau_1} x, s_{\sigma_p} \ldots s_{\sigma_1} y] \in A_{p+q}.$$

Now for each summand $w_{\sigma,\tau} := [s_\tau x, s_\sigma y]$ in (2.13), with (σ,τ) a (p,q)-shuffle, there are two cases to consider:

The first is that there exist ℓ, m such that $\tau_\ell = k$, $\sigma_m = k-1$ – in which case there is an associated (p,q)-shuffle (σ', τ'), differing from (σ,τ) only in that τ_ℓ and σ_m are switched, so that $d_k(w_{\sigma,\tau}) = d_k(w_{\sigma',\tau'})$ but $(-1)^{\varepsilon(\sigma)} = -(-1)^{\varepsilon(\sigma')}$, and these pairs thus cancel in the sum (2.13).

In the second case, $k, k-1 \in \{\sigma_1, \ldots, \sigma_p\}$, say, and then there is some $0 \leq \ell \leq q$ with $\tau_\ell < k-1$ and $\tau_{\ell+1} > k$. Since necessarily $k+1-p \leq \ell \leq k-1$, we find that $d_k s_\tau x = s_{\tau_q - 1} \cdots s_{\tau_{\ell+1}-1} s_{\tau_\ell} \cdots s_{\tau_1} d_{k-\ell} x = 0$. □

2.14. Corollary. *$[\![\ ,\ ]\!]$ restricts to a bracket $C_{p,s} \otimes C_{q,t} \to C_{p+q,s+t}$.*

Moreover, if we forget the Lie structure, the Moore chain complex functor N induces an equivalence between the categories of simplicial graded vector spaces and bigraded chain complexes (cf. [Do, Thm 1.9]), with the inverse functor Γ defined for such a chain complex $(A_{*,*}, \partial)$ by

$$(\Gamma A_{*,*})_{n,s} := \bigoplus_{0 \leq \lambda \leq n} \ \bigoplus_{I \in \mathfrak{I}_{n,\lambda}} A_{n-\lambda,s}$$

(where for each $n \geq 0$ and $0 \leq \lambda \leq n$, we let $\mathfrak{I}_{\lambda,n}$ denote the set of all sequences of λ non-negative integers $i_1 < \cdots < i_\lambda (< n)$), with the obvious face maps (induced by ∂) and degeneracies (see [Ma, p. 95]).

The left adjoint $N^* : \mathrm{db}\,\mathcal{L}_0 \to s\mathcal{L}_0$ to N is defined

$$N^*((L_{*,*}, \partial)) = L(\Gamma(L_{*,*}))/I(L_{*,*}),$$

where L is the free graded Lie algebra functor, and $I(L_{*,*})$ is the ideal generated by $[\![\Gamma(x), \Gamma(y)]\!] - \Gamma([x,y])$. The identities (2.7) follow from the corresponding ones in the singly-graded case and the simplicial identities. □

3. Simplicial resolutions

The proper algebraic setting for defining our higher homotopy operations is a suitable notion of a simplicial resolution of $\pi_*\mathbf{X}_{\mathbb{Q}}$:

3.1. Definition. Recall that a category of *universal graded algebras* (or variety of graded algebras, in the terminology of [Mc2, V,§6]) is a category $\mathcal{C}$ in which the objects are graded sets X_*, together with an action of a fixed set of n-ary graded operators $W = \{\omega : X_{k_1} \times X_{k_2} \times \cdots \times X_{k_n} \to X_m\}$, satisfying a set of identities E, and the morphisms are functions on the sets which commute with the operators. Such categories always come equipped with a "free graded algebra" functor $F : \mathrm{gr}\,\mathcal{S}et \to \mathcal{C}$, left adjoint to the "underlying graded set" functor $U : \mathcal{C} \to \mathrm{gr}\,\mathcal{S}et$. In all the examples we shall be concerned with, the objects X_* will be "underlying-abelian" (see [BS, §2.1.1]), and in fact will have the underlying structure of a graded vector space over $\mathbb{Q}$.

Examples include $\mathcal{L}$, and the categories of associative (resp. non-associative) graded algebras. Note that any ordinary ungraded category of universal algebras may be thought of as a CUGA with all objects concentrated in degree 0.

3.2. Definition. A *free simplicial resolution* of an object B in a CUGA $\mathcal{C}$ is a weak equivalence from a cofibrant object $A_\bullet \in s\mathcal{C}$ to the constant simplicial object associated to B (with respect to the closed model category structure on the category $s\mathcal{C}$ defined in [Q1, II, §4]). Such resolutions always exist, by [Q1, II, §4]; see Section 5 below for a specific construction.

3.3. Bisimplicial objects

We shall be interested in a particular type of simplicial resolution, which may be defined for an arbitrary CUGA $\mathcal{C}$ (cf. [DKS] and [BS]), though we shall only need it for the case where $\mathcal{C}$ is a category of ungraded universal algebras, such as $\mathcal{L}ie$ or $\mathcal{A}lg$:

Consider the category $ss\mathcal{C}$ of bisimplicial objects over $\mathcal{C}$. We think of an object $A_{\bullet\bullet} \in ss\mathcal{C}$ as having *internal* and *external* simplicial structures, with corresponding homotopy group objects $\pi^i_t A_{\bullet\bullet}$ and $\pi^e_s A_{\bullet\bullet}$ (each taking value in $s\mathcal{C}$ – see [BS, App.]). Let $sF : \mathrm{gr}\,\mathcal{S} \to s\mathcal{C}$ denote the free graded algebra functor, extended dimensionwise, and let $S^n(k)_\bullet$ be the graded simplicial set having the simplicial n-sphere $S^n_\bullet := \Delta[n]/\Delta[n]^{n-1}$ in degree k. We think of the simplicial graded algebras $F(S^n(k)_\bullet)$ as the $\mathcal{C}$-spheres, or *models*, for $s\mathcal{C}$ (cf. [BS, §3.1]). (In the ungraded case one can of course omit the extra degree k, and write simply $F(\mathbf{S}^n)$.) Similarly, if $D^n(k)_\bullet$ is the graded simplicial set having $\Delta[n]$ in degree k, we can think of $F(D^n(k)_\bullet)$ as the $\mathcal{C}$-disks for $s\mathcal{C}$. The full subcategory of $s\mathcal{C}$ whose objects are weakly equivalent to coproducts of such models will be denoted by $\mathcal{M}_{\mathcal{C}}$, or simply $\mathcal{M}$.

One can use these models to define the so-called "E^2-model category structure" for $ss\mathcal{C}$, as in [DKS, §5], in which a map $f : X_{\bullet\bullet} \to Y_{\bullet\bullet}$ is a weak equivalence if

(3.4) $\qquad f_\star : \pi_s \pi^i_t X_{\bullet\bullet} \to \pi_s \pi^i_t Y_{\bullet\bullet}$ is an isomorphism for each $s, t \geq 0$.

We shall not need an explicit description of the fibrations and cofibrations in $ss\mathcal{C}$, but only a particular type of cofibrant object, as follows:

3.5. Definition. A bisimplicial object $A_{\bullet\bullet} \in ss\mathcal{C}$ is called $\mathcal{M}$-*free* if for each $m \geq 0$ there are graded simplicial sets $X[m]_\bullet \simeq \bigvee_i \mathbf{S}^{n_i}(k_i)_\bullet$ such that $A_{\bullet,m} \cong F(X[m]_\bullet)$ (so that $A_{\bullet,m} \in \mathcal{M}$), and the external degeneracies of $A_{\bullet\bullet}$ are induced under F by maps $X[m]_\bullet \to X[m+1]_\bullet$ which are, up to homotopy, the inclusion of sub-coproduct summands. Any $X_\bullet \in s\mathcal{C}$ may be resolved by an $\mathcal{M}$-free bisimplicial algebra $A_{\bullet\bullet}$ (see [BS, §4.1]); this is called an $\mathcal{M}$-*free resolution* of $X_\bullet$.

3.6. Definition. The *diagonal* of a bisimplicial object $A_{\bullet\bullet} \in ss\mathcal{C}$ is a simplicial object $\operatorname{diag}(A_{\bullet\bullet}) \in s\mathcal{C}$ with $\operatorname{diag}(A_{\bullet\bullet})_n := A_{n,n}$, face maps $d_k = d^i_k \circ d^e_k$, and degeneracies $s_j = s^i_j \circ s^e_j$.

3.7. *Remark.* There is a first quadrant spectral sequence with

$$E^2_{s,t} = \pi^e_s(\pi^i_t A_{\bullet\bullet}) \Rightarrow \pi_{s+t} \operatorname{diag}(A_{\bullet\bullet})$$

(see [Q2], and compare [BF, Thm B.5]).

Thus in particular if $A_{\bullet\bullet} \to X_\bullet$ is a resolution (in the E^2-model category sense), we see that $\varepsilon : A_{0,\bullet} \to X_\bullet$ induces a weak equivalence $\operatorname{diag}(A_{\bullet\bullet}) \simeq X_\bullet$.

Moreover, the same is true if we disregard the degeneracies and consider only the Δ-bisimplicial resolution $A^{\Delta}_{\bullet\bullet} \to X_\bullet$.

4. Resolutions for rational spaces

Given a simply-connected space $\mathbf{X} \in \mathcal{T}_1$, the first approximation to an algebraic description of its rational homotopy type is given by its rational homotopy Lie algebra $\Pi^X_* := \pi_{*-1}\mathbf{X}_\mathbb{Q} \in \mathcal{L}$. If $\mathbf{X}_\mathbb{Q}$ were coformal (§2.4), then in particular all higher homotopy operations vanish in $\pi_*\mathbf{X}_\mathbb{Q}$, and no information beyond Π^X_* itself is needed to determine the rational homotopy type of $\mathbf{X}$. The higher homotopy operations we shall describe may thus be thought of as "obstructions to coformality", much in the spirit (though not the specific approach) of [HS].

4.1. Topological resolutions

To proceed further, we need some kind of a "topological" simplicial object $C_\bullet$ which realizes a suitable "algebraic" simplicial resolution $V_{\bullet,*} \to \Pi^X_*$ in $s\mathcal{L}$, in the sense that $V_{\bullet,*} = \pi_{*-1}C_\bullet$. The higher homotopy operations we want then arise as the obstructions to realizing the "algebraic" augmentation map $\pi_{*-1}C_\bullet \to \Pi^X_*$ topologically.

This can be done using actual topological spaces, as in the integral case (see [Bl2, §7], as simplified in [Bl3, §4.9]), but for rational spaces it is more convenient to use an algebraic model, in a category such as $d\mathcal{L}$. To allow us freedom in choosing this model, we give a general definition:

4.2. *Assumptions.* Let $\mathrm{gr}\,\mathcal{C}$ be a CUGA (which we may assume to have the underlying structure of a graded vector space), and $\mathcal{C}$ the category of (ungraded) universal algebras corresponding to objects of $\mathrm{gr}\,\mathcal{C}$ concentrated in degree 0. The cases we shall be interested in are $\mathcal{C} = \mathcal{L}ie$ (with $\mathrm{gr}\,\mathcal{C} = \mathcal{L}$) and $\mathcal{C} = \mathcal{A}lg$ (with $\mathrm{gr}\,\mathcal{C} = \mathcal{A}$).

As shown in [BS, App.], for each simplicial algebra $A_\bullet \in s\mathcal{C}$, the graded homotopy object $\pi_* A_\bullet$ actually takes value in $\mathrm{gr}\,\mathcal{C}$.

For a given $A_\bullet \in s\mathcal{C}$, let $C_{\bullet\bullet} \to A_\bullet$ be an $\mathcal{M}_\mathcal{C}$-free resolution (Definition 3.5). In particular, this implies that upon applying the functor π_* we obtain a free simplicial resolution $\pi^i_* C_{\bullet\bullet}$ (in the "external" direction!) of the graded algebra $\pi_* A_\bullet$. In fact, we only need a Δ-bisimplicial resolution (§3.7), but we shall nevertheless usually abuse notation by writing $C_{\bullet\bullet}$ for $C^\Delta_{\bullet\bullet}$.

Next, assume we are given another object $B_\bullet \in s\mathcal{C}$, together with an isomorphism $\varphi : \pi_* A_\bullet \cong \pi_* B_\bullet$ (in $\mathrm{gr}\,\mathcal{C}$). Define a sequence of morphisms $\psi_n : \pi_* C_{n,\bullet} \to \pi_* B_\bullet$ by $\psi_0 := \varphi \circ \varepsilon_\#$ and $\psi_{n+1} := \psi_n \circ d_0$ (which implies that $\psi_{n+1} = \psi_n \circ d_i$ for all $0 \le i \le n$, by the simplicial identities).

We choose once and for all a fixed map $f_0 : C_{0,\bullet} \to B_\bullet$ realizing ψ_0 (this is possible because $C_{\bullet\bullet} \to A_\bullet$ is $\mathcal{M}$-free) and define $f_n : C_{n,\bullet} \to B_\bullet$ inductively by setting $f_{n+1} := f_n \circ d_n$, so that $\pi_*(f_n) = \psi_n$ for all $n \ge 0$. It is usually most convenient to set $f|_{\mathcal{D}k} = 0$ for all $\mathcal{C}$-disks $\mathcal{D}k \hookrightarrow C_{0,\bullet}$ (cf. §3.3).

Note that, because $C_{\bullet\bullet}$ is $\mathcal{M}$-free, the maps $\{\psi_n\}_{n=0}^\infty$ define an augmented Δ-simplicial object ${}^h C^\Delta_{\bullet\bullet} \to B_\bullet$ *in the homotopy category* $\mathrm{ho}(s\mathcal{C})$ – or equivalently, an augmented Δ-simplicial object up-to-homotopy.

4.3. Definition. Let $D[n] \in \mathcal{S}_*$ denote the standard simplicial n-simplex, together with an indexing of its non-degenerate k-dimensional faces $D[k]^{(\gamma)}$ by the composite face maps

$$\gamma = d_{i_{n-k+1}} \circ \ldots \circ d_{i_n} : \mathbf{n} \to \mathbf{k}$$

in $\mathbf{\Delta}^{op}$ (cf. [Bl3, §4]). Its $(n-1)$-skeleton, which is a simplicial $(n-1)$-sphere, is denoted by $\partial D[n]$. We shall take $* := D[0]^{(d_0 d_1 d_2 \ldots d_{n-1})}$ as the base point of $D[n] \in \mathcal{S}_*$, and we choose once and for all a fixed isomorphism $\varphi^{(\gamma)} : D[k]^{(\gamma)} \to D[k]$ for each face $D[k]^{(\gamma)}$ of $D[n]$ (see, e.g., [Bl3, (4.5)]).

4.4. Definition. Given $Y_\bullet \in s\mathcal{C}$ and a simplicial set $K_\bullet \in \mathcal{S}$, we define their *half-smash* (in $s\mathcal{C}$) by:

$$Y_\bullet \rtimes K_\bullet := Y_\bullet \otimes K_\bullet / (\{0\} \otimes K_\bullet)$$

(where $(Y_\bullet \otimes K_\bullet)_n := \coprod_{x \in K_n} (Y_n)_{(x)}$ – cf. [Q1, II, §1, Prop. 2]).

Similarly, the *smash product* (in $s\mathcal{C}$) of $Y_\bullet$ with a *pointed* simplicial set $K_\bullet \in \mathcal{S}_*$ is defined $Y_\bullet \wedge K_\bullet := Y_\bullet \rtimes K_\bullet / (Y_\bullet \rtimes \{*\})$, and if $K_\bullet = \mathbf{S}^r$ (the simplicial sphere), we write $\Sigma^r Y_\bullet$ for $Y_\bullet \wedge \mathbf{S}^r$.

4.5. *Remark.* If $Y_\bullet = F(\mathbf{S}^n)$ is a $\mathcal{C}$-sphere (see §3.3), then $\Sigma^r Y_\bullet \cong F(\mathbf{S}^{n+r})$ is also a $\mathcal{C}$-sphere. In fact, many of the usual properties of spheres in $\mathrm{ho}\,\mathcal{T}$ also hold for $\mathcal{C}$-spheres – e.g., $\pi_r X_\bullet \cong [F(\mathbf{S}^n), X_\bullet]_{s\mathcal{C}}$ for any $X_\bullet \in s\mathcal{C}$ (cf. [Q1, I, §4]), and $Y_\bullet \simeq \coprod_i F(\mathbf{S}^{n_i}) \Rightarrow \Sigma^r Y_\bullet \simeq \coprod_i F(\mathbf{S}^{n_i+r})$ (cf. [Q1, I, §3]).

4.6. Definition. Under the assumptions of §4.2, for each $n \in \mathbb{N}$, we define a $\partial D[n]$-*compatible sequence* to be a sequence of maps $\{h_k : C_{k,\bullet} \rtimes D[k] \to B_\bullet\}_{k=0}^{n-1}$, such that $h_0 = f_0$ (under the natural identification $C_{0,\bullet} \rtimes D[0] = C_{0,\bullet}$), and for any iterated face maps $\delta = d_{i_{j+1}} \circ \cdots \circ d_{i_n}$ and $\gamma = d_{i_j} \circ \delta$ $(0 \leq j < n)$ we have

$$h_j \circ (d_{i_j} \rtimes id) = h_{j+1} \circ (id \rtimes \iota_\delta^\gamma) \text{ on} C_{j+1,\bullet} \rtimes D[j], \tag{4.7}$$

where $\iota_\delta^\gamma : D[j] \to D[j+1]$ is the composite $\iota_\delta^\gamma := \varphi^\delta \circ \iota \circ (\varphi^\gamma)^{-1}$. Here φ^γ and φ^δ are the isomorphisms of Definition 4.3, and $\iota : D[j]^{(\gamma)} \to D[j+1]^{(\delta)}$ is the inclusion (compare [Bl3, Def. 4.10]).

A sequence of maps $\{h_k : C_{k,\bullet} \rtimes D[k] \to B_\bullet\}_{k=0}^{\infty}$ satisfying condition (4.7) for all γ, δ, and n is called a $\partial D[\infty]$-*compatible sequence.*

4.8. Definition. Given such a $\partial D[n]$-compatible sequence $\{h_k : C_{k,\bullet} \rtimes D[k] \to B_\bullet\}_{k=0}^{n-1}$ the *induced map* $\bar{h} : C_{n,\bullet} \rtimes \partial D[n] \to B_\bullet$ is defined on the "faces" $C_{n,\bullet} \rtimes D[n-1]^{(d_i)}$ of $C_{n,\bullet} \rtimes D[n]$ by: $\bar{h}|_{C_{n,\bullet} \rtimes D[n-1]^{(d_i)}} = h_{n-1} \circ (d_i \rtimes id)$. The compatibility condition (4.7) above guarantees that $\bar{h}$ is well defined.

4.9. Definition. For each $n \geq 2$, the *n-th order homotopy operation* (associated to the choice of $C_{\bullet\bullet} \to A_\bullet$ in §4.2) is a subset $\langle\langle n \rangle\rangle$ of the track group $[\Sigma^{n-1} C_{n,\bullet}, B_\bullet]_{s\mathcal{C}}$ defined as follows:

Let $T_n \subseteq [C_{n,\bullet} \rtimes \partial D[n], B_\bullet]_{s\mathcal{C}}$ be the set of homotopy classes of maps $\bar{h} : C_{n,\bullet} \rtimes \partial D[n] \to B_\bullet$ induced as above by some $\partial D[n]$-compatible collection $\{h_k\}_{k=0}^{n-1}$. Since each $C_{n,\bullet}$ is a suspension, up to homotopy, by Remark 4.5, we have a splitting

$$C_{n,\bullet} \rtimes \partial D[n] \simeq (\mathbf{S}^{n-1} \wedge C_{n,\bullet}) \amalg C_{n,\bullet} \tag{4.10}$$

(as for topological spaces). We define $\langle\langle n \rangle\rangle \subseteq [\Sigma^{n-1} C_{n,\bullet}, B_\bullet]_{s\mathcal{C}}$ to be the image under the resulting projection of the subset $T_n \subseteq [C_{n,\bullet} \rtimes \partial D[n], B_\bullet]_{s\mathcal{C}}$.

Note that the projection of a class $[\bar{h}] \in T_n$ on the other summand $[C_{n,\bullet}, B_\bullet]_{s\mathcal{C}}$ coming from the splitting (4.10) is just the homotopy class of the map f_n of §4.2. On the other hand, since $C_{\bullet\bullet}$ was assumed to be $\mathcal{M}$-free, each $C_{n,\bullet} \simeq \coprod_{k=1}^{\infty} \coprod_{x \in K_{n,k}} F(\mathbf{S}_{(x)}^k)$ is weakly equivalent to a wedge of spheres over some indexing set $K_{*,*}$, so $\Sigma^{n-1} C_{n,\bullet} \simeq \coprod_{k=1}^{\infty} \coprod_{x \in K_{n,k}} F(\mathbf{S}_{(x)}^{k+n-1})$. Thus

$$[\Sigma^{n-1} C_{n,\bullet}, B_\bullet]_{s\mathcal{C}} \cong \prod_{k=1}^{\infty} \prod_{x \in K_{n,k}} [F(\mathbf{S}_{(x)}^{k+n-1}), B_\bullet]_{s\mathcal{C}}, \tag{4.11}$$

and we shall denote the components of $\langle\langle n \rangle\rangle$ under this product decomposition by $\langle\langle n, x \rangle\rangle \subseteq [F(\mathbf{S}_{(x)}^{k+n-1}), B_\bullet]_{s\mathcal{C}} = \pi_{k+n-1} B_\bullet$.

4.12. Definition. An operation $\langle\langle k \rangle\rangle$ *vanishes* if it contains the null class. We say that all the lower order operations $\langle\langle k \rangle\rangle$ $(2 \leq k < n)$ *vanish coherently* (cf. [Bl2, Def. 5.7]) if the $\partial D[m]$-compatible collections $\{h_k^\gamma\}_{k=0}^{m-1}$ for the various faces γ of $\partial D[n]$ can be chosen to agree on their intersections, so that they in fact fit together to form a $\partial D[n+1]$-compatible collection $\{h_k\}_{k=0}^{n}$.

4.13. Proposition. *A necessary condition for the subset* $\langle\langle n\rangle\rangle$ *to be non-empty that the lower order operations* $\langle\langle k\rangle\rangle$ *vanish for* $2 \leq k < n$*; a sufficient condition is that they vanish coherently.*

Proof. See [BM, Theorem 3.29]. □

4.14. *Remark.* The coherent vanishing of all the operations $\{\langle\langle n\rangle\rangle\}_{n=2}^{\infty}$ is equivalent, by [BV, Cor. 4.21 & Thm. 4.49] and [Bl2, §4.11], to the *rectifiability* of the augmented Δ-simplicial object up-to-homotopy ${}^{h}C_{\bullet\bullet}^{\Delta} \to B_{\bullet}$: that is, its replacement by augmented Δ-simplicial object $\hat{C}_{\bullet\bullet}^{\Delta} \to B_{\bullet}$ over $s\mathcal{C}$ (with the simplicial identities now holding precisely, in $s\mathcal{C}$, rather than just in $\mathrm{ho}(s\mathcal{C})$), such that $C_{n,\bullet} \simeq \hat{C}_{n,\bullet}$ for each n.

This in turn implies (by §3.7) that $\mathrm{diag}(\hat{C}_{\bullet\bullet}^{\Delta}) \simeq B_{\bullet}$; but since $\mathrm{diag}(\hat{C}_{\bullet\bullet}^{\Delta}) \simeq \mathrm{diag}(C_{\bullet\bullet}^{\Delta})$, and $\mathrm{diag}(C_{\bullet\bullet}^{\Delta}) \simeq A_{\bullet}$ by assumption, we conclude that $A_{\bullet} \simeq B_{\bullet}$ if and only if the higher homotopy operations $\{\langle\langle n\rangle\rangle\}_{n=2}^{\infty}$ vanish coherently.

4.15. Summary. This yields a first approximation to Theorem A, which may be described as follows:

We work in $\mathcal{C} = \mathcal{L}ie$ (and $\mathrm{gr}\,\mathcal{C} = \mathcal{L}$). Given a space $\mathbf{X} \in \mathcal{T}_1$ we consider the simplicial Lie algebra $B_{\bullet}$ corresponding to a DGL model $L_X \in d\mathcal{L}$ for $\mathbf{X}_{\mathbb{Q}}$ (under the functors of Proposition 2.2), and let $\Pi_*^X := \pi_{*-1}\mathbf{X}_{\mathbb{Q}} \in \mathcal{L}$ be its rational homotopy Lie algebra, with $A_{\bullet} \in s\mathcal{L}ie$ the simplicial Lie algebra corresponding to the trivial DGL $L^{(0)} := (\Pi_*^X, 0)$. Choose some $\mathcal{M}_{\mathcal{L}ie}$-free resolution $C_{\bullet\bullet} \in ss\mathcal{L}ie$ of $A_{\bullet}$.

$\mathbf{X}$ is coformal if and only if $A_{\bullet} \simeq B_{\bullet}$, and this happens if and only if all the higher homotopy operations $\{\langle\langle n\rangle\rangle\}_{n=2}^{\infty}$ associated to $C_{\bullet\bullet}$ vanish coherently, by Remark 4.14. If not, let n_0 denote the least $n \geq 2$ such that $0 \notin \langle\langle n\rangle\rangle$. Note that we can apply the above procedure to any DGL in $d\mathcal{L}(\mathbf{X})$ (Def. 2.4), not only to L_X; and the existence and vanishing or non-vanishing of the higher homotopy operation $\langle\langle n_0\rangle\rangle \subset \pi_*\mathbf{X}_{\mathbb{Q}}$ is a homotopy invariant. Denote by $\mathcal{H}^{(1)}$ the set of all homotopy types in $\mathrm{ho}\, d\mathcal{L}_0(\mathbf{X})$ for which $\langle\langle n_0\rangle\rangle$ is defined and has the same value as for $B_{\bullet}$ itself (i.e., those DGLs which are indistinguishable from L_X as far as the primary homotopy operations, and all the higher homotopy operations $\{\langle\langle n\rangle\rangle\}_{n=2}^{n_0}$ associated to $C_{\bullet\bullet}$, can see). For each $\alpha \in \mathcal{H}^{(1)}$, choose a representative DGL $L^{(1,\alpha)}$.

Next, choose a new $\mathcal{M}$-free resolution for the simplicial Lie algebra corresponding to $L^{(1,\alpha)}$, and repeat the above procedure, yielding a set of higher homotopy operations $\langle\langle n_{1,\alpha}\rangle\rangle \subset \pi_*\mathbf{X}_{\mathbb{Q}}$ which serve as obstructions to the existence of a homotopy equivalence $L^{(1,\alpha)} \xrightarrow{\simeq} L_X$. For each such higher operation $\langle\langle n_{1,\alpha}\rangle\rangle$, we denote by $\mathcal{H}^{(2,\alpha)}$ the set of all homotopy types in $\mathcal{H}^{(1)} \subseteq ho\, d\mathcal{L}_0(\mathbf{X})$ for which $\langle\langle n_{1,\alpha}\rangle\rangle$ has the same value as for L_X. Now choose representatives $L^{(2,\alpha,\alpha')}$ for each $\alpha' \in \mathcal{H}^{(2,\alpha)}$, and proceed as above.

In this way we obtain a tree T_X of rational homotopy types in $\mathrm{ho}\, d\mathcal{L}_0(\mathbf{X})$, which also indexes a collection of higher homotopy operations of the form $\langle\langle n_{k,\alpha_1,\alpha_2,\ldots,\alpha_k}\rangle\rangle \subseteq \pi_*\mathbf{X}_{\mathbb{Q}}$, and $\lim_{k\to\infty} n_k = \infty$ along any branch of the tree T_X, so that in fact this collection of operations determines the rational homotopy type of $\mathbf{X}$.

In [Bl5], we show how this tree of homotopy types in $\operatorname{ho} d\mathcal{L}_0(\mathbf{X})$, and thus the corresponding collection of higher homotopy operations, may be described more effectively in terms of a "Postnikov tower" for an $\mathcal{M}_{\mathcal{L}ie}$-free resolution for $\mathbf{X}$.

5. Minimal resolutions

We now explain how the bisimplicial theory described in Section 4 translates into a differential graded theory, when $\mathcal{C} = \mathcal{L}$. In particular, this allows an application of the Halperin-Stasheff-Félix perturbation theory to our context.

First, it is sometimes convenient to have *minimal* $\mathcal{M}$-free resolutions for a DGL, defined for any CUGA $\mathcal{C}$ as follows:

5.1. Definition. Any $B \in \mathcal{C}$ has a special kind of free simplicial resolution (see §3.2) $A_\bullet \to B$, called a *CW-resolution*, defined as follows (cf. [Bl1, §5.3]): Assume that for each $n \geq 0$, $A_n = F(T^n_*)$ is the free graded algebra on the graded set T^n_*, and that the degeneracies of $A_\bullet$ take T^n_* to T^{n+1}_*. Let $\bar{A}_n$ denote the sub-algebra of A_n generated by the non-degenerate elements in T^n_*. Then we require that $d_i|_{\bar{A}_n} = 0$ for $1 \leq i \leq n$. The sequence $\bar{A}_0 = A_0, \bar{A}_1, \ldots, \bar{A}_n, \ldots$ is called a *CW-basis* for $A_\bullet$, and $\bar{d}_0 = d_0|_{\bar{A}_n}$ is the *attaching map* for $\bar{A}_n$.

Such a $A_\bullet \to B$ will be called *minimal* if each $\bar{A}_{n+1}$ is minimal among those free algebras in $\mathcal{C}$ which map onto the Moore n-cycles $Z_n A_\bullet = \operatorname{Ker}(\partial_n)$ (see (2.11)).

5.2. Definition. When $\mathcal{C} = \mathcal{L}$, the category of graded Lie algebras, it will be more convenient at times to use of the adjoint functors of Proposition 2.9 to replace $A_{\bullet\bullet} \to X_\bullet$ by a simplicial DGL $L_{\bullet,*} \to X_*$. In this case the simplicial models are replaced by the corresponding DGLs, namely

1. A *$d\mathcal{L}$-n-sphere*, $\mathcal{S}^n_{(x)}$, is a DGL of the form $(\mathbb{L}\langle X_*\rangle, 0)$ where X_* is the graded set with $X_n = \{x\}$ and $X_i = \emptyset$ for $i \neq n$.
2. A *$d\mathcal{L}$-$(n+1)$-disk*, denoted $\mathcal{D}n+1_{(x)}$, is the DGL $(\mathbb{L}\langle X_*\rangle, \partial_L)$ where $X_{n+1} = \{x\}$, $X_n = \{\partial_L x\}$, and $X_i = \emptyset$ for $i \neq n, n+1$. Its *boundary* is the $d\mathcal{L}$-n-sphere $\partial\mathcal{D}n+1_{(x)} := \mathcal{S}^n_{(\partial_L x)}$.
3. A *two-stage DGL* is a DGL $(\mathbb{L}\langle X_*\rangle, \partial_L) \in d\mathcal{L}$, where for some $n \geq 0$ we have $X_i = \emptyset$ for $i \neq n, n+1$. Any coproduct (in $d\mathcal{L}$) of two-stage DGLs will be called a *free* DGL.

Evidently $d\mathcal{L}$-spheres and disks are free DGLs, and any free DGL may be described as the coproduct of $d\mathcal{L}$-spheres and disks – more precisely, as a coproduct of $d\mathcal{L}$-spheres, disks, and collections of disks with their boundaries identified to a single sphere.

5.3. Definition. Following Stover, we define a comonad $F : d\mathcal{L} \to d\mathcal{L}$ by setting

$$F(B) := (\coprod_{k=1}^{\infty} \coprod_{x \in B_k} \mathcal{D}k_{(x)})/\sim, \tag{5.4}$$

for any $B = (B_*, \partial_B) \in d\mathcal{L}$, where we set $\mathcal{D}k_{(x)} := \mathcal{S}^k_{(x)}$ if $\partial_B x = 0$, and let $\partial\mathcal{D}k+1_{(x)} \sim \mathcal{S}^k_{(\partial_B x)}$ if $\partial_B x \neq 0$.

Clearly $F(B)$ is a free DGL, and by iterating F we obtain a free simplicial DGL $W_{\bullet,*}$ with $W_n = F^{n+1}(B)$ (see [Gd, App., §3]), which we call the *canonical free simplicial DGL resolution* of $B = (B_*, \partial_B)$, which we denote by $W_{\bullet,*}(B)$. Observe that $W_{\bullet,*}$ (or equivalently, the corresponding bisimplicial Lie algebra $W_{\bullet\bullet}$) is an $\mathcal{M}$-free resolution of B.

5.5. *Remark.* Note that if $\partial_B \equiv 0$, by definition (5.4) $F(B_*, 0)$ has only spheres, and no disks, and thus the canonical resolution $W_{\bullet,*}(B)$ has $\partial_{W_n} = 0$ for all $n \geq 0$. Thus $W_{\bullet,*}$ may be identified with the usual canonical resolution of the graded Lie algebra B_* (coming from the "free graded Lie algebra on underlying graded set" comonad), which we shall denote by $V_{\bullet,*}(B_*)$. However – unlike the canonical resolution – the construction above can be mimicked topologically (cf. [St, §2.3]). Since we want to present our results in a manner which could be generalized (as far as possible) to the integral case, we have chosen the somewhat convoluted description of (5.4).

Note further that by (3.4), if we apply the functor H'_* to $W_{\bullet,*} \to B_*$ – or equivalently, the functor π^i_* to $W_{\bullet\bullet} \to B_\bullet$ – we obtain a free simplicial resolution of the graded Lie algebra $L_* := H'_*(B_*, \partial_B)$.

5.6. Notation. If we write $\langle x \rangle \in F(B)$ for the generator corresponding to an element $x \in B_*$, then recursively a typical DGL generator for $W_n = W_{n,*}$ (in the canonical resolution $W_{\bullet,*}(B)$) is $\langle \alpha \rangle$, for $\alpha \in W_{n-1}$, so an element of W_n is a sum of iterated Lie products of elements of B_*, arranged within $n+1$ nested pairs of brackets $\langle\langle \cdots \rangle\rangle$. With this notation, the i-th face map of $W_{\bullet,*}$ is "omit i-th pair of brackets", and the j-th degeneracy map is "repeat j-th pair of brackets". The operation of bracketing is defined to be linear $\langle - \rangle$ is linear – i.e., we set $\langle \alpha x + \beta y \rangle = \alpha\langle x \rangle + \beta\langle y \rangle$ for $\alpha, \beta \in \mathbb{Q}$ and $x, y \in B$.

In order to construct minimal $\mathcal{M}$-simplicial resolutions, first consider the coformal case:

5.7. The bigraded model

Any coformal DGL (§2.4), and in particular $L = (L_*, 0)$, has a *bigraded model* $A_{*,*} \to L_*$ – that is, a bigraded DGL $(A_{*,*}, \partial_A)$ (see §2.6) which is minimal in the sense of §2.5 (so in particular free as a graded Lie algebra), along with a quasi-isomorphism $A_{*,*} \to L_*$. The bigraded model is unique up to isomorphism. See [O, Part I] for an explicit construction.

This is just the Lie algebra version of the bigraded model of [HS, §3] (see also [F2]), which is in turn essentially the Tate-Jozefiak resolution (see [J]) of a graded commutative algebra.

$A = (A_*, \partial_A)$ will denote the DGL associated to $A_{*,*}$ (Definition 2.8); by construction A is the minimal model (§2.5) for L (which is not minimal itself, unless L_* happens to be a *free* graded Lie algebra).

5.8. Proposition. *Let $L = (L_*, 0) \in d\mathcal{L}$ be a coformal DGL, and $\phi : A_{*,*} \to L_*$ its bigraded model; then there is an $\mathcal{M}_{d\mathcal{L}}$-free simplicial resolution $C_{\bullet,*} \to L$, with a bijection $\theta : X_{**} \hookrightarrow C_{\bullet,*}$ between a bigraded set X_{**} of generators for $A_{*,*}$ and the set of non-degenerate (§5.1) $d\mathcal{L}$-spheres in $C_{\bullet,*}$. Moreover, $H'_*(C_{\bullet,*})$ is a minimal CW-resolution of $L_* = H'_*(A_{*,*})$, with CW basis generated by* $\mathrm{Im}\,(\theta)$.

Proof. By Proposition 2.9 there is a simplicial graded Lie algebra resolution $C_{\bullet,*} \to L_*$ corresponding to $A_{*,*}$, and thus a weak equivalence of simplicial graded Lie algebras $\psi : C_{\bullet,*} \to V_{\bullet,*} = V_{\bullet,*}(L_*)$ (see §5.5), which is one-to-one because $A_{*,*}$, and thus $C_{\bullet,*}$, are minimal (cf. [BL, §2]).

Now let $W_{\bullet,*}$ be the canonical free simplical DGL resolution of $A_{*,*}$; the fact that $\phi : A_{*,*} \to L_*$ is a quasi-isomorphism implies that there is a weak equivalence $\varphi : V_{\bullet,*} \to W_{\bullet,*}$ (as well as one in the other direction). The composite $\varphi \circ \psi : C_{\bullet,*} \to W_{\bullet,*}$ is again a one-to-one weak equivalence (by minimality); we may therefore think of $C_{\bullet,*}$ as a sub-simplicial object of $W_{\bullet,*}$.

Moreover, there is an embedding of bigraded vector spaces $\eta : A_{*,*} \to C_{\bullet,*}$ (see the proof of Proposition 2.9), and thus another such embedding $\theta : A_{*,*} \to W_{\bullet,*}$, which may be defined explicitly as follows (using the notation of §5.6):

For $x \in X_{0,*}$, set $\theta(x) = \langle x \rangle \in C_{0,*} = F(A_*)$. Since ϕ maps $X_{0,*}$ onto a (minimal) set of Lie algebra generators for $L_* = H'_*(A_{*,*})$, each $\theta(x)$ is a ∂_W-cycle, so $C^{(0)}_{0,*} := \coprod_{k=1}^{\infty} \coprod_{x \in X_{0,k}} \mathcal{S}^k_{(\theta(x))}$ is a sub free DGL of $W_{0,*}$.

By minimality of $A_{*,*}$, any $x \in X_{n,*}$ $(n \geq 1)$ is uniquely determined by $\partial_A(x) \in A_{n-1,*}$. Thus if we require θ to be multiplicative (with respect to the ordinary bracket in $A_{*,*}$, and with respect to the simplicial Lie bracket $[\![\ ,\ ]\!]$ of (2.10) in $W_{\bullet,*}$), we may define $\theta : A_{*,*} \to W_{\bullet,*}$ inductively by

$$\theta(x) = \langle \theta(\partial_A(x)) \rangle, \tag{5.9}$$

and we shall write $x^{(0)}$ for $\theta(x)$ if $x \in X_{**}$.

By definition (see Proposition 2.9), $d_0 \circ \theta = \theta \circ \partial_A$, so for $x \in X_{n,*}$ $(n \geq 2)$ we have $d_1(x^{(0)}) = d_1 \langle \theta(\partial_A(x)) \rangle = \langle d_0 \theta(\partial_A(x)) \rangle = \langle \theta(\partial_A^2(x)) \rangle = 0$, while $\varepsilon(d_1(x^{(0)}))$ is a ∂_A-boundary for $x \in X_{1,*}$ (where $\varepsilon : W_{0,*} \to A_*$ is the augmentation). Thus Lemma 2.12 below implies that for $x \in X_{n,*}$ $(n \geq 1)$ we have $d_i(x^{(0)}) = 0$ for all $1 \leq i \leq n-1$, while $d_n(x^{(0)})$ is a ∂_W-boundary.

Therefore, if we set $C^{(0)}_{n,*} := \coprod_{k=1}^{\infty} \coprod_{x \in X_{n,k}} \mathcal{S}^k_{(x^{(0)})}$ for all $n \geq 0$, we see $\{H'_*(C^{(0)}_{n,*})\}_{n=1}^{\infty}$ is an $\mathcal{L}$-CW basis for $H'_*(C_{\bullet,*})$.

In order to give an explicit description of $C_{\bullet,*}$ in terms of $C^{(0)}_{\bullet,*}$, we need to know the Lie disks in which $d_n(x^{(0)})$ (and their faces) lie. By a double induction on $n \geq 1$ and $1 \leq r \leq n$, we shall now define, for all $x \in X_{n,k}$, elements $x^{(r)} \in W_{n-r,k+r}$ such that $\partial_W(x^{(r)}) = d_{n-r}(x^{(r-1)})$:

Note that for each $x \in A_{n,*}$ we have

$$\partial_A(x) = \sum_t a_t \omega_t[y_{i_1}, \ldots, y_{i_{m_t}}],$$

where $\omega_t[\ldots]$ is some m_t-fold iterated Lie bracket, $y_{i_j} \in X_{n_j,*}$ with $\sum_{j=1}^{m_t} n_j = n$, and $a_t \in \mathbb{Q}$. Then

$$\theta(x) = \langle \theta(\partial_A(x)) \rangle = \langle \sum_t a_t \omega_t [\![y^{(0)}_{i_1}, \ldots, y^{(0)}_{i_{m_t}}]\!] \rangle, \tag{5.10}$$

where $\omega_t[\![\ldots]\!]$ is the same m_t-fold iterated Lie bracket as above, but now with respect to the simplicial Lie bracket $[\![\,,\,]\!]$, rather than $[\,,\,]$.

If we set $x^{(s)} = 0$ for $i > n$, we may define $x^{(s)}$ for $0 < s \leq n$ inductively by:

$$x^{(s)} = \langle \sum_t a_t \sum_{\substack{r_1+\cdots+r_{m_t}=s \\ 0 \leq r_j}} \omega_t [\![y^{(r_1)}_{i_1}, \ldots, y^{(r_{m_t})}_{i_{m_t}}]\!] \rangle \in C^{(s)}_{n-s,k-n+s}. \tag{5.11}$$

Thus if we assume by induction that we have chosen $y^{(r_j)}_{i_j}$ with $\partial_W(y^{(r_j)}_{i_j}) = d_{n_{j_0}}(y^{(r_j-1)}_{i_j})$, it follows from Lemma 2.12 below that indeed $\partial_W(x^{(s+1)}) = d_n(x^{(s)})$ and $d_i(x^{(s)}) = 0$ for $0 < i < n$.

For example, $y^{(0)} = \langle y \rangle$ and $\varepsilon(\langle y \rangle) = y \in A_k$ for any $y \in X_{0,k}$. Therefore, for $x \in X_{1,*}$ we have $\varepsilon d_1(x^{(0)}) = \varepsilon d_0(x^{(0)}) = \partial_A(x)$, so we may set $x^{(1)} = \langle x \rangle \in W_{1,k+1}$, with $\partial_W(x^{(1)}) = d_1(x^{(0)})$.

Now if we define by induction

$$C^{(r)}_{n,*} := C^{(r-1)}_{n,*} \amalg \coprod_{k=1}^{\infty} \coprod_{x \in X_{n+r,k}} \mathfrak{D}k + r_{(x^{(r)})},$$

then it is not hard to see that $C_{\bullet,*}$ is the sub-simplicial graded Lie algebra of $W_{\bullet,*}$ generated (under the degeneracies of $W_{\bullet,*}$) by $(C^{(r)}_{n,*})^n_{r=0}$ for all $n \in \mathbb{N}$. (In particular, this is closed under face maps and includes $\operatorname{Im}(\theta)$, and $\theta : A_{*,*} \to C_{\bullet,*}$ is a weak equivalence. The only non-degenerate Lie spheres in $C_{\bullet,*}$ are those of $C^{(0)}_{\bullet,*}$, as required. □

5.12. The filtered model

If $B = (B_*, \partial_B) \in d\mathcal{L}$ is an arbitrary DGL, it no longer has a bigraded model, as in §5.7 above. However, it does have a *filtered model*, constructed as follows:

Let $L_* := H'_*(B)$ be the homotopy Lie algebra of B_*, and $(A_{*,*}, \partial_A)$ the bigraded model for $(L_*, 0)$. The filtered model for B is the free graded Lie algebra $A_{*,*}$, equipped with an increasing filtration $0 = \mathfrak{F}^{-1}(A) \subset \mathfrak{F}^0(A) \subset \cdots \mathfrak{F}^r(A) \subset \mathfrak{F}^{r+1}(A) \subset \cdots$ (defined by $\mathfrak{F}^r(A) := \bigoplus_{i=0}^r A_{i,*}$), and a new differential $D_A =$

$\partial_A + \delta_A$ on $A_{*,*}$ such that $\delta_A : A_{n,*} \to \mathcal{F}^{n-2}(A)$. (Of course, D_A is still required to be a derivation.)

We may decompose $D_A : A_{n,*} \to A_{*,*}$ as $D_A = \partial_0 + \partial_1 + \cdots + \partial_{n-1}$, where $\partial_r : A_{n,*} \to A_{n-r-1,*}$ (and $\partial_0 = \partial_A$, the original differential of the bigraded model).

See [O, II] or [Har]; this is again the Lie algebra version of a construction of Halperin-Stasheff and Félix in [HS, §4],[F1].

Note that the filtered model is no longer unique, since its construction depends on choices; in particular, it is not necessarily minimal. One again has the associated DGL (A_*, D_A), which is quasi-isomorphic to the original DGL B, and $A_{*,*}$ is obtained by filtering A_*.

5.13. Proposition. *Let $B = (B_*, \partial_B)$ be a DGL, and $(A_{*,*}, D_A)$ a filtered model for B; then there is an $\mathcal{M}_{d\mathcal{L}}$-free simplicial resolution $E_{\bullet,*} \to B$, with a bijection $\theta : X_{**} \hookrightarrow E_{\bullet,*}$ between a bigraded set X_{**} of generators for $A_{*,*}$ and the set of non-degenerate $d\mathcal{L}$-spheres in $E_{\bullet,*}$.*

Proof. We start with the minimal $\mathcal{M}$-free resolution $C_{\bullet,*} \to L_*$ for $L_* = H'_*(B_*)$, constructed as in the proof of Proposition 5.8, and deform it into an $\mathcal{M}$-free resolution for B, using the filtered model $(A_{*,*}, D_A)$ as a guideline. This time we shall embed the resulting $\mathcal{M}$-free resolution in the canonical free DGL resolution $W_{\bullet,*}$ of (A_*, D_A), the DGL associated to the filtered model:

For each $x \in X_{n,k}$ (where X_{**} is a bigraded set of generators for the bigraded Lie algebra $A_{*,*}$, as above), set $x^{(n)} = \langle x \rangle \in W_{0,k}$, and let $D_A(x) = \partial_0(x) + \partial_1(x) + \cdots + \partial_{n-1}(x)$ as above, with

$$\partial_r(x) = \sum_t a_t^{(r)} \omega_t^{(r)} [y_{i_1}, \ldots, y_{i_{m_t}}] \in A_{n-r-1,*},$$

where $\omega_t^{(r)}[\ldots]$ is some m_t-fold iterated Lie bracket, as above, and each $y_{i_j} \in X_{n_j,*}$ with $\sum_{j=1}^{m_t} n_j = n - r - 1$.

If we set $x^{(s)} = 0$ for $i > n$, we may define $x^{(s)}$ for $0 < s \leq n$ inductively by:

$$(5.14) \quad x^{(s)} = \langle \sum_{r=0}^{s} \sum_t a_t^{(r)} \sum_{\substack{r_1 + \cdots + r_{m_t} = s - r \\ 0 \leq r_j}} \omega_t^{(r)} [\![y_{i_1}^{(r_1)}, \ldots, y_{i_{m_t}}^{(r_{m_t})}]\!] \rangle \in C_{n-s,k-n+s}^{(s)}$$

Using Lemma 2.12 and the fact that for any $A_\bullet \in s\mathcal{L}$, $x \in A_p$ and $y \in A_q$ we have $d_{p+q}([\![x, y]\!]) = [\![d_p(x), y]\!] + (-1)^q [\![x, d_q(y)]\!]$, one may then verify inductively that $d_{n-s}(x^{(s)}) = \partial_W(x^{(s+1)})$ and $d_i(x^{(s)}) = 0$ for $0 < i < n - s$, for all $0 \leq s < n$. The rest of the construction is as in the proof of Proposition 5.8. □

We have the following analogue of Definition 4.4:

5.15. Definition. Given a DGL $L = (\mathbb{L}\langle X_* \rangle, \partial_L) \in d\mathcal{L}$ and a simplicial set $A \in \mathcal{S}$, we define their *half-smash* $L \rtimes A = (\mathbb{L}\langle Y_* \rangle, \partial') \in d\mathcal{L}$ by setting $Y_n := \coprod_{k=0}^{n} X_k \times$

$\hat{A}_{n-k}$, where $\hat{A}_i$ denotes the set of non-degenerate i-simplices of A. For $a \in \hat{A}_k$ and $x \in X_m$, we set

$$\partial'(x,a) = \sum_{i=0}^{k}(-1)^{i+m}(x, d_i a) + (\partial_L x, a)$$

(and extend ∂' by requiring that it be a derivation).

5.16. *Remark.* In order to apply the obstruction theory described in §4.15, note that all the definitions of Section 4 pass over to the DGL setting in a straightforward manner. However, if we now start with the trivial DGL $A = L_*^{(0)} := (\Pi_*^X, 0)$, we may take $C_{\bullet,*} \to L_*^{(0)}$ to be the minimal $\mathcal{M}$-free resolution of Proposition 5.8, corresponding to the bigraded model for $(\Pi_*^X, 0)$, and let $B = (B_*, \partial_B)$ (corresponding to $B_\bullet$ in §4.15) be the filtered model for L_X. We assume that $A \not\simeq B$.

As explained in §4.15, there is a least $n_0 \geq 2$ such that $0 \notin \langle\langle n_0 \rangle\rangle \subseteq H'_*(B)$, and we write $\langle\langle n_0 \rangle\rangle = (\langle\langle n_0, x_i \rangle\rangle)_{i \in I}$, in the notation of 4.9, where $x_i \in X_{n_0,i}$ and $\mathcal{S}^{k_i}_{(x_i)}$ are corresponding DGL spheres in $C_{n_0,*}$ (we include in the index set I only those coordinates of (4.11) which do not vanish).

Again let $\mathcal{H}^{(1)}$ denote the set of all homotopy types in $\mathrm{ho}\, d\mathcal{L}(\mathbf{X})$ for which $\langle\langle n_0 \rangle\rangle$ has the same value as for L_X, and choose a representative $L^{(1,\alpha)} \in d\mathcal{L}(\mathbf{X})$ for each $\alpha \in \mathcal{H}^{(1)}$. By [HS, §3], we may assume $L^{(1,\alpha)}$ is obtained from B by perturbation of ∂_B. Proceeding as in §4.15 we obtain a tree of DGLs $L^{(k,\alpha_1,\alpha_2,\ldots,\alpha_k)} \in d\mathcal{L}(\mathbf{X})$, and by [Bl1, Theorem 3.1], we know that $L^{(k,\alpha_1,\alpha_2,\ldots,\alpha_k)}$ may be chosen to agree with L_X through degree $n+1$ at least, so $\mathrm{colim}_n\, L^{(n)} \simeq L_X$ along any branch of the tree.

Note also that because $H'_*(C_{\bullet,*})$ is a (minimal) CW resolution of $H'_*(B)$, in each case, the maps $\psi_n : H'_*(C_{n,*}) \to H'_*(B)$ are null for all $n \geq 1$ (see §4.2). Thus any $\partial D[n]$-compatible collection $\{h_k\}_{k=0}^{n-1}$ in §4.9 induce a map $C_{n,*} \wedge \partial D[n] \to B$ directly, without need of the splitting (4.10).

5.17. *Remark.* The second order operation described in the previous example is actually a secondary Whitehead product. Unlocalized higher order Whitehead products were defined by G. Porter in [P, 1.3], and the relation between this definition and the rational version has been studied by several authors – see [AA], [A1, A2], [R2, R1] and [T, V.1]. However, there are other higher order rational homotopy operations, too: For example, in the DGL $L_* = (\mathbb{L}\langle a_1, b_1, c_1, d_1, x_4, y_4, z_4, w_4\rangle, \partial)$, with $\partial(x) = [[b,a],c]$, $\partial(y) = [[b,a],d]$, $\partial(z) = [[d,c],a]$ and $\partial(w) = [[d,c],b]$, the cycle $[x,d] + [y,c] + [z,b] + [w,a]$ represents such an operation. There appears to be no general procedure for representing these as *integral* higher order operations in $\pi_*\mathbf{X}$; we shall offer a (partial) answer to this difficulty in Section 7.

6. Homology of DGLs

Obstructions in algebraic topology traditionally take values in suitable cohomology groups. In order to show that this holds in our setting, too, we recall Quillen's definition of homology and cohomology in model categories:

6.1. Definition. An object X in a category $\mathcal{C}$ is said to be *abelian* if it is an abelian group object – that is, if $\mathrm{Hom}_{\mathcal{C}}(Y,X)$ has a natural abelian group structure for any $Y \in \mathcal{C}$. When $\mathcal{C}$ is $\mathcal{L}ie$, $\mathcal{A}lg$, $s\mathcal{L}ie$, $s\mathcal{A}lg$, $\mathcal{L}$, or $d\mathcal{L}$, for example, this is equivalent to requiring that all products vanish in X (cf. [BS, §5.1.3]).

The full subcategory of abelian objects in $\mathcal{C}$ is denoted by $\mathcal{C}_{ab} \subset \mathcal{C}$. In the cases of interest to us, this will itself be an abelian category. It is equivalent to the category $\mathcal{V}ec$ of vector spaces if $\mathcal{C} = \mathcal{L}ie$ or $\mathcal{A}lg$, to $\mathcal{V}$ if $\mathcal{C} = \mathcal{L}$, to the category $s\mathcal{V}ec$ of simplicial vector spaces if $\mathcal{C} = s\mathcal{L}ie$ or $s\mathcal{A}lg$, and to the category $d\mathcal{V}$ if $\mathcal{C} = d\mathcal{L}$ (see §1.1).

In these cases, we have an *abelianization* functor $Ab : \mathcal{C} \to \mathcal{C}_{ab}$, along with a natural transformation $\theta : Id \to Ab$ having the appropriate universal property. In all the examples above, $Ab(X) = X/I(X)$, where $I(X)$ is the ideal in $X \in \mathcal{C}$ generated by all non-trivial products.

6.2. Definition. Let $\mathcal{C}$ be a category as above, which also has a closed model category structure: in [Q1, II, §5] (or [Q4, §2]), Quillen defines the *homology* of an object $X \in \mathcal{C}$ to be the total left derived functor $\mathbf{L}(Ab)$ of Ab, applied to X (cf. [Q1, I, §4]).

In more familiar terms, this means that we construct a resolution $A \to X$ (i.e., replace X by a weakly equivalent cofibrant object $A \in \mathcal{C}$), and then define the i-th homology group of X by $H_i X := \pi_i(Ab(A))$, for an appropriate concept of homotopy groups π_* in $\mathcal{C}_{ab}$ (see [Q1, II, §4]). One must verify, of course, that this definition is independent of the choice of the resolution $A \to X$.

Similarly, the *cohomology* of X with coefficients in $M \in \mathcal{C}_{ab}$ is defined:

$$H^i(X;M) := [\mathbf{L}(Ab)X, \Omega^{i+N}\Sigma^N M]_{\mathcal{C}_{ab}} \qquad \text{for } N \qquad \text{large enough}$$

(where the loop and suspension functors Ω and Σ are defined in [Q1, I, §2]).

Again, in the cases that interest us, Ω is essentially the shift operator Σ^{-1} of §1.1, and so the i-th cohomology group of X with coefficients in M is then $H^i(X;M) := [\Sigma^i Ab(A), M]_{\mathcal{C}_{ab}}$.

6.3. Definition. If $\mathcal{C}$ itself does not have a closed model category structure, one often defines the homology of $\mathbf{X} \in \mathcal{C}$ by embedding $\mathcal{C}$ in some category which does have such a structure, which in many cases may be taken to be $s\mathcal{C}$, the category of simplicial objects over $\mathcal{C}$ (see [Q1, II, §4]). Thus, if $\iota : \mathcal{C} \hookrightarrow s\mathcal{C}$ is the embedding of categories defined by taking $\iota(C)$ to be the constant simplicial object equal to C in all dimensions, then $H_i(C) := \pi_i(\mathbf{L}(Ab \circ \iota)C)$.

This is the approach usually taken for $\mathcal{C} = \mathcal{L}ie$, $\mathcal{A}lg$, or $\mathcal{L}$: to define the homology of a graded Lie algebra $L_* \in \mathcal{L}$, say, one chooses a free simplicial resolution $A_{\bullet,*} \to L_*$ (such as the canonical resolution – cf. §5.3), and then calculate the homotopy groups of the simplicial graded vector space $Ab(A_{\bullet,*}) \in s\mathcal{V}$ (or the homology groups of the bigraded chain complex in $\mathrm{db}\,\mathcal{V}$ corresponding to $Ab(A_{\bullet,*})$ – see the proof of Proposition 2.9).

6.4. *Remark.* Note that if we apply Definition 6.2 as is to a DGL $L = (L_*, \partial_L) \in d\mathcal{L}$, we may take the resolution A to be the minimal model $\hat{L} = (\hat{L}_*, \hat{\partial})$ for L (cf. §2.5), and since its abelianization is just the graded vector space $Q(\hat{L})$ of indecomposables, and $Q(\hat{\partial}) = 0$ by definition, $H_i(L)$ would be isomorphic to the vector space spanned by a set of generators for $\hat{L}$ in dimension i.

If we want cohomology with coefficients in an object $M_* \in d\mathcal{L}_{ab} \approx d\mathcal{V}$ with *trivial* differential – i.e., M_* is just a graded vector space – we find

$$H^i(X; M_*) \cong \prod_{j=1}^{\infty} \mathrm{Hom}_{\mathcal{V}ec}(H_j(X),\ M_{i+j}),$$

by the universal coefficients theorem.

However, since L is itself graded, we would like H_*L to be bigraded (with a "homological" degree, as well as a "topological" one). This requires a combination of Definitions 6.2 and 6.3, as follows:

6.5. Definition. The *homology* $H_{*,*}(L_\bullet)$ of a simplicial Lie algebra $L_\bullet \in s\mathcal{L}ie$ is defined to be the left derived functors of the abelianization, with respect to the E^2-closed model category structures (§3.3) on $ss\mathcal{L}ie$ and $ss\mathcal{L}ie_{ab} \approx dd\mathcal{V}$ respectively. More precisely,

$$H_{s,t}(L_\bullet) := \pi_s(\mathbf{L}(Ab \circ \iota)L_\bullet)_t = \pi_s\pi^i_t(AbA_{\bullet\bullet}), \tag{6.6}$$

where $A_{\bullet\bullet} \to L_\bullet$ is some $\mathcal{M}$-free bisimplicial resolution of $L_\bullet$.

Similarly, for any DGL $L \in d\mathcal{L}$ we may define $H_{s,t}(L) := \pi_s H^{\prime}_t(Ab(A_{\bullet,*}))$, for a $\mathcal{M}_{d\mathcal{L}}$-free simplicial resolution $A_{\bullet,*} \to L$; and these two definitions agree under the equivalence of homotopy categories $\mathrm{ho}(s\mathcal{L}ie) \approx \mathrm{ho}(d\mathcal{L})$ of Proposition 2.2.

The bigraded cohomology of a DGL L with coefficients in the abelian DGL (i.e., chain complex) M is defined analogously as

$$H^s_t(L; M) := \pi^s(\mathrm{Hom}_{d\mathcal{L}_{ab}}(Ab(A_{\bullet,*}), M)_t).$$

We note that the homology and cohomology of differential graded (commutative) algebras have been defined by Goodwillie (in [Go]) and Burghelea & Vigué-Poirrier (in [BV]), in a manner analogous to the traditional definitions of Hochschild homology. See [Lo, §5.3].

6.7. Proposition. *For any DGL $L \in d\mathcal{L}$, there is a monomorphism of graded vector spaces $H_{n,t}(L) \hookrightarrow H_{n,t}(L')$, where $L' \simeq (H^{\prime}_*(L), 0)$ is the coformal model for L; the same holds for cohomology with trivial coefficients.*

Proof. If $A_{*,*}$ is the bigraded model for L', and $C_{\bullet,*} \to L'$ the simplicial resolution of Proposition 5.8, then the non-degenerate spheres $\mathcal{S}^k_{(x^{(0)})} \subset C_{n,t}$, which correspond to a vector space basis for $H_{n,t}(L')$, are in bijective correspondence with the generators $x \in X_{n,t}$ for $A_{*,*}$.

Now let $B_{*,*}$ be a filtered model for L obtained by perturbing $(A_{*,*}, \partial_A)$, and $E_{\bullet,*} \to L$ the associated simplicial resolution of Proposition 5.13: since $B_{*,*}$ need no longer be minimal (§5.12), a vector space basis for $H_{n,t}(L)$ now corresponds to a *subset* of the collection of non-degenerate spheres $\mathcal{S}^k_{(x^{(0)})} \subset E_{n,t}$, (which are still in bijective correspondence with the generators $x \in X_{n,t}$ for $A_{*,*}$ or $B_{*,*}$). □

Note that this description of the homology implies that $H_{*,*}(L)$ is indeed just a bigraded version of the DGL homology defined in §6.3.

6.8. Proposition. *The collection of higher operations $\langle\langle n_{k,\alpha_1,\alpha_2,\ldots,\alpha_k}, x\rangle\rangle$ which determine the rational homotopy type of $\mathbf{X} \in \mathcal{T}_1$ (described in §4.15 above) are indexed by elements $x \in H_{n_{k,\alpha_1,\alpha_2,\ldots,\alpha_k},t}(L^{(k,\alpha_1,\alpha_2,\ldots,\alpha_k)})$ in the homology of the DGLs of §5.16, and take value in the cohomology of these DGLs, with*

$$\langle\langle n_{k,\alpha_1,\alpha_2,\ldots,\alpha_k}, x\rangle\rangle \subseteq H^{n_{k,\alpha_1,\alpha_2,\ldots,\alpha_k}}_{t+n_k-1}(L^{(k,\alpha_1,\alpha_2,\ldots,\alpha_k)}; \pi_*\mathbf{X}_{\mathbb{Q}}).$$

Proof. We may construct a simplicial resolution $E_{\bullet,*}$ for each successive DGL $L^{(k)} = L^{(k,\alpha_1,\alpha_2,\ldots,\alpha_k)}$, corresponding to the filtered models obtained as perturbations $(A_{*,*}, D_A)$ of the bigraded model $(A_{*,*}, \partial_A)$ for $L^{(0)}$, as above. The non-degenerate spheres $\mathcal{S}^m_{(x)} = \mathcal{S}^m_{(x^{(0)})} \subset E_{m,t}$ which index the higher homotopy operations $\langle\langle m, x\rangle\rangle$ are thus in bijective correspondence with the generators $x \in X_{m,t}$ for $A_{*,*}$. However, if x is not minimal – in the sense that $D_A(x) \notin [A, A]$, or $x + \alpha = D_a(y)$ for some $\alpha \in A_{*,*}$ and $y \in X_{m+1,t}$ – then we can construct a new simplicial resolution $E'_{\bullet,*}$ of $L^{(k)}$ in which $\mathcal{S}^m_{(x)}$ has been eliminated (though of course new spheres may appear in higher simplicial dimensions). By the universal property of resolutions (i.e., of cofibrant objects in the E^2 model category for $sd\mathcal{L}$ – see §3.3) there is a map of resolutions $E_{\bullet,*} \to E'_{\bullet,*}$, and there can be no non-vanishing higher operation $\langle\langle n_{k,\alpha_1,\ldots,\alpha_k}, x\rangle\rangle$ which serves as an obstruction to rectifying the augmentation up-to-homotopy $\varphi : E_{\bullet,*} \to L^X$, since $\varphi_n|_{\mathcal{S}^m_{(x)}}$ can be factored through $0 \in E'_{\bullet,*} \to L^X$. Thus the only homotopy operations which can appear are those corresponding to *non-trivial* homology classes in $H_*(L^{(k,\alpha_1,\alpha_2,\ldots,\alpha_k)})$. These yield the requisite cohomology classes by Universal Coefficients. □

Proposition 6.7 thus implies that we may if we like think of all the higher homotopy operations described in §4.15 (associated to the various deformations of $L^{(0)}$) as lying in one fixed bigraded group $H^*_*(L^{(0)}; \pi_*\mathbf{X}_{\mathbb{Q}})$, which is of course just the usual cohomology of a graded Lie algebra, and is easier to compute than the cohomology of a non-trivial DGL.

7. Non-associative algebra models

The DGL higher homotopy operations are unsatisfactory from a topological point of view because there is no obvious way to translate them, in general, into *unlocalized* topological homotopy operations. We now describe an algebraic model for rational homotopy theory which may serve to answer this objection.

7.1. Non-associative graded algebras

Let $\mathcal{A}$ denote the category of *non-associative graded algebras*: an object $A_* \in \mathcal{A}$, is just a graded vector space equipped with a bilinear graded product $A_p \otimes A_q \to A_{p+q}$. Let $\mathbb{A}\langle X\rangle$ denote the free non-associative graded algebra generated by a graded set X_*. As in §2.1, the functor $\mathbb{A} : gr\mathcal{S}et \to \mathcal{A}$ factors through $A : \mathcal{V} \to \mathcal{A}$.

$d\mathcal{A}$ will denote the category of differential graded non-associative algebras (A_*, ∂_A), called *DGNAs*; the differential ∂_A must satisfy $\partial_A \circ \partial_A = 0$ and $\partial_A(x \cdot y) = \partial_A x \cdot y + (-1)^{|x|} x \cdot \partial_A y$, as for DGLs.

For simplicity we assume each $A_* \in \mathcal{A}$, $d\mathcal{A}$ is connected – that is, $A_0 = \{0\}$. Again, we have a category $\mathrm{db}\,\mathcal{A}$ of differential bigraded non-associative algebras (*DBGNA*s), as in §2.6.

As for any CUGA (§3.1), one can define a closed model category structure on $s\mathcal{A}$ (see [Q1, II, §4]) and thus by [Bl4, §4] on $\mathrm{db}\,\mathcal{A}$, and one has the expected analogues of Propositions 2.2 and 2.9:

7.2. Proposition. *There are adjoint functors* $s\mathcal{A}lg \underset{N^*}{\overset{N}{\rightleftharpoons}} d\mathcal{A}$, *which induce equivalences of the corresponding homotopy categories* $\mathrm{ho}(s\mathcal{A}lg) \approx \mathrm{ho}(d\mathcal{A})$.

7.3. Proposition. *There are adjoint functors* $s\mathcal{A} \underset{N^*}{\overset{N}{\rightleftharpoons}} \mathrm{db}\,\mathcal{A}$, *which induce equivalences of the corresponding homotopy categories* $\mathrm{ho}(s\mathcal{A}) \approx \mathrm{ho}(\mathrm{db}\,\mathcal{A})$; *and* N^* *takes free DBGNAs to free simplicial non-associative algebras.*

7.4. Notation. For any $(X_*, \cdot) \in \mathcal{A}$, let $[x, y]$ denote $\frac{1}{2}(x \cdot y + (-1)^{|x||y|+1} y \cdot x)$. We then have $[y, x] = (-1)^{|x||y|+1}[x, y]$, so $(X_*, [\,,\,])$ is now a non-associative graded algebra with a graded-commutative (or: graded skew-symmetric) multiplication. Moreover, any graded derivation ∂ on $(X_*, \cdot)$ is also a derivation with respect to $[\,,\,]$, and any morphism of algebras from $(X_*, \cdot)$ to a graded-commutative algebra will also respect $[\,,\,]$. Therefore we can (and will) assume that our non-associative algebras are all graded-commutative, and denote the product by $[\,,\,]$.

Moreover, if $A_\bullet \in s\mathcal{A}$ is a simplicial graded algebra, we shall also use the notation

$$[\![x, y]\!] = \sum_{(\sigma,\tau)\in S_{p,q}} (-1)^{\varepsilon(\sigma)+p|y|} [s_{\tau_q} \dots s_{\tau_1} x, s_{\sigma_p} \dots s_{\sigma_1} y]$$

for the corresponding simplicial bracket (compare (2.10)).

7.5. Definition. Any simplicial Lie algebra $L_\bullet \in s\mathcal{L}ie$ is in particular an object in $s\mathcal{A}lg$; let $\iota : d\mathcal{L} \hookrightarrow d\mathcal{A}$ be the inclusion functor. Note that even if each L_n is free as a Lie algebra, it is not free as a non-associative algebra: a free simplicial resolution $J_\bullet \to \iota(L_\bullet)$ in the category $s\mathcal{A}lg$ (see §3.2) will be called a *$s\mathcal{A}lg$-model* for $L_\bullet$. Such models can be constructed functorially, for example by a variant of §5.5.

There is also the analogous concept of a *$d\mathcal{A}$-model* $J_* \in d\mathcal{A}$ of a DGL L; we can of course translate back and forth between these two types of models using Proposition 7.2.

Since the DGL $L = (L_*, \partial_L)$ has an internal grading, and its $d\mathcal{A}$-model $J = (J_*, \partial_J)$ is constructed as a resolution of L, it is natural to define a second "homological" degree on J_*, so as to have a *filtered* $d\mathcal{A}$-model (cf. §5.12). If the DGL is trivial (i.e., $\partial_L = 0$), a $d\mathcal{A}$-model for $(L_*, 0)$ will be a differential bigraded non-associative algebra (DBGNA) $J = (J_{*,*}, \partial_J)$ (cf. §5.7).

7.6. *Remark.* Define a *Jacobi algebra* to be a DGNA $J = (J_*, \partial_J) \in d\mathcal{A}$ such that $H'_* J \in \mathcal{L}$. In particular, any $d\mathcal{A}$-model J of a DGL L is a Jacobi algebra, since $H'_* J \cong H'_* L$. We denote by $\mathcal{J} \subset d\mathcal{A}$ the full subcategory of Jacobi algebras. These algebras are clearly related to the strongly homotopy Lie algebras of [SS] (see also [LM]), though in general a Jacobi algebra is just a "Lie algebra up to homotopy".

Note that even though $d\mathcal{A}$ itself is a CUGA, $\mathcal{J}$ apparently is not, and it does not inherit many desirable properties from $d\mathcal{A}$: for example, $\mathcal{J}$ is not closed under the coproduct in $d\mathcal{A}$. However, one still has *free* Jacobi algebras, in the following sense:

7.7. Lemma. *There is a functor $J : d\mathcal{V} \to d\mathcal{A}$, and a natural transformation $\theta : J \to L$ such that:*

(a) *JV_* is free as an algebra, for any chain complex V_*;*
(b) *θ_{V_*} is a surjective quasi-isomorphism;*
(c) *any chain map $\varphi : V_* \to K_*$ (where $V_* \in d\mathcal{V}$ and $K_* \in \mathcal{J}$) extends to a $d\mathcal{A}$ map $\hat{\varphi} : JV_* \to K_*$.*

Proof. As noted above, we may define $J_{*,*} = J(V_*)$ by induction on the homological filtration:

Start with $J_{0,*} = A(V_*)$ (the free non-associative algebra on the differential graded vector space (V_*, ∂_V), with ∂_J extending ∂_V as a derivation), and let $\theta_0 : J_{0,*} \to L(V_*)$ be the obvious surjection, with $K_{0,*} = \mathrm{Ker}(\theta_0)$ a two-sided $d\mathcal{A}$-ideal of $J_{0,*}$.

Choose once and for all some collection of generators $M_0 = \{\mu_i\}_{i\in I}$ for $K_{0,*}$ as a $J_{0,*}$-bimodule: for each choice of a $d\mathcal{V}$-basis $\{x_\gamma\}_{\gamma\in\Gamma}$ for V_* – that is, of a graded vector space basis of the form $\{x_\alpha, \partial_V x_\alpha, x_\beta\}$, with $\partial_V x_\beta = 0$ – we may write each μ_i as some expression $\mu_i(x_{\gamma_{i_1}}, \dots, x_{\gamma_{i_n}})$. If we then choose some other $d\mathcal{V}$-basis $\{x'_{\gamma'}\}_{\gamma'\in\Gamma'}$ for V_*, again we will have $\mu'_i := \mu_i(x'_{\gamma'_{i_1}}, \dots, x'_{\gamma'_{i_n}}) \in K_{0,*}$; define $J_{1,*}$ to be the free non-associative algebra on the DG vector subspace of $K_{0,*}$ spanned by all such "canonical operations" μ'_i $(i \in I)$, for all possible choices of $d\mathcal{V}$-bases $\{x'_{\gamma'}\}_{\gamma'\in\Gamma'}$ for V_*. Again one has the obvious augmentation $J_{1,*} \to J_{0,*}$ to serve as ∂_J (with $\partial_J \circ \partial_J = 0$ by construction), and one takes the kernel $K_{1,*}$ of this augmentation for the next step.

Proceeding in this way we may define the functor J by induction on the homological filtration; if the collections of operations M_n are chosen canonically, the functor itself will be canonical. Properties (a)–(c) are readily verified. □

7.8. Example. Let $L = (\mathbb{L}\langle X_*\rangle, 0)$ be the trivial free DGL on a graded set X_*; in this case the canonical DBGNA model $(J_{*,*}, \partial_J) = J(X_*, 0)$ for L may be described in part as follows:

Let $J_{0,*} = \mathbb{A}\langle X_*\rangle$ (the free non-associative algebra on X_*). Since the Jacobi identity holds in $\mathbb{L}\langle X_*\rangle$, but not in $\mathbb{A}\langle X_*\rangle$, we have $\mu(x,y,z) = [x,[y,z]] - [[x,y],z] + (-1)^{qr}[[x,z],y] \in K_{0,*}$ for all $x, y, z \in X_*$. Thus $J_{1,*}$ will be generated as a $J_{0,*}$-bimodule by the image of $(J_{0,*})^{\otimes 3}$ under the Σ_3-equivariant multilinear map $\lambda_3 : J_{i,p} \otimes J_{j,q} \otimes J_{k,r} \to J_{i+j+k+1,p+q+r}$. Here the symmetric group Σ_n acts on $J_{*,p_1} \otimes \cdots \otimes J_{*,p_n}$ by permutations, and on $J_{*,p_1+\cdots+p_n}$ via the *Koszul sign* homomorphism $\varepsilon_I : \Sigma_n \to \{1,-1\}$ (defined by letting $\varepsilon_I((k,k+1)) = (-1)^{i_k i_{k+1}+1}$ for any adjacent transposition $(k,k+1) \in \Sigma_n$). We set

$$\partial_J(\lambda_3(x \otimes y \otimes z)) = [x,[y,z]] - [[x,y],z] + (-1)^{qr}\ [[x,z],y].$$

However, there are relations among these elements $\lambda_3(x \otimes y \otimes z)$, so we define a Σ_4-equivariant multilinear map $\lambda_4 : J_{i,p} \otimes J_{j,q} \otimes J_{k,r} \otimes J_{\ell,s} \to J_{i+j+k+\ell+2,p+q+r+s}$,

$$\begin{aligned}\partial_J(\lambda_4(x \otimes y \otimes z \otimes w)) = {} & [x, \lambda_3(y \otimes z \otimes w)] + [\lambda_3(x \otimes y \otimes z), w] \\ & - (-1)^{|z||w|}\ [\lambda_3(x \otimes y \otimes w), z] \\ & + (-1)^{|y|(|z|+|w|)}[\lambda_3(x \otimes z \otimes w), y] \\ & - (\lambda_3([x,y] \otimes z \otimes w)\ +\ \lambda_3(x \otimes y \otimes [z,w])) \\ & + (-1)^{|y||z|}\ (\lambda_3([x,z] \otimes y \otimes z) \\ & + \lambda_3(x \otimes z \otimes [y,w])) \\ & - (-1)^{|w|(|y|+|z|)}\ (\lambda_3([x,w] \otimes y \otimes z) \\ & + \lambda_3(x \otimes w \otimes [y,z])).\end{aligned}$$

In fact, one can define a sequence of "higher Jacobi relations" $\lambda_n(x_1 \otimes \cdots \otimes x_n)$, for all $n \geq 3$, which yield an explicit construction of $J(X_*)$ for a the free (graded) Lie algebra LX_*. See [AB], and compare [LM, 2.1].

7.9. $\mathcal{A}$-homotopy operations

One can now apply the theory of Section 4 verbatim to any space $\mathbf{X} \in \mathcal{T}_1$ with $\mathcal{C} = \mathcal{A}lg$ rather than $\mathcal{L}ie$, to obtain a sequence of higher homotopy operations as in §4.15 which determining the rational homotopy type of $\mathbf{X}$ – the only difference being that the simplicial resolutions $C_{\bullet\bullet}$ of the successive simplicial Lie algebras $L_\bullet^{(k)}$ are now $\mathcal{M}_{\mathcal{A}lg}$ free resolutions of $L_\bullet^{(k)}$ in $s\mathcal{A}lg$.

This is the reason that the theory of Section 3 was stated for an arbitrary CUGA, rather than specifically for $\mathcal{C} = \mathcal{L}ie$. The reason that our general theory was stated for *simplicial* rather than *differential graded* universal algebras is that there seems to be no reasonable version of Proposition 7.2 for an arbitrary CUGA.

7.10. Minimal resolutions

To make the construction more accessible, it is again useful to have *minimal* $\mathcal{M}_{Alg}$ resolutions, as in Section 5. For this purpose, we consider a variant of the above approach:

Even though $\mathcal{J}$ does not inherit a closed model category structure from $d\mathcal{A}$, one may define *models* for $\mathcal{J}$, in the sense of §3.3, by letting a *Jacobi sphere* be any $d\mathcal{A}$-model of a $\mathcal{L}$-sphere (§5.2), and more generally let $\mathcal{M}_{\mathcal{J}}$ denote the full subcategory of $\mathcal{J}$ consisting of DGNAs weakly equivalent to objects in $\mathcal{M}_{d\mathcal{L}}$ – i.e., Jacobi models of DGLs which are (up to homotopy) coproducts of $d\mathcal{L}$-spheres.

An $\mathcal{M}_{\mathcal{J}}$ resolution of a DGL L, which we shall call simply a *Jacobi resolution*, is then defined to be a free simplicial resolution of DGNAs $A_{\bullet,*} \to \iota(L)$ (Def. 3.2), with each $A_{n,*} \in \mathcal{M}_{\mathcal{J}}$. Note that such an $A_{\bullet,*} \to \iota(L)$ is at the same time also an $\mathcal{M}_{\mathcal{J}}$-Jacobi resolution of the $d\mathcal{A}$-model J_* of L, and it is usually more convenient to think of it as such.

There is a comonad $F : d\mathcal{A} \to d\mathcal{A}$ as in (5.4), which yields the *canonical Jacobi resolution* $U_{\bullet,*}$ for any $C \in \mathcal{J}$, as in §5.3. Again we may use the notation of §5.6.

One also has an analogue of Propositions 5.8 and 5.13, as follows:

7.11. Proposition. *Let $B = (B_*, \partial_B) \in d\mathcal{L}$ be any DGL, and $(A_{*,*}, D_A)$ a filtered model for B: then there is a Jacobi resolution $J_{\bullet,*} \to \iota(B)$, with a bijection $\theta : X_{**} \hookrightarrow J_{\bullet,*}$ between a bigraded set X_{**} of generators for $A_{*,*}$ and the set of non-degenerate $d\mathcal{A}$-spheres in $J_{\bullet,*}$.*

Proof. Let $G = G_B$ be a $d\mathcal{A}$-model for the DGL B, and $U_{\bullet,*} \to G$ the canonical Jacobi resolution. As in the proof of Proposition 5.8, we may define a map $\theta : A_{*,*} \to U_{\bullet,*}$ inductively by the equation $\theta(x) = \langle\theta(\partial_A(x))\rangle$ (compare (5.9)), and we shall again write $x^{(0)}$ for $\theta(x)$ if $x \in X_{**}$ (a set of generators for $A_{*,*}$), and let $V_{*,*}$ be the bigraded vector space spanned by $\theta(X_{**})$. For simplicity of notation we consider first the case where B has trivial differential and $A_{*,*}$ is bigraded (with $D_A = \partial_A$).

For each $n \in \mathbb{N}$, define the sub-DGNA $J^{(0)}_{n,*}$ of $U_{n,*}$ to be $J(V_{n,*})$, in the notation of Lemma 7.7 – that is, $J^{(0)}_{n,*}$ is the coproduct, in $\mathcal{J}$, of a set of Jacobi spheres $\mathcal{S}^k_{(x^{(0)})}$, one for each generator $x \in X_{n,k}$ of $A_{*,*}$. By Lemma 7.7(c), it is enough to define the face and degeneracy maps of $J_{\bullet,*}$ on each x – where we may use the description of §5.6.

Once again, we want $d_i(x^{(0)})$ to be a ∂_U-boundary for each $1 \leq i \leq n$; but the analogues of the elements $x^{(s)}$ of Propositions 5.8 and 5.13 are more complicated, so we need some definitions:

For each $0 \leq s \leq n$, let $\mathcal{K}_{n,s}$ denote the set of all sequences $I = (i_1, \ldots, i_s)$ of integers $1 \leq i_1 < \cdots < i_s \leq n$, corresponding to the s-fold face map $d_I = d_{i_1} \circ \cdots \circ d_{i_s} : \mathbf{n} \to \mathbf{n-s}$ in $\mathbf{\Delta}^{op}$ (compare Definition 4.3 and the proof of Proposition 5.8). Given $I = (i_1, \ldots, i_s) \in \mathcal{K}_{n,s}$, for each $1 \leq j \leq s$ let $I(\hat{j}) := (i_1, \ldots, \widehat{i_j}, \ldots, i_s) \in \mathcal{K}_{n,s-1}$ be obtained from I by omitting the j-th entry. By repeatedly using the identity $d_k d_m = d_{m-1} d_k$ $(k < m)$, we can find a unique $\kappa(j) \in \{1, 2, \ldots, n\}$ such that $d_{\kappa(j)} \circ d_{I(\hat{j})} = d_I$. For each $x \in X_{n,k}$, $0 \leq s \leq n$, and $I \in \mathcal{K}_{n,s}$, we want to choose an element $x^{(s;I)} \in J^{(s)}_{n-s,k-n+s} \subset U_{n-s,k-n+s}$ by induction on $n-s$, starting with $x^{(0,\emptyset)} := x^{(0)} = \theta(x)$, so that:

$$\partial_U(x^{(s;I)}) = \sum_{j=1}^{s} (-1)^j d_{\kappa(j)}(x^{(s-1;I(\hat{j}))}) \tag{7.12}$$

for $s \geq 1$. (The index s is not really needed, since $s = |I|$, but it is useful for keeping the analogy with the notation of (5.11) in mind.)

Note that since $d_0 \circ \theta = \theta \circ \partial_A$ no longer holds here (because d_0 is a morphism in $d\mathcal{A}$, not in $d\mathcal{L}$), it is not generally true that $d_1(x^{(0)}) = 0$ for all $x \in X_{**}$. However, since applying H'_* still yields a CW resolution $H'_* J^{(0)}_{\bullet,*} \to H'_* C = H'_* B$, (where C is the $d\mathcal{A}$-model for the DGL B), we know that $d_i(x^{(0)})$ must be a ∂_U-boundary for each $1 \leq i \leq n$. Thus we can choose an element $x^{(1;1)} \in U_{n-1,k-n+1}$ with $\partial_U(x^{(1;1)}) = d_1(x^{(0)})$ (a special case of (7.12)) – and in fact $x^{(1;1)}$ may be expressed in terms of the "canonical operations" μ_i of Lemma 7.7.

Now let $\partial_A(x) = \sum_t a_t \omega_t[y_{i_1}, \ldots, y_{i_{m_t}}]$ for $y_{i_j} \in A_{n_j,*}$, so

$$\begin{aligned} x^{(0)} &= \langle \sum_t a_t \omega_t [\![y^{(0)}_{i_1}, \ldots, y^{(0)}_{i_{m_t}}]\!] \rangle \\ &= \langle \sum_t a_t \sum_{(J_1,\ldots,J_{m_t})} (-1)^{\varepsilon_t} \omega_t[s_{J_1}(y^{(0)}_{i_1}), \ldots, s_{J_{m_t}}(y^{(0)}_{i_{m_t}})] \rangle \end{aligned} \tag{7.13}$$

as in (5.10), where the $(n-n_j)$-multi-index $J_j \subseteq \{0, 1, \ldots, n-1\}$ is obtained by repeated shuffles, which also determine the sign $(-1)^{\varepsilon_t}$ (see (2.10) ff.). Therefore,

$$d_k(x^{(0)}) = \langle \sum_t a_t \sum_{(I_1,\ldots,I_{m_t})} (-1)^{\varepsilon_t} \omega_t[d_{k-1} s_{J_1}(y^{(0)}_{i_1}), \ldots, d_{k-1} s_{J_{m_t}}(y^{(0)}_{i_{m_t}})] \rangle.$$

The proof of Lemma 2.12 (which is valid in $d\mathcal{A}$, too) implies by induction on $t \geq 2$ that for each summand

$$v_t = \omega_t[s_{J_1}(y^{(0)}_{i_1}), \ldots, s_{J_{m_t}}(y^{(0)}_{i_{m_t}})],$$

there is exactly one $1 \leq j \leq t$ and $0 \leq \ell \leq k-1$ such that

$$\begin{aligned} d_{k-1}(v_t) = \omega_t[& s_{J'_1}(y^{(0)}_{i_1}), \ldots, s_{J'_{j-1}}(y^{(0)}_{i_{j-1}}), \\ & s_{J''_j} d_\ell(y^{(0)}_{i_j}), s_{J'_{j+1}}(y^{(0)}_{i_{j+1}}), \ldots, s_{J'_{m_t}}(y^{(0)}_{i_{m_t}})], \end{aligned}$$

for suitable multi-indices $J'_1, \ldots, J'_{j-1}, J'_{j+1}, \ldots, J'_{m_t}$ and J''_j.

Since $\ell < k$ and $n_j < n$, we may assume by induction that we have defined $y_{i_j}^{(1;\ell)} \in U_{n_j-1,*}$ such that $\partial_U(y_{i_j}^{(1;\ell)}) = d_\ell(y_{i_j}^{(0)})$, and then let $x^{(1;k)}$ be

$$\langle \sum_t \sum_{(J_1,\ldots,J_{m_t})} (-1)^{\varepsilon t}\omega_t[s_{J'_1}(y_{i_1}^{(0)}),\ldots,s_{J'_{j-1}}(y_{i_{j-1}}^{(0)}),$$

$$s_{J''_j}(y_{i_j}^{(1;\ell)}), s_{J'_{j+1}}(y_{i_{j+1}}^{(0)}),\ldots,s_{J'_{m_t}}(y_{i_{m_t}}^{(0)})]\rangle.$$

The rest of the construction of the elements $x^{(s;I)}$, as well as the generalization to the filtered case, is similar to that in the proofs of Propositions 5.8 and 5.13. □

We can now summarize the main result of this paper in the following

7.14. Theorem. *Let* $\mathbf{X}$ *be a simply connected space, and* $\Pi_*^X := \pi_{*-1}\mathbf{X}_{\mathbb{Q}} \in \mathcal{L}_0$ *its rational homotopy Lie algebra. There is a tree* T_X *of DGLs* $L^{(k,\alpha_1,\ldots,\alpha_k)}$*, starting with* $L^{(0)} \simeq (\Pi_*^X, 0)$*, and for each branch* $\alpha_1,\ldots,\alpha_k,\ldots$ *of* T_X*, an increasing sequence of positive integers (or* ∞*)* $(n_k = n_{k,\alpha_1,\ldots,\alpha_k})_{k=1}^{\infty}$ *such that*

(a) $H'_*(L^{(k,\alpha_1,\ldots,\alpha_k)}) \cong \Pi_*^X$;

(b) *The higher homotopy operations* $\langle\langle m\rangle\rangle \subset H_*^m(L^{(k,\alpha_1,\ldots,\alpha_k)};\Pi_*^X)$ *associated to a minimal Jacobi resolution of* $L^{(k,\alpha_1,\ldots,\alpha_k)}$ *as in §4.9, vanish for* $m < n_k$.

(c) *The operation* $\langle\langle n_k\rangle\rangle = \langle\langle n_{k,\alpha_1,\ldots,\alpha_k}\rangle\rangle \subset H_*^{n_k}(L^{(k)};\Pi_*^X)$ *does not vanish (unless* $n_k = \infty$*).*

(d) *For any* α_{k+1} *along the branch of* $(\alpha_1,\ldots,\alpha_k)$*, the DGL* $L^{(k+1,\alpha_1,\ldots,\alpha_k,\alpha_{k+1})}$ *may be chosen to agree with* $L^{(k,\alpha_1,\ldots,\alpha_k)}$ *in degrees* $\le n_k+1$*, so the sequential colimit* $L^{(\infty)} = \operatorname{colim}_k L^{(k)}$*, along any branch, is well defined, and is a DGL model for* $\mathbf{X}$.

The main difference between the construction described here (in $s\mathcal{J}$) and that of Proposition 5.13 is that the "higher order information" in $J_{\bullet,*}$ is no longer concentrated in the last face map $d_n : J_{n,*} \to J_{n-1,*}$. As a result, the higher homotopy operations associated to the (minimal) Jacobi resolution (as in Section 4) are true simplicial operations, which can be translated more directly into topological ones.

References

[AB] R. Adin & D. Blanc, "Resolutions of Associative and Lie algebras", *Can. Math. Bull.* **43** (2000) No. 1, pp. 3–16.

[A1] C. Allday, "Rational Whitehead products and a spectral sequence of Quillen", *Pac. J. Math.* **46** (1973) No. 2, pp. 313–323.

[A2] C. Allday, "Rational Whitehead products and a spectral sequence of Quillen, II", *Houston J. Math.* **3** (1977), No. 3, pp. 301–308.

[AA] P.G. Andrews & M. Arkowitz, "Sullivan's minimal models and higher order Whitehead products", *Can. J. Math.* **30** (1978) No. 5, pp. 961–982.

[BL] H.J. Baues & J.-M. Lemaire. "Minimal models in homotopy theory", *Math. Ann.* **225** (1977), pp. 219–242.

[Bl1] D. Blanc, "Derived functors of graded algebras", *J. Pure Appl. Alg.* **64** (1990) No. 3, pp. 239–262.

[Bl2] D. Blanc, "Higher homotopy operations and the realizability of homotopy groups", *Proc. Lond. Math. Soc. (3)* **70** (1995), pp. 214–240.

[Bl3] D. Blanc, "Homotopy operations and the obstructions to being an H-space", *Manus. Math.* **88** (1995) No. 4, pp. 497–515.

[Bl4] D. Blanc, "New model categories from old", *J. Pure Appl. Math.* **109** (1996) No. 1, pp. 37–60.

[Bl5] D. Blanc, "Algebraic invariants for homotopy types", *Math. Proc. Camb. Phil. Soc.* **27** (1999), No. 3, pp. 497–523.

[Bl6] D. Blanc, "CW simplicial resolutions of spaces, with an application to loop spaces", *Topology & Appl.* **100** (2000), No. 2-3, pp. 151–175.

[Bl7] D. Blanc, "Realizing coalgebras over the Steenrod algebra", *Topology* **40** (2001), No. 5, pp. 993–1016.

[BDG] D. Blanc, W.G. Dwyer, & P.G. Goerss, "The realization space of a Π-algebra: a moduli problem in algebraic topology", to appear in *Topology*.

[BG] D. Blanc & P.G. Goerss, "Realizing simplicial Π-algebras", preprint 2002.

[BM] D. Blanc & M. Markl, "Higher homotopy operations", to appear in *Math. Z.*.

[BS] D. Blanc & C.S. Stover, "A generalized Grothendieck spectral sequence", in N. Ray & G. Walker, eds., *Adams Memorial Symposium on Algebraic Topology, Vol. 1*, Lond. Math. Soc. Lec. Notes Ser. **175**, Cambridge U. Press, Cambridge, 1992, pp. 145–161.

[BV] J.M. Boardman & R.M. Vogt, *Homotopy Invariant Algebraic Structures on Topological Spaces*, Springer-Verlag *Lec. Notes Math.* **347**, Berlin-New York, 1973.

[BF] A.K. Bousfield & E.M. Friedlander, "Homotopy theory of Γ-spaces, spectra, and bisimplicial sets", in M.G. Barratt & M.E. Mahowald, eds., *Geometric Applications of Homotopy Theory, II*, Springer-Verlag *Lec. Notes Math.* **658**, Berlin-New York, 1978, pp. 80–130.

[BV] D. Burghelea & M. Vigué-Poirrier, "Cyclic homology of commutative algebras, I", in Y. Félix, ed., *Algebraic Topology – Rational Homotopy 1986*, Springer-Verlag *Lec. Notes Math.* **1318**, Berlin-New York, 1987, pp. 51–72.

[Do] A. Dold, "Homology of symmetric products and other functors of complexes", *Ann. Math. (2)* **68** (1958), pp. 54–80.

[DKS] W.G. Dwyer, D.M. Kan, & C.R. Stover, "An E^2 model category structure for pointed simplicial spaces", *J. Pure & Appl. Alg.* **90** (1993) No. 2, pp. 137–152.

[F1] Y. Félix, *Dénombrement des types de k-homotopie: théorie de la déformation*, Mem. Soc. Math. France **3**, Paris, 1980.

[F2] Y. Félix, "Modèles bifiltrés: une plaque tournant en homotopie rationelle", *Can. J. Math.* **33** (1981) No. 6, pp. 1448–1458.

[GHT] A. Gómez-Tato, S. Halperin & D. Tanré, "Rational homotopy theory for non simply connected spaces", *preprint* 1996.

[Gd] R. Godement, *Topologie algébrique et théorie des faisceaux*, Act. Sci. & Ind. No. 1252, Publ. Inst. Math. Univ. Strasbourg **XIII**, Hermann, Paris 1964.

[Go] T.G. Goodwillie, "Cyclic homology, derivations, and the free loopspace", *Topology* **24** (1985) No. 2, pp. 187–216.

[HS] S. Halperin & J.D. Stasheff, "Obstructions to homotopy equivalences", *Adv. in Math.* **32** (1979) No. 3, pp. 233–279.

[Har] Y. Haralambous, "Perturbation d'algèbres de Lie différentielles et coformalité modérée", *Bull. Soc. Math. Belg. Sér. B* **43** (1991) No. 1, pp. 59–67.

[J] T. Jozefiak, "Tate resolutions for commutative graded algebras over a local ring", *Fund. Math.* **74** (1972) No. 3, pp. 209–231.

[LM] T.J. Lada & M. Markl, "Strongly homotopy Lie algebras", *Comm. in Alg.* **23** (1995) No. 6, pp. 2147–2161.

[LS] J.-M. Lemaire & F. Sigrist, "Dénombrement de types d'homotopie rationelle", *C. R. Acad. Sci., Paris, I* **287** (1978) No. 3, pp. 109–112.

[Lo] J.-L. Loday, *Cyclic Homology*, Springer-Verlag *Grund. math. Wissens.* **301**, Berlin-New York, 1992.

[Mc1] S. Mac Lane, *Homology*, Springer-Verlag *Grund. math. Wissens.* **114**, Berlin-New York, 1963.

[Mc2] S. Mac Lane, *Categories for the Working Mathematician*, Springer-Verlag *Grad. Texts Math.* **5**, Berlin-New York, 1971.

[Ma] J.P. May, *Simplicial Objects in Algebraic Topology*, U. Chicago Press, Chicago-London, 1967.

[Me] W. Meier, "Minimal models for nonnilpotent spaces", *Comm. Math. Helv.* **55** (1980), No. 4, pp. 622–633.

[MN] T.J. Miller & J A. Neisendorfer. "Formal and coformal spaces", *Ill. J. Math.* **22** (1978) No. 4, pp. 565–580.

[N] J.A. Neisendorfer, "Lie algebras, coalgebras and rational homotopy theory for nilpotent spaces", *Pac. J. Math.* **74** (1978) No. 2, pp. 429–460.

[O] A. Oukili, *Sur l'homologie d'une algèbre différentielle de Lie*, Ph.D. thesis, Univ. de Nice, 1978.

[P] G.J. Porter. "Higher order Whitehead products", *Topology* **3** (1965) pp. 123–165.

[Q1] D.G. Quillen, *Homotopical Algebra*, Springer-Verlag *Lec. Notes Math.* **20**, Berlin-New York, 1963.

[Q2] D.G. Quillen, "Spectral sequences of a double semi-simplicial group", *Topology* **5** (1966), pp. 155–156.

[Q3] D.G. Quillen, "Rational homotopy theory", *Ann. Math.* **90** (1969) No. 2, pp. 205–295.

[Q4] D.G. Quillen, "On the (co-)homology of commutative rings", *Applications of Categorical Algebra*, Proc. Symp. Pure Math. **17**, AMS, Providence, RI, 1970, pp. 65–87.

[R1] V.S. Retakh, "Massey operations in Lie superalgebras and differentials of the Quillen spectral sequence", *Funk. Ann. i Pril.* **12** (1978) No. 4, pp. 91–92 [transl. in *Funct. Anal. & Applic.* **12** (1978) No. 4, pp. 319–321].

[R2] V.S. Retakh, "Lie-Massey brackets and n-homotopically multiplicative maps of differential graded Lie algebras", *J. Pure & Appl. Alg.* **89** (1993) No. 1–2, pp. 217–229.

[SS] M. Schlessinger & J.D. Stasheff. "Deformation theory and rational homotopy type", to appear in *Pub. Math. Inst. Hautes Et. Sci.*.

[St] C.R. Stover, "A Van Kampen spectral sequence for higher homotopy groups", *Topology* **29** (1990), pp. 9–26.

[T] D. Tanré, *Homotopie Rationelle: Modèles de Chen, Quillen, Sullivan*, Springer-Verlag *Lec. Notes Math.* **1025**, Berlin-New York, 1983.

David Blanc
Dept. of Mathematics
University of Haifa
Haifa 31905, Israel
e-mail: `blanc@math.haifa.ac.il`

Progress in Mathematics, Vol. 215
Algebraic Topology: Categorical Decomposition Techniques, 77–92

The Loop Homology Algebra of Spheres and Projective Spaces

Ralph L. Cohen, John D.S. Jones, and Jun Yan

Abstract. In [3] Chas and Sullivan defined an intersection product on the homology $H_*(LM)$ of the space of smooth loops in a closed, oriented manifold M. In this paper we will use the homotopy theoretic realization of this product described by the first two authors in [2] to construct a second quadrant spectral sequence of algebras converging to the loop homology multiplicatively, when M is simply connected. The E_2 term of this spectral sequence is $H^*(M; H_*(\Omega M))$ where the product is given by the cup product on the cohomology of the manifold $H^*(M)$ with coefficients in the Pontryagin ring structure on the homology of its based loop space $H_*(\Omega M)$. We then use this spectral sequence to compute the ring structures of $H_*(LS^n)$ and $H_*(L\mathbb{CP}^n)$.

Introduction

The loop homology of a closed orientable manifold M^d of degree d is the ordinary homology of the free loop space $LM = \mathrm{Map}(S^1, M^d)$, with degree shifted by $-d$, i.e.,

$$\mathbb{H}_*(LM; \mathbb{Z}) = H_{*+d}(LM; \mathbb{Z}).$$

In [3], Chas and Sullivan defined a type of intersection product on the chains of LM, yielding an algebra structure on $\mathbb{H}_*(LM)$.

Roughly, the loop product is defined as follows. Let $\alpha : \Delta^p \to LM$ and $\beta : \Delta^q \to LM$ be singular simplices in LM. The evaluation at $1 \in S^1 \subset \mathbb{C}$ defines a map $ev : LM \to M$. Assume that $ev \circ \alpha$ and $ev \circ \beta$ define a map

$$\Delta^p \times \Delta^q \to LM \times LM \to M \times M$$

that is transverse to the diagonal. At each point $(s,t) \in \Delta^p \times \Delta^q$ where $ev \circ \alpha$ intersects $ev \circ \beta$, one can define a single loop by first traversing the loop $\alpha(s)$ and then traversing the loop $\beta(t)$. This then defines a chain $\alpha \circ \beta \in C_{p+q-d}(LM)$. In [3] Chas and Sullivan showed that this procedure defines a chain map

$$C_p(LM) \otimes C_q(LM) \to C_{p+q-d}(LM)$$

which induces an associative, commutative algebra structure on the loop homology, $\mathbb{H}_*(LM)$. Chas and Sullivan also described other structures this pairing induces,

The first author was partially supported by a grant from the NSF.

such as a Lie algebra structure on the equivariant homology of the loop space. In [2], the first two authors used the Pontryagin–Thom construction to show that the loop product is realized on the homotopy level on Thom spaces (spectra) of bundles over the loop space. In particular let TM denote the tangent bundle of M, and $-TM$ denotes its inverse as a virtual bundle in K-theory. Let M^{-TM} denote the Thom spectrum of this bundle, and LM^{-TM} the Thom spectrum of $ev^*(-TM)$. Then in [2] it was shown that LM^{-TM} is a homotopy commutative ring spectrum with unit, whose product realizes the Chas–Sullivan product in homology, after applying the Thom isomorphism, $H_q(LM^{-TM}) \cong H_{q+d}(LM) \cong \mathbb{H}_q(LM)$.

The goal of this paper is to describe a spectral sequence of algebras converging to the loop homology algebra of a manifold, and to use it to compute the loop homology algebra of spheres and projective spaces. More specifically we shall prove the following theorems.

Theorem 1. *Let M be a closed, oriented, simply connected manifold. There is a second quadrant spectral sequence of algebras $\{E^r_{p,q}, d^r : p \leq 0, q \geq 0\}$ such that*

1. *$E^r_{*,*}$ is an algebra and the differential $d^r : E^r_{*,*} \to E^r_{*-r,*+r-1}$ is a derivation for each $r \geq 1$.*
2. *The spectral sequence converges to the loop homology $\mathbb{H}_*(LM)$ as algebras. That is, $E^\infty_{*,*}$ is the associated graded algebra to a natural filtration of the algebra $\mathbb{H}_*(LM)$.*
3. *For $m, n \geq 0$,*
$$E^2_{-m,n} \cong H^m(M; H_n(\Omega M)).$$
Here ΩM is the space of base point preserving loops in M. Furthermore the isomorphism $E^2_{-*,*} \cong H^*(M; H_*(\Omega M))$ is an isomorphism of algebras, where the algebra structure on $H^*(M; H_*(\Omega M))$ is given by the cup product on the cohomology of M with coefficients in the Pontrjagin ring $H_*(\Omega M)$.
4. *The spectral sequence is natural with respect to smooth maps between manifolds.*

We then use this spectral sequence to do the following calculations. Let $\Lambda[x_1, \ldots, x_n]$ denote the exterior algebra (over the integers) generated by $x_1, \ldots, x_n$, and let $\mathbb{Z}[a_1, \ldots, a_m]$ denote the polynomial algebra generated by $a_1, \ldots, a_m$.

Theorem 2. *There exist isomorphisms of graded algebras,*

1. $$\mathbb{H}_*(LS^1) \cong \Lambda[a] \otimes \mathbb{Z}[t, t^{-1}]$$
where $a \in \mathbb{H}_{-1}(LS^1)$ and $t, t^{-1} \in \mathbb{H}_0(LS^1)$.
2. *For $n > 1$,*
$$\mathbb{H}_*(LS^n) = \begin{cases} \Lambda[a] \otimes \mathbb{Z}[u], & \text{for } n \text{ odd} \\ (\Lambda[b] \otimes \mathbb{Z}[a, v])/(a^2, ab, 2av), & \text{for } n \text{ even,} \end{cases}$$
where $a \in \mathbb{H}_{-n}(LS^n)$, $b \in \mathbb{H}_{-1}(LS^n)$, $u \in \mathbb{H}_{n-1}(LS^n)$, and $v \in \mathbb{H}_{2n-2}(LS^n)$.

Theorem 3. *There is an isomorphism of algebras,*

$$\mathbb{H}_*(L\mathbb{CP}^n) \cong (\Lambda[w] \otimes \mathbb{Z}[c,u])/(c^{n+1}, (n+1)c^n u, wc^n)$$

where $w \in \mathbb{H}_{-1}(L\mathbb{CP}^n)$, $c \in \mathbb{H}_{-2}(L\mathbb{CP}^n)$, *and* $u \in \mathbb{H}_{2n}(L\mathbb{CP}^n)$.

The organization of this paper is as follows. In Section 1 we will review the construction of the loop product that was used in [2] and describe it on the chain level. In Section 2 we will construct the spectral sequence and prove Theorem 1. In Section 3 we use this spectral sequence to do the calculations presented in Theorems 2 and 3.

1. The loop product

For the remainder of the paper, let M be a closed, connected, oriented manifold of dimension d. The goal of this section is to give a description of the loop product on $\mathbb{H}_*(LM)$ at the chain level. Of course this was done originally by Chas and Sullivan in [3], but our approach will be slightly different. It will be more amenable to the construction and the analysis of the loop homology spectral sequence to be done in the next section.

We begin by recalling a chain level description of the intersection product arising from Poincaré duality on the homology of the manifold,

$$\langle , \rangle : H_q(M) \otimes H_p(M) \to H_{p+q-d}(M).$$

For pairs of spaces (X, A) with $A \subset X$, denote by $C_*(X, A)$ and $C^*(X, A)$ the groups of singular chains and cochains.

Recall that the normal bundle of the diagonal embedding $\Delta : M \to M \times M$ is naturally isomorphic to the tangent bundle TM. Let M^{TM} denote the Thom space of this bundle. The Thom–Pontryagin map for the diagonal embedding is therefore a map

$$\tau : M \times M \to M^{TM}, \tag{1.1}$$

which collapses everything outside a tubular neighborhood of the diagonal to the base point of M^{TM}.

Choose a Riemannian metric on TM and let D_TM and S_TM be the unit disk and sphere bundles of TM. Since S_TM has a collar, $S_TM \times I \hookrightarrow D_TM$, there is a chain equivalence

$$\theta_\sharp : C_*(D_TM/S_TM, *) \to C_*(D_TM \cup cS_TM, cS_TM) \to C_*(D_TM, S_TM)$$

where $*$ is the base point of D_TM/S_TM, cS_TM is the cone on S_TM, and $D_TM \cup cS_TM$ is the mapping cone of the inclusion $S_TM \hookrightarrow D_TM$. Let $t \in C^d(D_TM, S_TM)$ be a cochain that represents the Thom class $[t] \in H^d(D_TM, S_TM)$. Then taking the cap product with t at the chain level followed by the projection from D_TM to M,

$$\sigma_\sharp : C_*(D_TM, S_TM) \xrightarrow{\cap t} C_{*-d}(D_TM) \xrightarrow{\pi_*} C_{*-d}(M) \tag{1.2}$$

induces the Thom isomorphism

$$\tilde{H}_*(M^{TM}) \xrightarrow{\theta_*} H_*(D_TM, S_TM) \xrightarrow{\sigma_*} H_{*-d}(M)$$

in homology.

Let $\epsilon_\sharp : C_*(D_TM/S_TM) \to C_*(D_TM/S_TM, *)$ be the projection map onto the quotient complex, and denote by $u_\sharp$ the composition of the chain maps

$$(1.3) \quad u_\sharp : C_*(M^{TM}) \xrightarrow{\epsilon_\sharp} C_*(D_TM/S_TM, *) \xrightarrow{\theta_\sharp} C_*(D_TM, S_TM) \xrightarrow{\sigma_\sharp} C_{*-d}(M).$$

Now let A be a graded ring. Recall that the intersection product structure on $H_*(M; A)$ is defined so that the Poincaré duality isomorphism

$$D : H^*(M; A) \to H_{d-*}(M; A)$$
$$\alpha \to \alpha \cap [M]$$

is an isomorphism of graded algebras. Here $[M] \in H_d(M; A)$ is the fundamental class determined by the orientation of M. The commutativity of the following diagram describes the well-known relation between the intersection product, the Thom–Pontryagin map and the Thom isomorphism:

$$\begin{array}{ccccc} H_{p+q}(M \times M) & \xrightarrow{\tau_*} & H_{p+q}(M^{TM}) & \xrightarrow{u_*} & H_{p+q-d}(M) \\ \times \uparrow & & & & \| \\ H_p(M) \otimes H_q(M) & & \xrightarrow{(-1)^{d(d-p)}\langle\cdot\rangle} & & H_{p+q-d}(M), \end{array}$$

where $\langle\cdot\rangle$ is the intersection product on $H_*(M)$, and $\times$ denotes the cross product. We therefore have the following chain description of the intersection pairing:

Proposition 4. *Let M be as above. Then the following composition of chain maps*

$$C_p(M) \otimes C_q(M) \xrightarrow{\times} C_{p+q}(M \times M) \xrightarrow{\tau_\sharp} C_{p+q}(M^{TM}) \xrightarrow{u_\sharp} C_{p+q-d}(M)$$

is a representative of the intersection product

$$(-1)^{d(d-p)}\langle\cdot\rangle : H_p(M) \otimes H_q(M) \to H_{p+q-d}(M).$$

We next turn our attention to the loop space and the loop product. Let $LM = C^\infty(S^1, M)$ be the free loop space of M, where S^1 denotes the unit circle in the complex line, parametrized by $\exp : [0,1] \to S^1$ with $1 \in S^1$ chosen as the basepoint. We now recall some constructions in [2].

The loop evaluation map

$$ev : LM \to M$$
$$\gamma \to \gamma(1),$$

is a Serre fibration. Let $LM \times_M LM$ denote the pull back of the product of this fibration with itself along the diagonal embedding $\Delta : M \to M \times M$, denoted by

$LM \times_M LM$:

$$\begin{array}{ccc} LM \times_M LM & \xrightarrow{\tilde{\Delta}} & LM \times LM \\ {\scriptstyle \mathrm{ev}_\infty}\downarrow & & \downarrow{\scriptstyle ev \times ev} \\ M & \xrightarrow[\Delta]{} & M \times M. \end{array}$$

Notice that $LM \times_M LM$ is the space of pairs of loops having the same basepoint. The map ev_∞ in this fiber square is given by $\mathrm{ev}_\infty(\alpha, \beta) = \alpha(1) = \beta(1)$. The map $\tilde{\Delta}$ is an embedding of a codimension d infinite-dimensional submanifold of the infinite-dimensional manifold $LM \times LM$.

As shown in [2] this pullback square allows for a Thom–Pontryagin map $\tilde{\tau}: LM \times LM \to (LM \times_M LM)^{TM}$, where $(LM \times_M LM)^{TM}$ denotes the Thom space of the pull-back bundle $\mathrm{ev}_\infty^*(TM)$, which is the normal bundle of the embedding $\tilde{\Delta}$. Notice that we have a commutative diagram of Thom–Pontryagin maps:

$$\begin{array}{ccc} LM \times LM & \xrightarrow{\tilde{\tau}} & (LM \times_M LM)^{TM} \\ {\scriptstyle ev \times ev}\downarrow & & \downarrow{\scriptstyle \mathrm{ev}_\infty} \\ M \times M & \xrightarrow{\tau} & M^{TM}. \end{array}$$

In this diagram the map ev_∞ is the induced map on the Thom spaces.

Loop composition $\alpha\beta$ is defined for two loops α and β having the same base point by first traversing the loop α, then the loop β, i.e.,

$$\alpha\beta(t) = \begin{cases} \alpha(2t) & \text{if } 0 \le t \le \frac{1}{2}, \\ \beta(2t-1) & \text{if } \frac{1}{2} \le t \le 1. \end{cases} \tag{1.4}$$

Denote this operation by

$$\gamma : LM \times_M LM \to LM$$
$$(\alpha, \beta) \to \alpha\beta.$$

Notice that γ preserves the base points of loops in $LM \times_M LM$ and LM, thus the composition

$$LM \times_M LM \xrightarrow{\gamma} LM \xrightarrow{ev} M$$

coincides with ev_∞. Therefore γ induces a map of bundles, $\gamma : \mathrm{ev}_\infty^*(TM) \to ev^*(TM)$, and therefore an induced map of Thom spaces,

$$\tilde{\gamma} : (LM \times_M LM)^{TM} \to LM^{TM}.$$

Putting the above maps together, we obtain the following commutative diagram

$$\begin{array}{ccccc} LM \times LM & \xrightarrow{\tilde{\tau}} & (LM \times_M LM)^{TM} & \xrightarrow{\tilde{\gamma}} & LM^{TM} \\ {\scriptstyle ev \times ev}\downarrow & & \downarrow{\scriptstyle \mathrm{ev}_\infty} & & \downarrow{\scriptstyle ev} \\ M \times M & \xrightarrow{\tau} & M^{TM} & \xrightarrow{=} & M^{TM}. \end{array} \tag{1.5}$$

Let $D_L M$ and $S_L M$ be the pull-backs of $D_T M$ and $S_T M$ via ev. They are the unit disk and sphere bundles of $ev^*(TM)$. Recall that the cochain t in $C^d(D_T M, S_T M)$ represents the Thom class and therefore its pull-back $\tilde{t} = ev^*(t)$ is a cochain in $C^d(D_L M, S_L M)$ representing the Thom class of $ev^*(TM)$. Similar to (1.2), capping with $\tilde{t}$ at the chain level followed by the projection of $D_L M$ to LM

$$\tilde{\sigma}_\sharp : C_*(D_L M, S_L M) \xrightarrow{\cap \tilde{t}} C_{*-d}(D_L M) \xrightarrow{\tilde{\pi}_*} C_{*-d}(LM)$$

induces the Thom isomorphism

$$\tilde{H}_*(LM^{TM}) \xrightarrow{\cong} H_*(D_L M, S_L M) \xrightarrow{\tilde{\sigma}_*} H_{*-d}(LM)$$

in homology. Similar to the argument used for (1.3), there is a chain map

(1.6)

$$\tilde{u}_\sharp : C_*(LM^{TM}) \xrightarrow{\tilde{\epsilon}_\sharp} C_*(D_L M/S_L M, *) \xrightarrow{\tilde{\theta}_\sharp} C_*(D_L M, S_L M) \xrightarrow{\tilde{\sigma}_\sharp} C_{*-d}(LM),$$

such that the following diagram commutes:

$$\begin{array}{ccc} H_*(LM^{TM}) & \xrightarrow{\tilde{u}_*} & H_{*-d}(LM) \\ {\scriptstyle ev_*}\downarrow & & \downarrow{\scriptstyle ev_*} \\ H_*(M^{TM}) & \xrightarrow{u_*} & H_{*-d}(M). \end{array}$$

In [2] the first two authors proved that the composition of $\tilde{\tau}_*$, $\tilde{\gamma}_*$, and $\tilde{u}_*$ realizes the Chas–Sullivan loop product. That is, the following diagram commutes:

(1.7)

$$\begin{array}{ccccc} H_p(LM) \otimes H_q(LM) & & \xrightarrow{(-1)^{d(d-p)}(\circ)} & & H_{p+q-d}(LM) \\ \times\downarrow & & & & \uparrow{\scriptstyle \tilde{u}_*} \\ H_{p+q}(LM \times LM) & \xrightarrow{\tilde{\tau}_*} & H_{p+q}(LM \times_M LM)^{TM} & \xrightarrow{\tilde{\gamma}_*} & H_{p+q}(LM^{TM}) \end{array}$$

where $\circ : H_p(LM) \otimes H_q(LM) \to H_{p+q-d}(LM)$ is the Chas–Sullivan loop product. We therefore have the following proposition, which should be viewed as the analogue of Proposition 4.

Proposition 5. *The composition of the four chain maps*

(1.8) $\times : \quad C_*(LM) \otimes C_*(LM) \to C_*(LM \times LM),$

(1.9) $\tilde{\tau}_\sharp : \quad C_*(LM \times LM) \to C_*(LM \times_M LM)^{TM},$

(1.10) $\tilde{\gamma}_\sharp : \quad C_*(LM \times_M LM)^{TM} \to C_*(LM^{TM}),$ *and*

(1.11) $\tilde{u}_\sharp : \quad C_*(LM^{TM}) \to C_{*-d}(LM).$

gives a chain representative of the Chas–Sullivan loop product up to sign

$$H_p(LM) \otimes H_q(LM) \to H_{p+q-d}(LM)$$
$$\alpha \otimes \beta \to \alpha \circ \beta.$$

2. The loop algebra spectral sequence

In this section we will describe the loop algebra spectral sequence and prove Theorem 1. To do this we describe a filtration of simplicial sets arising from the fibration $\Omega M \hookrightarrow LM \to M$, which will induce the Serre spectral sequence for this fibration. We then analyze how the loop product behaves with respect to this spectral sequence, using the chain level description of the product given in the last section. We then apply the Poincaré duality to regrade the spectral sequence (and in particular change a first quadrant spectral sequence into a second quadrant one), and prove Theorem 1.

Given a topological space X, let $\mathcal{S}.X$ denote its singular simplicial set. The p-simplices are given by singular simplices $S_p(X) = \{\sigma : \Delta^p \to X\}$, where Δ^p is the standard p-simplex, $\Delta^p = \{(x_0, \ldots, x_p) \in \mathbb{R}^{p+1} : x_i \geq 0 \text{ and } \sum x_i = 1\}$. $\mathcal{S}.X$ has the usual face and degeneracy operations, and it is well known that its geometric realization $|\mathcal{S}.X|$ has the weak homotopy type of X. See [4] for details.

Number the vertices of Δ^p $\{0, 1, \cdots, p\}$. Then for a non-decreasing sequence of integers $0 \leq i_0 \leq \ldots \leq i_r \leq p$, define a map of simplices

$$(i_0, i_1, \ldots, i_r) : \Delta^r \to \Delta^p,$$

by requiring that $(i_0, i_1, \ldots, i_r)$ be a linear map that sends the vertex k of Δ^r to the vertex i_k of Δ^p. Given a singular p-simplex $\sigma : \Delta^p \to X$, the composition of σ with $(i_0, i_1, \ldots, i_r)$ defines an r-simplex

$$\sigma(i_0, i_1, \ldots, i_r) : \Delta^r \to \Delta^p \to X.$$

Now let $F \hookrightarrow E \xrightarrow{\pi} B$ be a fibration. There is a filtration of the simplicial set $\mathcal{S}.E$ defined as follows.

Definition 1. Let $F_p(\mathcal{S}.E) \subset \mathcal{S}.E$ be the subsimplicial set whose r simplices are given by

$$F_p(\mathcal{S}_r(E)) = \{T : \Delta^r \to E : \pi \circ T = \sigma(i_0, \ldots, i_r), \text{ for some } \sigma \in \mathcal{S}_q(B),\ q \leq p, \\ \text{and some sequence } 0 \leq i_0 \leq \ldots \leq i_r \leq q\}.$$

Given a simplicial set $Y.$, let $C_*(Y.)$ be the associated simplicial chain complex, whose q-chains $C_q(Y.)$ are the free abelian group on the q-simplices Y_q, and whose boundary homomorphisms are given by the alternating sum of the face maps. Again, see [4] for details. In particular for a space X, we have $C_*(\mathcal{S}.X)$ is the singular chain complex which we previously denoted simply by $C_*(X)$. The following is verified in [6].

Proposition 6. *Let $F \hookrightarrow E \to B$ be a fibration as above. Consider the filtration of chain complexes,*

$$\{0\} \hookrightarrow \cdots \hookrightarrow F_{p-1}(C_*(E)) \hookrightarrow F_p(C_*(E)) \hookrightarrow \cdots \hookrightarrow C_*(E)$$

defined by $F_p(C_*(E)) = C_*(F_p(\mathcal{S}.E))$ where $F_p(\mathcal{S}.E)$ is the pth filtration of the singular simplicial set defined above. Then this filtration induces the Serre spectral sequence converging to $H_*(E)$.

We will study this spectral sequence in the examples of the fibrations

$$\begin{aligned} ev : & LM \to M, \\ \mathrm{ev}_\infty : & LM \times_M LM \to M, \\ ev : & D_L M \to D_T M, \text{ and} \\ ev : & S_L M \to S_T M \end{aligned}$$

described in the last section. In particular notice that by taking the filtration of pairs

$$F_p(C_*(D_L M, S_L M)) = F_p(C_*(D_L M))/F_p(C_*(S_D M))$$

we get a spectral sequence (the relative Serre spectral sequence) converging to

$$H_*(D_L M, S_L M) = \tilde{H}_*(LM^{TM})$$

and whose E_2 term is

$$E^2_{p,q} = H_p(D_T M, S_T M; H_q(\Omega M)) = \tilde{H}_p(M^{TM}; H_q(\Omega M)).$$

Using the fibration

$$\mathrm{ev}_\infty : LM \times_M LM \to M$$

there is a similar filtration

$$F_p(C_*(\mathrm{ev}^*_\infty(D_T M), \mathrm{ev}^*_\infty(S_T M)))$$

which yields a relative Serre spectral sequence converging to

$$\tilde{H}_*((LM \times_M LM)^{TM})$$

and whose E_2 term is

$$\tilde{H}_*(M^{TM}; H_*(\Omega M \times \Omega M)).$$

The following is an immediate observation based on the chain descriptions in the last section.

Proposition 7. *The chain maps*

$$\begin{aligned} \times : &\quad C_*(LM) \otimes C_*(LM) \to C_*(LM \times LM) \quad (1.8), \\ \tilde{\tau}_\sharp : &\quad C_*(LM \times LM) \to C_*(LM \times_M LM)^{TM} \quad (1.9), \quad \textit{and} \\ \tilde{\gamma}_\sharp : &\quad C_*(LM \times_M LM)^{TM} \to C_*(LM^{TM}) \quad (1.10) \end{aligned}$$

described in the last section all preserve the above filtrations:

$$\begin{aligned} \times : &\quad F_p(C_*(LM)) \otimes F_q(C_*(LM)) \to F_{p+q}(C_*(LM \times LM)), \\ \tilde{\tau}_\sharp : &\quad F_m(C_*(LM \times LM)) \to F_m(C_*(ev^*_\infty(D_T M), ev^*_\infty(S_T M))), \\ \tilde{\gamma}_\sharp : &\quad F_m(C_*(ev^*_\infty(D_T M), ev^*_\infty(S_T M))) \to F_m(C_*(D_L M, S_L M)), \end{aligned}$$

and therefore induce maps of the associated Serre spectral sequences.

The following is a bit more delicate.

Theorem 8. *The chain map*

$$\tilde{u}_\sharp : C_*(LM^{TM}) \to C_{*-d}(LM) \ (1.11)$$

induces a map of filtered chain complexes that lowers the filtration by d,

$$\tilde{u}_\sharp : F_p(C_*(D_LM, S_LM)) \to F_{p-d}(C_*(LM)).$$

and therefore induces a map of the associated Serre spectral sequences that shifts grading,

$$\tilde{u}_* : E^r_{p,q}(D_LM, S_LM) \to E^r_{p-d,q}(LM).$$

In particular on the E_2*-level* $\tilde{u}_*$ *is the Thom isomorphism,*

$$H_p(M^{TM}; H_q(\Omega M)) \to H_{p-d}(M; H_q(\Omega M)).$$

Proof. By the definition of the chain map $\tilde{u}_\sharp$, to prove this theorem it suffices to show that taking the cap product with the Thom class $\tilde{t} \in C^d(D_LM, S_LM)$ induces a map of filtered chain complexes that lowers the filtration by d,

$$F_p(C_*(D_LM, S_LM)) \xrightarrow{\cap \tilde{t}} F_{p-d}(C_*(D_LM)).$$

To verify this, recall that the cap product has the following chain level description. Consider the operations on singular n-simplices,

$$\rfloor_p \quad \text{and} \quad {}_p\lfloor : \mathcal{S}_n(X) \to \mathcal{S}_p(X)$$

defined by

$$\sigma\rfloor_p = (d_{p+1})^{n-p}(\sigma), \qquad {}_p\lfloor\sigma = (d_0)^{n-p}(\sigma).$$

That is, $\sigma\rfloor_p$ is the restriction of σ to the "front" p-face, and ${}_p\lfloor\sigma$ is the restriction of σ to the "back" p-face.

Now we can choose our cochain $\tilde{t} \in C^*(D_LM, S_LM)$ to represent the Thom class so that if we view it as an element in $\mathrm{Hom}(C_d(D_LM); \mathbb{Z})$ it satisfies

(2.1) $\qquad C_d(S_LM) \subset \mathrm{Ker}(\tilde{t}),$

(2.2) $\qquad \mathrm{Im}(s_j : \mathcal{S}_{d-1}(D_LM) \to \mathcal{S}_d(D_LM) \subset C_d(D_LM)) \subset \mathrm{Ker}(\tilde{t}),$

That is, $\tilde{t}$ vanishes on chains on S_LM and degenerate chains on D_LM. Let $\sigma \in \mathcal{S}_p(D_LM)$ for some $p \geq d$, then the cap product can be described as

$$\sigma \cap \tilde{t} = (-1)^{d(p-d)}\tilde{t}({}_d\lfloor\sigma) \cdot \sigma\rfloor_{p-d}. \tag{2.3}$$

This gives a well-defined chain map $\cap\tilde{t} : C_p(D_LM, S_LM) \to C_{p-d}(D_LM)$ that represents capping with the Thom class in cohomology.

The map $\cap\tilde{t}$ on $C_*(D_LM, S_LM)$ is completely determined by its composition with the projection $C_*(D_LM) \to C_*(D_LM, S_LM)$. So to prove the lemma it now suffices to prove that $\cap\tilde{t}$ maps $F_p(C_*(D_LM))$ to $F_{p-d}(C_*(D_LM))$.

Let $T \in F_p(\mathcal{S}_r(D_LM))$ be a singular r-simplex in filtration p. Then, by definition,

$$ev(T) = \sigma(i_0, i_1, \ldots, i_r)$$

for some q-simplex $\sigma \in \mathcal{S}_q(D_T M)$, and some sequence $i_0 \leq \cdots \leq i_r \leq q \leq p$. By the above formula for the cap product, we then have

$$ev(T \cap \tilde{t}) = ev(T) \cap t = \pm t(\sigma(i_{r-d}, \ldots, i_r))\sigma(i_0, \ldots, i_{r-d}),$$

where, as above, $t \in C^d(D_T M, S_T M)$ represents the Thom class of the tangent bundle $TM \to M$. By (2.2) t vanishes on degenerate simplices. Therefore this expression can only be nonzero if $i_{r-d} < \cdots < i_r \leq q$, and therefore $r - d \leq q - d \leq p - d$. Hence $T \cap \tilde{t} \in F_{p-d}(C_*(D_L M))$ as claimed. □

Notice that if we compose the chain maps in Proposition 7 and Theorem 8, we have a map of filtered chain complexes

$$\mu : F_p(C_*(LM)) \otimes F_q(C_*(LM)) \longrightarrow F_{p+q-d}(C_*(LM))$$

which, by Proposition 5 induces the loop product in homology. Therefore μ induces a map of spectral sequences

$$\mu : E^r_{p,s}(LM) \otimes E^r_{q,t}(LM) \to E^r_{p+q-d,s+t}(LM). \tag{2.4}$$

For simply connected M, on the E_2-level, μ defines a map

$$\mu : H_p(M; H_s(\Omega M)) \otimes H_q(M; H_t(\Omega M)) \longrightarrow H_{p+q-d}(M; H_{s+t}(\Omega M))$$

which we claim is given up to sign, by the intersection product with coefficients on the Pontryagin ring $H_*(\Omega M)$. More explicitly,

$$\mu((a \otimes g) \otimes (b \otimes h)) = \pm(a \cdot b) \otimes (gh)$$

where $a \in H_p(M)$, $g \in H_s(\Omega M)$, $b \in H_q(M)$, and $h \in H_t(\Omega M)$, and where $a \cdot b$ is the intersection product and gh is the Pontryagin product.

To see this, notice that the composition of chain maps used to define μ is given by a composition of chain maps of fibrations (and pairs of fibrations). On the base space level this is given by the composition of maps described in Proposition 4 realizing the intersection product. On the fiber level, the fact that this chain map induces the Pontryagin product comes from the fact that map $\gamma : LM \times_M LM \to LM$ as defined in (1.4) is a map of fibrations

$$\begin{array}{ccc}
\Omega M \times \Omega M & \xrightarrow{\rho} & \Omega M \\
\downarrow & & \downarrow \\
LM \times_M LM & \xrightarrow{\gamma} & LM \\
{\scriptstyle \mathrm{ev}_\infty}\downarrow & & \downarrow{\scriptstyle ev} \\
M & \xrightarrow[=]{} & M
\end{array}$$

where ρ is the Pontryagin product.

Thus μ defines a multiplicative structure on the Serre spectral sequence for the fibration $\Omega M \to LM \to M$ which converges to the loop homology algebra structure on $H_*(LM)$, and on the E^2-level is given (up to sign) by the intersection product on M with coefficients in the Pontryagin ring $H_*(\Omega M)$. However the grading is shifted in a way that is confusing for calculational purposes. To remedy this, define a second quadrant spectral $\{E^r_{s.t}(\mathbb{H}_*(LM));\ d_r : E^r_{s,t} \to E^r_{s-r,t+r-1}\}$ with $s \leq 0$, $t \geq 0$, which converges to the loop homology $\mathbb{H}_{s+t}(LM)$ by regrading the Serre spectral spectral sequence in the following way.

$$E^r_{s.t}(\mathbb{H}_*(LM)) = E^r_{s+d,t}(LM), \tag{2.5}$$

where d is the dimension of M, and the right-hand side is the Serre spectral sequence for the fibration $\Omega M \to LM \to M$ we have been considering. Notice that $E^r_{s.t}(\mathbb{H}_*(LM))$ can only be nonzero for $-d \leq s \leq 0$. Moreover with the new indexing the spectral sequence converges to the loop homology in a grading preserving way, $E^r_{s.t}(\mathbb{H}_*(LM)) \rightrightarrows \mathbb{H}_{s+t}(LM)$. We also see that with this new indexing, the loop multiplication in the spectral sequence (2.4) preserves the bigrading,

$$\mu : E^r_{s,t}(\mathbb{H}_*(LM)) \otimes E^r_{p,q}((\mathbb{H}_*(LM)) \longrightarrow E^r_{s+p,t+q}((\mathbb{H}_*(LM)).$$

Finally notice that the E^2-term is given by

$$E^2_{s,t}(\mathbb{H}_*(LM)) = H_{s+d}(M; H_t(\Omega M))$$

for $-d \leq s \leq 0$. By applying the Poincaré duality we have

$$E^2_{s,t}(\mathbb{H}_*(LM)) = H^{-s}(M; H_t(\Omega M)).$$

Since under the Poincaré duality the intersection pairing in homology coincides with the cup product in cohomology, on the E^2-level the multiplication μ is given (up to sign) by cup product with coefficients in the Pontryagin ring $H_*(\Omega M)$.

$$\mu : H^{-s}(M; H_t(\Omega M)) \otimes H^{-p}(M; H_q(\Omega M)) \to H^{-(s+p)}(M; H_{t+q}(\Omega M))$$

for $-d \leq s, p \leq 0$ and $t, q \geq 0$. This completes the proof of Theorem 1.

3. The Loop Product on S^n and $\mathbb{CP}^n$

Our goal in this section is to use the loop homology spectral sequence constructed in the last section to perform the calculations described in Theorems 2 and 3.

We first prove Theorem 2 by calculating the ring structure of $\mathbb{H}_*(LS^n)$. If $n = 1$, the fibration $\Omega S^1 \to LS^1 \to S^1$ is trivial. Since the components of the based loop space ΩS^1 are all contractible, and $\pi_0(\Omega S^1) \cong \mathbb{Z}$, we have a homotopy equivalence

$$LS^1 \simeq S^1 \times \Omega S^1 \simeq S^1 \times \mathbb{Z}.$$

Similarly, there is a homotopy equivalence $LS^1 \times_{S^1} LS^1 \simeq S^1 \times \mathbb{Z} \times \mathbb{Z}$. With respect to these equivalences, it is clear that the map $\gamma : LS^1 \times_{S^1} LS^1 \to LS^1$ is given by

$$\begin{aligned} S^1 \times \mathbb{Z} \times \mathbb{Z} &\longrightarrow S^1 \times \mathbb{Z} \\ (x, m, n) &\longrightarrow (x, m+n). \end{aligned} \tag{3.1}$$

It is also clear that with respect to these equivalences, the Thom–Pontryagin map $\tilde{\tau} : LS^1 \times LS^1 \to (LS^1 \times_{S^1} LS^1)^{TS^1}$ is given by

$$\tau \times 1 : S^1 \times S^1 \times \mathbb{Z} \times \mathbb{Z} \longrightarrow (S^1)^{TS^1} \wedge (\mathbb{Z} \times \mathbb{Z})_+$$

where $\tau : S^1 \times S^1 \to (S^1)^{TS^1}$ is the Thom–Pontryagin construction for the diagonal map $\Delta : S^1 \to S^1 \times S^1$. Thus by (1.7) the loop homology algebra structure on $\mathbb{H}_*(LS^1)$ is, with respect to the equivalence $LS^1 \simeq S^1 \times \mathbb{Z}$, given by the tensor product of the intersection ring structure on $H_*(S^1)$ with the group algebra structure, $H_0(\mathbb{Z}) \cong \mathbb{Z}[t, t^{-1}]$. Using the Poincaré duality, we then have an algebra isomorphism

$$\mathbb{H}_{-*}(LS^1) \cong H^*(S^1) \otimes H_0(\mathbb{Z}) \cong \Lambda[a] \otimes \mathbb{Z}[t, t^{-1}] \tag{3.2}$$

where $a \in \mathbb{H}_{-1}(LS^1)$ corresponds to the generator in $H^0(S^1)$.

We now proceed with a calculation of $\mathbb{H}_*(LS^n)$ for $n > 1$. Consider the loop homology spectral sequence in this case. For dimension reasons, the only nontrivial differentials occur at the E^n level. Recall that there is an isomorphism of algebras, $H_*(\Omega S^n) \cong \mathbb{Z}[x]$, where x has degree $n-1$. It then follows that

$$E^2_{-p,q}(\mathbb{H}_*(LS^n)) \cong \cdots \cong E^n_{-p,q} \cong H^p(S^n) \otimes H_q(\Omega S^n).$$

The differentials d^n can be computed using the results in [1] and [7]. An exposition of this calculation (for the Serre spectral sequence of the fibration $\Omega S^n \to LS^n \to S^n$) is given in [5]. Inputting the change of grading used to define the loop homology spectral sequence, we have the following picture of the differentials:

E^n terms:

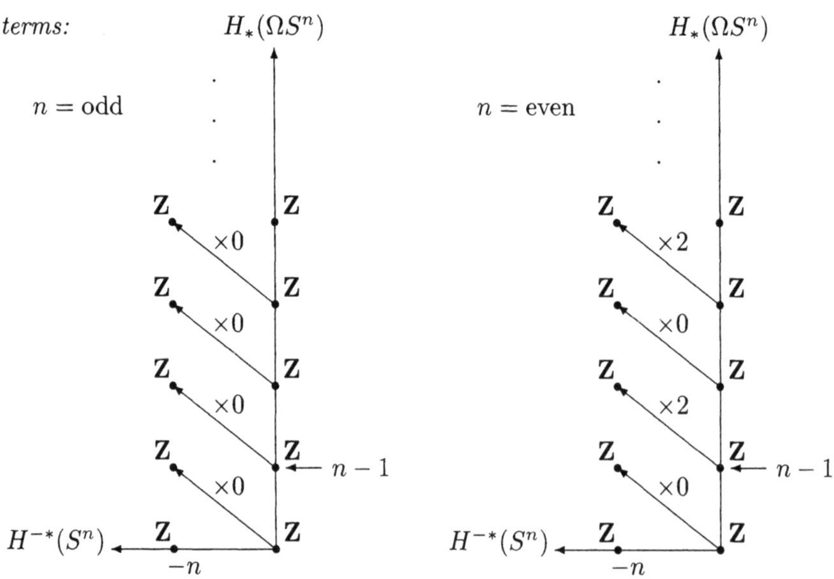

The Spectral Sequences for LS^n

Denote by ι, σ the generators of $H^n(S^n)$ and $H^0(S^n)$, respectively. Let 1_Ω be the unit of $H_*(\Omega S^n)$. In both spectral sequences $\sigma \otimes 1_\Omega$ is an infinite cycle and therefore represents a class in $\mathbb{H}_0(LS^n)$. This class is the unit in the algebra $\mathbb{H}_*(LS^n)$, which we denote by 1. Similarly $a = \iota \otimes 1_\Omega$ represents a class in $\mathbb{H}_{-n}(LS^n)$, and for dimension reasons a^2 must vanish in $\mathbb{H}_*(LS^n)$.

Let $u = \sigma \otimes x \in E^n_{0,n-1}$. In the case when n is odd, the spectral sequence collapses. Since $n > 1$, for dimensional reasons there can be no extension problems, and so we have an isomorphism of algebras

$$\mathbb{H}_*(LS^n) \cong E^2_{*,*} \cong H^{-*}(S^n) \otimes H_*(\Omega S^n) \cong \Lambda[a] \otimes \mathbb{Z}[u].$$

If n is even, the u^{2k} terms survive to E^∞, and

$$d_n(u^{2k+1}) = 2au^{2k+2}$$

for all nonnegative integers k. Let $v = u^2 = \sigma \otimes x^2$. Being an infinite cycle it represents a class in $\mathbb{H}_{2(n-1)}(LS^n)$. Then v^k represents a generator of a subgroup of $\mathbb{H}_{2k(n-1)}(LS^n)$ represented by classes in $E^\infty_{0,2k(n-1)}$. Notice that the ideal $(2av)$ vanishes in E^∞. Thus $\mathbb{Z}[a,v]/(a^2, 2av)$ is a subalgebra in $E^\infty_{*,*}$. Let $b = \iota \otimes x \in E^\infty_{-n,n-1}$. Then $E^\infty_{*,*}$ is generated by $\mathbb{Z}[a,v]/(a^2, 2av)$ and b. Again for dimension reasons $b^2 = ab = 0 \in E^\infty$, and so

$$E^\infty_{*,*} \cong (\Lambda[b] \otimes \mathbb{Z}[a,v])/(a^2, ab, 2av), \qquad \text{if } n \text{ is even.}$$

When $n > 2$, for dimensional reasons there can be no extension issues, so

$$\mathbb{H}_*(LS^n) \cong (\Lambda[b] \otimes \mathbb{Z}[a,v])/(a^2, ab, 2av) \quad \text{for } n \text{ even and } n > 2.$$

For $n = 2$ we consider the potential extension problem. For filtration reasons, there are unique classes in $\mathbb{H}_*(LS^2)$ represented by a and b in $E^\infty_{*,*}$. The ambiguity in the choice of class represented by $v \in E^\infty_{0,2}$ lies in $E^\infty_{-2,4} \cong \mathbb{Z}$ generated by av. Since $a^2 = 0$ in $\mathbb{H}_*(LS^2)$, any choice $\tilde{v} \in \mathbb{H}_2(LS^2)$ will satisfy $2a\tilde{v} = 0$. Thus any choice of $\tilde{v}$ together with a and b will generate the same algebra, namely $\mathbb{H}_*(LS^2) \cong (\Lambda[b] \otimes \mathbb{Z}[a,\tilde{v}])/(a^2, ab, 2a\tilde{v})$.

This completes the proof of Theorem 2.

We now proceed with the proof of Theorem 3 by calculating the ring structure of $\mathbb{H}_*(\mathbb{CP}^n)$. Notice that if $n = 1$, $\mathbb{H}_*(L\mathbb{CP}^n) \cong \mathbb{H}_*(LS^2)$, and this case was already discussed above. So for what follows we assume $n > 1$. The E^2-term of the loop homology spectral sequence is $H^*(\mathbb{CP}^n; H_*(\Omega\mathbb{CP}^n))$. The cohomology ring $H^*(\mathbb{CP}^n) \cong \mathbb{Z}[c_1]/(c_1^{n+1})$ is generated by the first Chern class c_1. We now recall the Pontryagin ring structure of $H_*(\Omega\mathbb{CP}^n)$.

Consider the homotopy fibration

$$\Omega S^{2n+1} \xrightarrow{\Omega\eta} \Omega\mathbb{CP}^n \to \Omega\mathbb{CP}^\infty \simeq S^1,$$

where $\eta : S^{2n+1} \to \mathbb{CP}^n$ is the Hopf map. Since this is a fibration of loop spaces that has a section (because $\pi_1(\Omega\mathbb{CP}^n) \cong \pi_1(\Omega\mathbb{CP}^\infty)$) we can conclude the following.

1. The fibration is homotopically trivial, $\Omega\mathbb{CP}^n \simeq \Omega S^{2n+1} \times S^1$.
2. The Serre spectral sequence of this fibration is a spectral sequence of algebras.

Therefore we have that in the Serre spectral sequence,

$$
\begin{aligned}
E^{\infty}_{*,*} &\cong H_*(S^1) \otimes H_*(\Omega S^{2n+1}) \\
&\cong \Lambda[t] \otimes \mathbb{Z}[x]
\end{aligned}
\tag{3.3}
$$

where $t \in H_1(S^1)$ and $x \in H_{2n}(\Omega S^{2n+1})$ are generators. This isomorphism is one of algebras. Indeed it is clear that there are no extension issues in this spectral sequence so that

$$H_*(\Omega \mathbb{CP}^n) \cong H_*(S^1) \otimes H_*(\Omega S^{2n+1}) \cong \Lambda[t] \otimes \mathbb{Z}[x].$$

Now consider the loop homology spectral sequence for $L\mathbb{CP}^n$. We therefore have

$$
\begin{aligned}
E^2_{-*,*}(\mathbb{H}_*(L\mathbb{CP}^n)) &\cong H^*(\mathbb{CP}^n; H_*(\Omega \mathbb{CP}^n)) \\
&\cong \mathbb{Z}[c_1]/(c_1^{n+1}) \otimes \Lambda[t] \otimes \mathbb{Z}[x]
\end{aligned}
$$

as algebras.

It was computed in [8] that

$$
H_k(L\mathbb{CP}^n) = \begin{cases} \mathbf{Z} & \text{if } k = 0, 1, \ldots, k \neq 2mn,\ m \geq 1, \\ \mathbf{Z} \oplus \mathbf{Z}_{n+1} & \text{if } k = 2mn,\ m \geq 1. \end{cases}
$$

This calculation implies the following pattern of differentials in the loop homology spectral sequence:

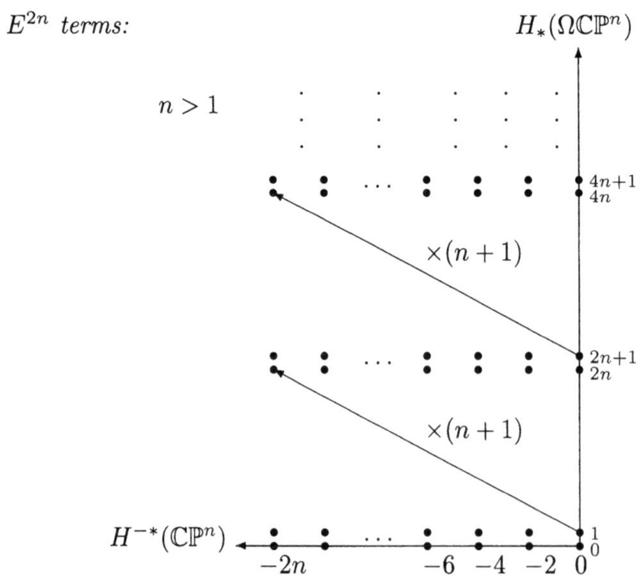

The loop homology spectral sequence for $L\mathbb{CP}^n$

That is, the only nonzero differentials are d_{2n} and

$$d_{2n}(y) = (n+1)c^n u,$$

where $c = c_1 \otimes 1$, $u = \sigma \otimes x$, and $y = \sigma \otimes t$. Here σ is a generator of $H^0(\mathbb{CP}^n) \cong \mathbb{Z}$. For dimension reasons $d_{2n}(u) = 0$, and since d_{2n} is a derivation,

$$d_{2n}(yu^k) = d_{2n}(y)u^k = (n+1)c^n u^{k+1},$$

and these are all the nonzero differentials.

The spectral sequence collapses beyond E^{2n} level, therefore c and u represent homology classes in $\mathbb{H}_{-2}(L\mathbb{CP}^n)$ and $\mathbb{H}_{2n}(L\mathbb{CP}^n)$ respectively. Notice also that the ideal $(c^{n+1}, (n+1)c^n u)$ vanishes in $E^{\infty}_{*,*}$. Therefore we have a subalgebra

$$\mathbf{Z}[c,u]/(c^{n+1}, (n+1)c^n u) \subset E^{\infty}_{*,*}(\mathbb{H}_*(L\mathbb{CP}^n)).$$

Let $w = yc \in E^2_{*,*}$. w is an infinite cycle in this spectral sequence and represents a class in $\mathbb{H}_{-1}(\mathbb{CP}^n)$. Notice that $w^2 = 0$ in $E^{\infty}_{-4,2}$. Similarly wc^n also vanishes in $E^{\infty}_{*,*}$. Thus the E^{∞}-term of the loop homology spectral sequence can be written as follows:

$$E^{\infty}_{*,*}(\mathbb{H}_*(L\mathbb{CP}^n)) \cong (\Lambda[w] \otimes \mathbf{Z}[c,u])/(c^{n+1}, (n+1)c^n u, wc^n).$$

The extension issues are handled just as they were for $\mathbb{H}_*(LS^2) = \mathbb{H}_*(L\mathbb{CP}^1)$ above. We then conclude that

$$\mathbb{H}_*(L\mathbb{CP}^n) \cong (\Lambda[w] \otimes \mathbf{Z}[c,u])/(c^{n+1}, (n+1)c^n u, wc^n)$$

as algebras. This completes the proof of Theorem 3.

References

[1] E.H. Brown, *Twisted tensor products*, Annals of Math. 1(1959), 223–246.

[2] R.L. Cohen and J.D.S. Jones, *A Homotopy Theoretic Realization Of String Topology*, preprint: math.GT/0107187.

[3] M. Chas and D. Sullivan, *String Topology*, preprint: math.GT/9911159, 1999.

[4] J.P. May, *Simplicial Objects in Algebraic Topology*. Van Nostrand Co., New York, 1967.

[5] J. McCleary, *Homotopy Theory and Closed Geodesics*, Homotopy Theory and Related Topics, Proceedings, Kinosaki 1988, Springer-Verlag, 1990.

[6] J. McCleary, *A User's Guide to Spectral Sequences*, 2nd edition, Cambridge, 2001.

[7] W. Shih, *Homologie des espaces fibre*, Publ. I.H.E.S. 3(1962), 93–176.

[8] W. Ziller, *The Free Loop Space of Globally Symmetric Spaces*, Inv. Math. 41, 1–22 (1977).

Ralph L. Cohen
Dept. of Mathematics
Stanford University
Stanford, California 94305, USA
e-mail: `ralph@math.stanford.edu`

John D.S. Jones
Department of Mathematics
University of Warwick
Coventry, CV4 7AL England
e-mail: `jdsj@maths.warwick.ac.uk`

Jun Yan
Dept. of Mathematics
Stanford University
Stanford, California 94305
e-mail: `jyan@math.stanford.edu`

Progress in Mathematics, Vol. 215
Algebraic Topology: Categorical Decomposition Techniques, 93–105

On Braid Groups, Free Groups, and the Loop Space of the 2-sphere

F.R. Cohen* and J. Wu†

Abstract. The purpose of this article is to describe a connection between the single loop space of the 2-sphere, Artin's braid groups, a choice of simplicial group whose homotopy groups are given by modules called $Lie(n)$, as well as work of Milnor [17, 18], and Habegger-Lin [11, 15] on "homotopy string links". The novelty of the current article is a description of connections between these topics.

1. A tale of two groups

In 1924 E. Artin [1, 2] defined the n-th braid group B_n together with the n-th pure braid group P_n, the kernel of the natural map of B_n to the n-th symmetric group. It is the purpose of this article to derive some additional connections of these groups to homotopy theory, as well as some overlaps with other subjects.

This article gives certain new relationships between free groups on n generators F_n, and braid groups which serve as a bridge between different structures. These connections, at the interface of homotopy groups of spheres, braids, knots, and links, and homotopy links, admit a common thread given by a simplicial group.

Recall that a simplicial group Γ_* is a collection of groups

$$\Gamma_0, \Gamma_1, \dots, \Gamma_n, \dots$$

together with face operations

$$d_i : \Gamma_n \to \Gamma_{n-1},$$

and degeneracy operations

$$s_i : \Gamma_n \to \Gamma_{n+1},$$

for $0 \leq i \leq n$. These homomorphisms are required to satisfy the standard simplicial identities.

One example is Milnor's free group construction $F[K]$ for a pointed simplicial set K. In case K is reduced (K is reduced provided K consists of a single point in degree 0), Milnor proved that the geometric realization of $F[K]$, denoted $|F[K]|$,

---

Key words and phrases. braid group, simplicial group, homotopy groups homotopy string links.
*Partially supported by the NSF.
†Partially supported by a grant from the National University of Singapore.

is homotopy equivalent to $\Omega\Sigma|K|$ [17]. The first theorem below addresses one connection concerning the simplicial group given by $F[\Delta[1]]$ (which is not reduced) where $\Delta[1]$ is the simplicial 1-simplex, and a simplicial group given in terms of the pure braid groups described next.

A second example is given by the simplicial group which in degree n is given by $\Gamma_n = P_{n+1}$, the $n+1$-st pure braid group, and which is elucidated in [6]. The face operations are given by deletion of a strand, while the degeneracies are gotten by "doubling" of a strand. This simplicial group is denoted AP_*.

Theorem 1.1. *The (simplicial) loop space of AP_* is isomorphic to $F[\Delta[1]]$ and is thus contractible. Hence $\pi_n AP_*$ is the trivial group for all n.*

A connection between Artin's braid group and the loop space of the 2-sphere is as follows. The second pure braid group is isomorphic to the integers with a choice of generator denoted $A_{1,2}$. Notice that the simplicial circle S^1 has a single non-degenerate point in degree 1 given by $\langle 0,1\rangle$. Thus there exists a unique map of simplicial groups

$$\Theta : F[S^1] \to AP_*$$

such that $\Theta(<0,1>) = A_{1,2}$. One of the theorems of [6] is as follows.

Theorem 1.2. *The morphism of simplicial groups*

$$\Theta : F[S^1] \to AP_*$$

is an embedding. Hence the homotopy groups of $F[S^1]$ are natural sub-quotients of AP_*, and the geometric realization of quotient simplicial set $AP_*/F[S^1]$ is homotopy equivalent to the 2-sphere. Furthermore, the smallest sub-simplicial group of AP_* which contains the element $\Theta(<0,1>) = A_{1,2}$ is isomorphic to $F[S^1]$.

This theorem gives that the homotopy groups of $F[S^1]$, those of the loop space of the 2-sphere, are given as "natural" sub-quotients of the braid groups, a result first given in work of the second author [21] by other methods. The proof of the above theorem given in [6] relies heavily on the structure of a Lie algebra arising from the "infinitesimal braid relations" arising in the Vassiliev invariants of braids by work of T. Kohno [12, 13, 14] as well as work of V. Drinfel'd [7] on the KZ equations. A specific computation is given here.

A naturality argument provides a bridge between the above theorems, various quotients of the braid groups as well as the modules $Lie(n)$ given by the representation $H_{n-1}P_n$ tensored with the sign representation, homotopy classes of self-maps of the loop space of a double suspension [4], and "homotopy string links" as considered in Milnor [17, 18], Habegger-Lin [11, 15]. This bridge will be described in Section 7 here.

This article gives a proof of Theorem 1.1 as well as a proof of a special case of Theorem 1.2. The remainder of this article gives the simplicial structure for AP_* together with computations concerning simplicial groups, and the Lie algebra which arises from the descending central series of the pure braid groups. Theorems 7.1, and 7.2 give connections to topics similar to some at this conference.

The authors thank the organizers of this conference for providing an extremely stimulating mathematical environment.

2. On looping AP_*

A proof of the following theorem is given next while a detailed development is given in [6].

Theorem 2.1. *The loop space (regarded as a simplicial group) of the simplicial group AP_*, $\Omega(AP_*)$, is isomorphic to $F[\Delta[1]]$ as a simplicial group. Thus AP_* is contractible, and the realization of the simplicial set $AP_*/F[S^1]$ is homotopy equivalent to S^2.*

Remarks: The proof given next exhibits the utility of the long exact homotopy sequence for the fundamental fibration sequence due to Fadell, and Neuwirth [8] in determining the homotopy type of the loop space for the simplicial group obtained from the pure braid groups. This sequence is essentially giving the structure of the loop space $\Omega(AP_*)$.

Proof. John Moore gave a definition for the loop space $\Omega\Gamma_*$ of a reduced simplicial group Γ_* (a simplicial group for which Γ_0 consists of a single element) [20]. This procedure corresponds to the topological notion of looping in the sense that the loop space of the geometric realization of Γ_* is homotopy equivalent to the geometric realization of $\Omega\Gamma_*$. This process of looping a simplicial group is described next where it is convenient to use a standard, but slightly different convention than that of [20].

Define a simplicial group $E\Gamma_*$ where the group $E\Gamma_n$ in degree n is given by the group Γ_{n+1} with face, and degeneracies given by the first $n+1$ operators for Γ_{n+1}. Then $\Omega\Gamma_n$, the looping of Γ_* in degree n, is defined to be the kernel of the map

$$d_{n+1} : \Gamma_{n+1} \to \Gamma_n,$$

and so

$$\Omega\Gamma_n = ker[d_{n+1} : \Gamma_{n+1} \to \Gamma_n].$$

In the special case that Γ_n is P_{n+1}, then $E\Gamma_n$ is P_{n+2}. Recall that P_{n+1} is the fundamental group of the configuration space of ordered $n+1$-tuples of distinct points in $\mathbb{R}^2$, $Conf(\mathbb{R}^2, n+1)$ [8, 3]. In addition P_{n+1} is generated by symbols $A_{i,j}$ for $1 \le i < j \le n+1$. A complete set of relations is given in Section 3 here. Furthermore, the map $d_{n+1} : \Gamma_{n+1} \to \Gamma_n$ is induced by the map $p_* : \pi_1(Conf(\mathbb{R}^2, n+2)) \to \pi_1(Conf(\mathbb{R}^2, n+1))$ where $p : Conf(\mathbb{R}^2, n+2) \to Conf(\mathbb{R}^2, n+1)$ is the map given by projection to the first $n+1$ coordinates.

The fibre of the map p is the plane punctured $(n+1)$-times, $\mathbb{R}^2 - Q_{n+1}$ [8]. The kernel of p_* is the free group F_{n+1} with generators $A_{j,n+2}$ for $1 \leq j < n+2$ [16, 8, 3]. The face, and degeneracy operations of the simplicial group AP_* are listed in the next section. Furthermore, the kernel of p_* is exactly the kernel of $d_{n+1} : P_{n+2} \to P_{n+1}$ by inspection. Hence $\Omega(AP_*)$ in degree n is the free group with basis $\{A_{j,n+2,}|1 \leq j \leq n+1\}$.

Recall that the simplicial 1-simplex in degree n is given by $\{\langle 0^i, 1^{n+1-i}\rangle\}$ where $0 \leq i \leq n+1$ with the single relation that $\langle 0^{n+1}\rangle = \mathit{identity}$ in the simplicial group $F[\Delta[1]]$. By Lemma 2.2 below, there is a dimension-wise morphism of simplicial groups

$$\Psi : F[\Delta[1]] \to \Omega(AP_*)$$

defined by sending

$$\Psi(\langle 1\rangle) = A_{1,2},$$

and

$$\Psi(\langle 0,1\rangle) = A_{2,3}.$$

That Ψ is an isomorphism of simplicial groups is given by Lemma 2.2, and the first part of the theorem follows.

To prove the second part of the theorem, notice that by Theorem 1.2, there is a morphism of simplicial groups

$$\Theta : F[S^1] \to AP_*$$

which is a dimension-wise monomorphism. Since Θ is multiplicative, there is a fibration

$$F[S^1] \longrightarrow AP_* \longrightarrow AP_*/F[S^1].$$

Hence, the geometric realization of $AP_*/F[S^1]$ is homotopy equivalent to the 2-sphere as AP_* is contractible. The theorem follows. □

Lemma 2.2. *There is a morphism of simplicial groups*

$$\Psi : F[\Delta[1]] \to \Omega(AP_*)$$

such that $\Psi(\langle 0,1\rangle) = A_{2,3}$, *and thus* $\Psi(\langle 1\rangle) = A_{1,2}$. *Furthermore,* Ψ *is a dimension-wise isomorphism, and is the unique morphism of simplicial groups satisfying the condition* $\Psi(\langle 0,1\rangle) = A_{2,3}$, *and thus* $\Psi(\langle 1\rangle) = A_{1,2}$.

Proof. Before considering the map Ψ, define a map of sets

$$\Phi : \Delta[1] \to \Omega(AP_*)$$

on points in simplicial degree n by the following formulas:

1. $\Phi(\langle 0^j, 1^{n+1-j}\rangle) = A_{j+1,n+2} \cdot A_{j+2,n+2} \cdots A_{n+1,n+2}$ in case $n+1-j > 0$, and
2. $\Phi(\langle 0^{n+1}\rangle) = \mathit{identity}$ in degree n for $\Omega(AP_*)$.

By a direct check of the simplicial identities for $\Omega(AP_*)$, the sets of elements in $\Omega(AP_*)$ given by

$$\{\Phi(\langle 0^j, 1^{n+1-j}\rangle) | 0 \leq j \leq n+1\}$$

in each degree n gives a simplicial subset of $\Omega(AP_*)$ with "base-point" $\langle 0^{n+1}\rangle$. Furthermore, the map Φ restricts to an isomorphism of simplicial sets onto the image of Φ.

Another direct inspection gives that the sets $\{\Phi(\langle 0^j, 1^{n+1-j}\rangle) | 0 \leq j < n+1\}$ freely generate $\Omega(AP_*)$ in degree n. Thus Φ admits a unique extension to a morphism of simplicial groups $\Psi : F[\Delta[1]] \to \Omega(AP_*)$ satisfying the conditions $\Psi(\langle 1\rangle) = A_{1,2}$, and $\Psi(\langle 0,1\rangle) = A_{2,3}$.

In addition, notice that Ψ is a dimension-wise isomorphism by a natural change of basis. Furthermore, any morphism of simplicial groups $\zeta : F[\Delta[1]] \to \Omega(AP_*)$ which satisfies $\zeta(\langle 0,1\rangle) = A_{2,3}$ must agree with Ψ as the map is multiplicative, and agrees on all points $\langle 0^j, 1^{n+1-j}\rangle$ by application of degeneracies. Thus Ψ is unique.

The lemma follows. □

3. The simplicial structure for AP_*

Recall that P_{n+1} is generated by symbols $A_{i,j}$ for $1 \leq i < j \leq n+1$. Artin's relations are listed in [16] while a reformulated complete set of relations is as follows:

1. $[A_{r,s}, A_{i,k}] = 1$ for either $r < s < i < k$ or $i < k < r < s$,
2. $[A_{k,s}, A_{i,k}] = [A_{i,s}^{-1}, A_{i,k}]$ for $i < k < s$,
3. $[A_{r,k}, A_{i,k}] = [A_{i,k}^{-1}, A_{i,r}^{-1}]$ for $i < r < k$, and
4. $[A_{r,s}, A_{i,k}] = [[A_{i,s}^{-1}, A_{i,r}^{-1}], A_{i,k}]$ for $i < r < k < s$.

The origin of the face, and degeneracy maps below is gotten by omitting a strand in pure braids for the case of face maps, and doubling a strand in pure braids for the case of degeneracy maps. The results are stated, but the computations are omitted. The face operations in the simplicial group AP_* are defined as follows:

$$d_t(A_{i,j}) = \begin{cases} A_{i-1,j-1} & \text{if } t+1 < i, \\ 1 & \text{if t +1 = i}, \\ A_{i,j-1} & \text{if } i < t+1 < j\ , \\ 1 & \text{if t + 1 = j}, \\ A_{i,j} & \text{if } t+1 > i\ . \end{cases}$$

The degeneracy operations are defined as follows:

$$s_t(A_{i,j}) = \begin{cases} A_{i+1,j+1} & \text{if } t+1 < i\ , \\ A_{i,j+1} \cdot A_{i+1,j+1} & \text{if t+1} = \text{i}\ , \\ A_{i,j+1} & \text{if } i < t+1 < j\ , \\ A_{i,j} \cdot A_{i,j+1} & \text{if t + 1} = \text{j}\ , \\ A_{i,j} & \text{if } t+1 > j. \end{cases}$$

These operations give a convenient method for describing the behavior of Θ_n in Section 6.

4. On embeddings of free groups

This section is expository. Let $\rho : \Pi \to G$ be a homomorphism between discrete groups. The i-th stage of the descending central series for Π is the subgroup of Π generated by commutators of weight at least i, and is denoted $\Gamma^i(\Pi)$. The associated graded is given by

$$E_0^j(\Pi) = \Gamma^j(\Pi)/\Gamma^{j+1}(\Pi).$$

A group homomorphism ρ preserves the stages of the descending central series. There is an induced morphism of associated graded Lie algebras

$$E_0^*(\rho) : E_0^*(\Pi) \to E_0^*(G).$$

Recall that a discrete group Γ is said to be residually nilpotent group if

$$\bigcap_{i \geq 1} \Gamma^i(\Pi) = \{identity\}.$$

Proposition 4.1.

1. *Assume that Π is a residually nilpotent group. Let*

$$\rho : \Pi \to G$$

be a homomorphism of discrete groups such that the morphism of associated graded Lie algebras

$$E_0^*(\rho) : E_0^*(\Pi) \to E_0^*(G)$$

is a monomorphism. Then ρ is a monomorphism.

2. *If Π is a free group, and $E_0^*(\rho)$ is a monomorphism, then ρ is a monomorphism.*

Proof. Let x denote a non-identity element in the kernel of ρ. Since Π is residually nilpotent, there exists a natural number n such that the element x is in $\Gamma^n(\Pi)$ and not in $\Gamma^{n+1}(\Pi)$. But then x projects to a non-identity element in $E_0^n(\Pi)$, and thus has non-trivial image in $E_0^n(G)$ contradicting the fact that x is a non-identity element in the kernel of ρ.

To finish the second part, observe that free groups are residually nilpotent [16]. □

The approach above for showing that a group homomorphism is an injection is suited for groups Π which are residually nilpotent such as P_n. However, the application given in Section 6 is restricted to free groups.

5. The Lie algebra associated to the descending central series for P_k

The method of proof for Theorem 1.2, that Θ_n is an embedding, relies heavily on the structure of the associated graded Lie algebra obtained from the descending central series for both F_n, and P_{n+1}. Thus the structure Lie algebra associated to the descending central series for P_k is recalled here as occurring in work of T. Kohno [12, 13], Falk-Randall [9], and others. Let $B_{i,j}$ denote the projections of the $A_{j,i}$ to $E_0^*(P_k)$. (Caution: The order of indices is intentionally reversed here.)

Theorem 5.1. *The Lie algebra obtained from the descending central series for P_k is given by $\mathcal{L}_k$ the free Lie algebra generated by elements $B_{i,j}$ with $k \geq i > j \geq 1$, modulo the infinitesimal braid relations:*

(i) $[B_{i,j}, B_{s,t}] = 0$ *if* $\{i,j\} \cap \{s,t\} = \phi$,
(ii) $[B_{i,j}, B_{i,t} + B_{t,j}] = 0$ *if* $1 \leq j < t < i \leq k$, *and*
(iii) $[B_{t,j}, B_{i,j} + B_{i,t}] = 0$ *if* $1 \leq j < t < i \leq k$.

Furthermore there is a split short exact sequence of Lie algebras

$$0 \to E_0^*(F_n) \xrightarrow{E_0^*(i)} E_0^*(P_{n+1}) \xrightarrow{E_0^*(d_n)} E_0^*(P_n) \to 0$$

where $E_0^*(F_n)$ is the free Lie algebra generated by $B_{n+1,j}$ for $1 \leq j < n+1$. In addition, $E_0^*(P_{n+1})$ is additively isomorphic to $E_0^*(P_n) \oplus E_0^*(F_n)$.

6. On $\Theta_n : F_n \to P_{n+1}$

The proof that Θ_n is an embedding is given in the first non-trivial case with $n = 2$ below while the complete proof is given in [6]. The other cases arise via induction on n together with choices of elements in $E_0^*(P_n)$ which centralize certain sub-Lie algebras of $E_0^*(P_{n+1})$.

The method here is to consider the homomorphism $\Theta_n : F_n \to P_{n+1}$ on the level of associated graded Lie algebras

$$E_0^*(\Theta_n) : E_0^*(F_n) \to E_0^*(P_{n+1}),$$

and to appeal to Proposition 4.1 to deduce that Θ_n is an embedding.

Proposition 6.1. *If $n \geq 1$, $E_0^*(\Theta_n) : E_0^*(F_n) \to E_0^*(P_{n+1})$ is a monomorphism of Lie algebras. Hence, by Proposition 4.1, $\Theta_n : F_n \to P_{n+1}$ is a monomorphism.*

Before giving the proof, a preparatory computation is required. Given two elements x, and y in a Lie algebra, define $ad^k(x)(y)$ inductively by the formulas

1. $ad^1(x)(y) = [y, x]$, and
2. $ad^{k+1}(x)(y) = [ad^k(x)(y), x]$.

Next, notice that

1. $\Theta_1 : F_1 \to P_2$ is an isomorphism with $\Theta_1(x_1) = A_{1,2}$,
2. $d_2(A_{1,2}) = A_{1,2}$,
3. $d_2(A_{1,3}) = 1$, and
4. $d_2(A_{2,3}) = 1$.

Furthermore, by the simplicial identities, it follows that

1. $\Theta_2(x_1) = A_{1,3} \cdot A_{2,3}$,
2. $\Theta_2(x_2) = A_{1,2} \cdot A_{1,3}$,
3. $d_2(\Theta_2(x_1)) = 1$, and
4. $d_2(\Theta_2(x_2)) = A_{1,2}$.

Proposition 6.2. *If $k \geq 0$, then*

$$E_0^*(\Theta_2)(ad^k(x_2)(x_1)) = (-1)^k ad^k(B_{3,2})(B_{3,1}).$$

Proof. That $E_0^*(\Theta_2)(ad^k(x_2)(x_1)) = (-1)^k ad^k(B_{3,2})(B_{3,1})$ is satisfied on the level of Lie algebras is given by the following induction on k.

1. $E_0^*(\Theta_2)(x_1) = E_0^*(\Theta_2)(s_0(x_1)) = s_0 E_0^*(\Theta_1)(x_1) = s_0(B_{2,1}) = B_{3,1} + B_{3,2}$
2. $E_0^*(\Theta_2)(x_2) = E_0^*(\Theta_2)(s_1(x_1)) = s_1\Theta_1(x_1) = s_1(B_{2,1}) = B_{2,1} + B_{3,1}$.

On the level of Lie algebras,

1. $E_0^*(\Theta_2)([x_1, x_2]) = [B_{3,1} + B_{3,2}, B_{3,1} + B_{2,1}]$,
2. $E_0^*(\Theta_2)([x_1, x_2]) = [B_{3,1} + B_{3,2}, B_{3,1}]$ by the infinitesimal braid relations of Theorem 5.1, and thus $E_0^*(\Theta_2)([x_1, x_2]) = [B_{3,2}, B_{3,1}]$.

Thus the proposition is correct for the case $k = 1$.

Assume that the formula $E_0^*(\Theta_2)(ad^k(x_2)(x_1)) = (-1)^k ad^k(B_{3,2})(B_{3,1})$ holds, and check it for $k+1$.

1. $E_0^*(\Theta_2)(ad^{k+1}(x_2)(x_1)) = [E_0^*(\Theta_2)(ad^k(x_2)(x_1)), E_0^*(\Theta)(x_2)]$,
2. $E_0^*(\Theta_2)(ad^{k+1}(x_2)(x_1)) = [(-1)^k ad^k(B_{3,2})(B_{3,1}), B_{3,1} + B_{2,1}]$,
3. $E_0^*(\Theta_2)(ad^{k+1}(x_2)(x_1)) = X + Y + Z$ where $X = (-1)^k([ad^k(B_{3,2})(B_{3,1}), B_{3,1}]$, $Y = -([ad^k(B_{3,2})(B_{3,1}), B_{3,1}]$, $Z = -[ad^k(B_{3,2})(B_{3,1}), B_{3,1}])$ by the infinitesimal braid relations of Theorem 5.1, and
4. $E_0^*(\Theta_2)(ad^{k+1}(x_2)(x_1)) = (-1)^{k+1} ad^{k+1}(B_{3,2})(B_{3,1})$.

Proposition 6.2 follows. □

The proof of Proposition 6.1 in the case of $n = 2$ will be given next.

Proof. There is a morphism of Lie algebras

$$\begin{array}{ccc} E_0^*(F_2) & \xrightarrow{E_0^*(d_2)} & E_0^*(F_1) \\ \downarrow{\scriptstyle E_0^*(\Theta_2)} & & \downarrow{\scriptstyle E_0^*(\Theta_1)} \\ E_0^*(P_3) & \xrightarrow{E_0^*(d_2)} & E_0^*(P_2). \end{array}$$

The first step is to determine the Lie algebra kernel of the surjection

$$E_0^*(d_2) : E_0^*(F_2) \to E_0^*(F_1).$$

To avoid confusion of notation, define

$$y = x_1 \qquad \text{as an element of} \qquad F_1.$$

Thus $d_2(x_1) = 1$, and $d_2(x_2) = y$.

Recall P. Hall's classical result [16] that $E_0^*(F_n) = E_0^*(F[x_1, x_2, \ldots, x_n])$ is isomorphic to the free Lie algebra over the integers $\mathbb{Z}$, $L[x_1, x_2, \ldots, x_n]$ where the x_i's in the free Lie algebra are the projections of the analogous elements in the group. Furthermore, the kernel of the projection map $L[x_1, x_2] \to L[x_2]$ which sends x_1 to zero is given by $L[S]$ where S is the set $\{x_1, ad^k(x_2)(x_1) | k > 0\}$[5].

Since $d_2(x_1) = 1$, and $d_2(x_2) = y$,

$$ker(E_0^*(d_2)) = L[T]$$

with

$$T = \{x_1, ad^k(x_2)(x_1) | k > 0\}.$$

Hence there is a split short exact sequence of Lie algebras

$$L[T] \longrightarrow L[x_1, x_2] \xrightarrow{E_0^*(d_2)} L[y].$$

Theorem 5.1 gives that the Lie algebra kernel of $E_0^*(d_2) : E_0^*(P_3) \to E_0^*(P_2)$ is $L[B_{1,3}, B_{2,3}]$, and that there is a split short exact sequence of Lie algebras

$$L[B_{1,3}, B_{2,3}] \longrightarrow E_0^*(P_3) \xrightarrow{E_0^*(d_2)} E_0^*(P_2).$$

Combining these two facts gives a morphism of exact sequences of Lie algebras in the following commutative diagram:

$$\begin{array}{ccccc} L[T] & \longrightarrow & L[x_1, x_2] & \xrightarrow{E_0^*(d_2)} & L[y] \\ \downarrow & & \downarrow E_0^*(d_2) & & \downarrow E_0^*(d_1) \\ L[B_{3,1}, B_{3,2}] & \longrightarrow & E_0^*(P_3) & \xrightarrow{E_0^*(d_2)} & E_0^*(P_2) \end{array}$$

Since $\Theta_1 : F_1 \to P_2$ is an isomorphism, so is $E_0^*(\Theta_1) : E_0^*(L[y]) \to E_0^*(P_2)$. Thus it suffices to show that $E_0^*(\Theta_2)$ restricted to $L[T]$ is a monomorphism.

The final step is to consider a sub-Lie algebra of $L[B_{1,3}, B_{2,3}]$. By a change of basis, consider the Lie algebra $L[W]$ where W is the 2-element set given by

$$W = \{u = B_{3,1} + B_{3,2}, v = B_{3,2}\}.$$

Thus

$$L[B_{3,1}, B_{3,2}] = L[u, v].$$

Define a morphism of Lie algebras

$$\mu : L[u, v] \to L[z]$$

by

1. $\mu(u) = 0$, and
2. $\mu(v) = z$.

Hence μ is a surjection of Lie algebras with kernel $L[u, ad^k(v)(u) | k > 0]$. By Proposition 6.2,

1. $E_0^*(\Theta_2)(x_1) = B_{3,1} + B_{3,2}$,
2. $E_0^*(\Theta_2)(x_2) = B_{2,1} + B_{3,1}$,
3. $E_0^*(\Theta_2)(ad(x_2)(x_1)) = ad(B_{3,1} + B_{3,2})(B_{3,1} + B_{2,1}) = ad(B_{3,1})(B_{3,2})$, and
4. $E_0^*(\Theta_2)(ad^k(x_2)(x_1)) = (-1)^k ad^k(B_{3,2})(B_{3,1})$for $k > 0$.

Hence

1. $E_0^*(\Theta_2)(x_1) = u$, and
2. if $k > 0$, then $E_0^*(\Theta_2)(ad^k(x_2)(x_1)) = (-1)^k ad^k(v)(u)$ as $[B_{3,1}+B_{3,2}, B_{3,2}] = [B_{3,1}, B_{3,2}]$.

Notice that $E_0^*(\Theta_2)$ preserves basis elements up to sign, and is an isomorphism. The proposition follows. □

7. Relationship to "homotopy string links" and other constructions

The morphism of simplicial groups

$$\Theta : F[S^1] \to AP_*$$

passes to several natural quotients. One such quotient arises in an overlap between certain groups which are generic homotopy classes of self-maps for the loop space of a double suspension [4]. On the other hand, these quotients admit a geometric interpretation as follows.

Consider the "reduced free group" due to Milnor[17, 18] used to analyze "homotopy string links" in Habegger-Lin [11], [15]. The "reduced free group" K_n is defined as the quotient of F_n modulo the relations

$$[x_i, gx_ig^{-1}] = 1$$

where x_i is any generator of F_n, g is any element in F_n, and $[x, y]$ denotes the commutator $x \cdot y \cdot x^{-1} \cdot y^{-1}$.

The notion of link homotopy is described crudely in this paragraph. Consider the space of ordered n-tuples of smooth maps of S^1 in $\mathbb{R}^3$ having disjoint images. Milnor defines an equivalence relation of "link homotopy" by allowing a strand in any component of a link to pass through itself. Lin describes the infinitesimal link-homotopy relations on page 5 of [15].

For example, consider the trivial n-component link $\mathcal{L}_n$ in $\mathbb{R}^3$. Thus the fundamental group of the complement $\mathbb{R}^3 - \mathcal{L}_n$ is isomorphic to F_n (by some, non-canonical, choice of isomorphism). Let β denote a simple closed curve in the complement which represents an element b in F_n.Then both Milnor, and Habbeger-Lin prove that the link $\mathcal{L}_n \cup \beta$ is link homotopically trivial if and only if the element b projects to the identity in K_n.

The "reduced free group" K_n was rediscovered in both a different form as well as different context in [4] in order to analyze Barratt's finite exponent conjecture. There K_n is the quotient of the free group on n generators modulo the smallest normal subgroup containing the simple commutators where at least one generator appears twice. More precisely, let Λ_n denote the smallest normal subgroup of F_n

containing all commutators of the form $[\cdots[x_{i_1}, x_{i_2}]x_{i_3}], \cdots]x_{i_t}]$ where $x_{i_j} = x_{i_k}$ for some $j < k$. That the "reduced free group" is isomorphic to the quotient

$$K_n = F_n/\Lambda_n$$

is an exercise which is omitted. The groups K_n are used to construct natural quotients of the pure braid groups by using the previous theorem. An application will be made to "reduced free groups", and the loop space of the 2-sphere. A theorem proven in [4] is as follows.

Theorem 7.1. *The natural quotient map $\rho : F_n \to K_n$ prolongs to a morphism of simplicial groups*

$$\rho : F[S^1] \to K[S^1]$$

for which $K[S^1]$ in degree n is K_n. The kernel of ρ, $\Gamma[S^1]$, is free in each degree.

1. *The map ρ is a surjection of simplicial groups, and is thus a fibration with fibre given by a simplicial group which is free in each degree.*
2. *The n-th homotopy group of $K[S^1]$, $\pi_n K[S^1]$, is isomorphic to $\oplus_{(n-1)!}\mathbb{Z}$, the $n-1$-st homology group of P_n given by Lie(n).*
3. *The induced map $\pi_*(\rho) : \pi_*(F[S^1]) \to \pi_*(K[S^1])$ is an isomorphism in dimension 1, and 2. This map is zero in degrees greater than 2.*
4. *If $n > 2$, there is a short exact sequence of homotopy groups*

$$0 \to \pi_{n+1}K[S^1] \to \pi_n\Gamma[S^1] \to \pi_n(F[S^1]) \to 0.$$

Naturality in Theorem 7.1 is used to give a simplicial group KAP_*, and to compare the map $\Theta : F[S^1] \to AP_*$ with a construction arising in Habegger-Lin [11], [15].

Theorem 7.2. *There is a simplicial group KAP_* which in degree n is given by the "reduced pure braid group", rP_{n+1}, a quotient of P_{n+1} obtained by replacing F_j with K_j for all $j \leq n$ in the extension $1 \to F_j \to P_{j+1} \to P_j \to 1$. There is an induced map $\theta : K[S^1] \to KAP_*$ together with a commutative diagram of simplicial groups:*

$$\begin{array}{ccc} F[S^1] & \xrightarrow{\Theta} & AP_* \\ \downarrow{\rho} & & \downarrow{\rho} \\ K[S^1] & \xrightarrow{\Theta} & KAP_* \end{array}$$

where each vertical map is an epimorphism of simplicial groups.

The previous theorem provides a comparison of link homotopy together with chains, and cycles for the loop space of the 2-sphere where an invariant arises by considering the coset representatives of $F[S^1]$ in AP_*. In addition, there are analogous quotients of the pure braid group arising from other families of characteristic subgroups of finitely generated free groups. This construction will be considered elsewhere [6].

References

[1] E. Artin, *Theorie der Zöpfe*, Hamburg Abhandlung (1924).

[2] E. Artin, *The theory of braids*, Ann. of Math., **48**(1947), 101–126.

[3] J. Birman, *Braids, Links and Mapping Class Groups*, Ann. of Math. Studies, **82**(1975), Princeton Univ. Press, Princeton, N.J..

[4] F.R. Cohen, *On combinatorial group theory in homotopy*, Cont. Math., **188**(1995), 57–63, and *Combinatorial group theory in homotopy theory I*, on website.

[5] F.R. Cohen, J.C. Moore and J.A. Neisendorfer, *Torsion in homotopy groups*, Ann. of Math., **109**(1979), 121–168.

[6] F.R. Cohen, and J. Wu, in preparation.

[7] V.G. Drinfel'd, *On the structure of quasi-triangular quasi-Hopf algebras*, Funct. Anal. Appl., **26**(1992), 63–65.

[8] E. Fadell and L. Neuwirth, *Configuration Spaces*, Math. Scand., **10**(1962), 111–118.

[9] M. Falk, and R. Randall, *The lower central series of a fiber-type arrangement*, Invent. Math., **82**(1985), 77–88.

[10] F. Gonzalez-Acuña, Lecture, (December 2001), Cinvestav.

[11] N. Habegger, X.-S. Lin, *Classification of links up to link homotopy*, J.A.M.S. **3**(1990), 389–419.

[12] T. Kohno, *Linear representations of braid groups and classical Yang-Baxter equations*, Cont. Math., **78**(1988), 339–363.

[13] T. Kohno, *Vassiliev invariants and de Rham complex on the space of knots*, in: Symplectic Geometry and Quantization, Contemp. Math., **179**(1994), Amer. Math. Soc., Providence, RI, 123–138.

[14] T. Kohno, *Elliptic KZ system, braid groups of the torus and Vassiliev invariants, Topology and its Applications*, **78**(1997), 79–94.

[15] X.-S. Lin, *Power series expansions and invariants of links*, 1993 Georgia Topology Conference.

[16] W. Magnus, A. Karras, and D. Solitar, *Combinatorial group theory: Representations of groups in terms of generators and relations*, Pure and Applied Mathematics, **13**(1966), Interscience Publishers.

[17] J. Milnor, *Link groups*, Ann. of Math., **78**(1954), 177–195.

[18] J. Milnor, *Isotopy of links*, in *Algebraic Geometry and Topology: A Symposium in Honor of S. Lefschetz, ed. R.H. Fox, D.C. Spencer, and A.W. Tucker* , Princeton Unversity Press, (1957), 280–306.

[19] J. Milnor, *On the construction F[K]*,In: A student's Guide to Algebraic Topology, J.F. Adams, editor, Lecture Notes of the London Mathematical Society, **4**(1972), 119–136.

[20] J.C. Moore, *Homotopie des complexes monoïdaux*, Séminaire H. Cartan, (1954/55).

[21] J. Wu, *Combinatorial descriptions of the homotopy groups of certain spaces*, Math. Proc. Camb. Philos. Soc., to appear.

F.R. Cohen
Department of Mathematics
University of Rochester,
Rochester, New York 14627, USA
e-mail: cohf@math.rochester.edu
URL: www.math.rochester.edu/~cohf

J. Wu
Department of Mathematics
National University of Singapore
Singapore
e-mail: matwuj@nus.edu.sg

Progress in Mathematics, Vol. 215
Algebraic Topology: Categorical Decomposition Techniques, 107–124

The K-completion of E_6

Donald M. Davis

Abstract. We compute the 2-primary v_1-periodic homotopy groups of the K-completion of the exceptional Lie group E_6. This is done by computing the Bendersky-Thompson spectral sequence of E_6. We conjecture that the natural map from E_6 to its K-completion induces an isomorphism in v_1-periodic homotopy, and discuss issues related to this conjecture.

1. Introduction

The p-primary v_1-homotopy groups of a topological space X, denoted $v_1^{-1}\pi_*(X;p)$ and defined in [11], are a localization of the portion of the homotopy groups of X detected by p-local K-theory. In [10], the author completed the determination of the odd-primary v_1-periodic homotopy groups of all compact simple Lie groups. The groups $v_1^{-1}\pi_*(X;2)$ have been determined for $X = SU(n)$ ([3]), $Sp(n)$ ([5]), G_2 ([12]), and F_4 ([4]). Joint work of the author and Bendersky is very close to completing the computation of $v_1^{-1}\pi_*(SO(n);2)$.[1] This will leave E_6, E_7, and E_8 to be determined, which would complete a program suggested to the author by Mimura in 1989. In this paper, we determine $v_1^{-1}\pi_*(\widehat{E}_6;2)$, where $\widehat{X}$ denotes the K-completion of X, as defined in [6]. We conjecture that the natural map $E_6 \to \widehat{E}_6$ induces an isomorphism in $v_1^{-1}\pi_*(-;2)$.[2]

In [6], Bendersky and Thompson defined the K-completion $\widehat{X}$ of a space X to be the homotopy limit of a certain tower of spaces under X. The space X is said to satisfy the Completion Telescope Property (CTP) at the prime p if $X \to \widehat{X}$ induces an isomorphism of p-primary v_1-periodic homotopy groups. They also constructed a spectral sequence (BTSS) which, for many spaces X, converges to $v_1^{-1}\pi_*(\widehat{X};p)$. It was shown in [7] and [4] that, localized at any prime, if X is K_*-strongly spherically resolved (KSSR), then X satisfies the CTP. This condition

2000 *Mathematics Subject Classification.* 55T15,55Q52,57T20.

Key words and phrases. K-theory, homotopy groups, exceptional Lie groups.

[1]Note added in proof: The determination of $v_1^{-1}\pi_*(SO(n);2)$ was completed and submitted for publication in July 2002.

[2]Note added in proof: In a January 2002 e-mail to the author, Bousfield outlined a proof that $v_1^{-1}\pi_*(E_6) \to v_1^{-1}\pi_*(\widehat{E}_6)$ is an isomorphism. In a September 2002 e-mail, he wrote that he had checked the details of that proof and was confident in it.

means that X can be built by fibrations from spheres S^{2n_i+1} such that K_*X is built as a K_*K-coalgebra as an extension of the $K_*S^{2n_i+1}$.

In [4], it was proved that F_4 satisfies the CTP at 2, and a general result ([4, 5.8, 5.11, 5.15]) was proved which implies that if E_6/F_4 satisfies the CTP, then so does E_6. Many standard functors of algebraic topology would lead one to expect that there is a fibration

$$S^9 \to E_6/F_4 \to S^{17}, \tag{1.1}$$

which would imply that E_6/F_4 is KSSR and then that E_6 satisfies the CTP at the prime 2. However, it was proved by Cohen and Selick ([9]) that there can be no such fibration. A fibration

$$\Omega S^9 \to \Omega(E_6/F_4) \to \Omega S^{17}$$

would also imply the CTP for E_6. It is not known whether such a fibration exists.

It was proved in [7] that, localized at an odd prime p, if X is K-algebraically spherically resolved and K-durable, then it satisfies the CTP. The first condition (KASR) means that $K_*(X)$ has the structure that it would if X were KSSR, and the second that X and its K-localization have isomorphic v_1-periodic homotopy groups. Both E_6/F_4 and E_6 are KASR and K-durable at the prime 2. However, it is not known whether, at the prime 2, this is enough to insure that the CTP is satisfied.

Our main result is the following determination of $v_1^{-1}\pi_*(\widehat{E}_6;2)$. As discussed above, the expectation is that this equals $v_1^{-1}\pi_*(E_6;2)$.

Theorem 1.2. *Let* $e = \min(12, \nu(\ell-18)+5)$ *and* $f = \min(12, 2\nu(\ell-3)+8)$. *Then*

$$v_1^{-1}\pi_{8\ell+d}(\widehat{E}_6;2) \approx \begin{cases} \mathbf{Z}/2^f & d=-3 \\ \mathbf{Z}/2^f \oplus \mathbf{Z}/2 & d=-2 \\ \mathbf{Z}/2^e \oplus \mathbf{Z}/2 \oplus \mathbf{Z}/2 & d=-1,0 \\ \mathbf{Z}/2^5 \oplus \mathbf{Z}/2 \oplus \mathbf{Z}/2 & d=1 \\ \mathbf{Z}/2^5 \oplus \mathbf{Z}/2 & d=2 \\ \mathbf{Z}/2^3 & d=3,4 \end{cases}$$

Here and throughout, $\nu(-)$ denotes the exponent of 2 in an integer. The picture of the BTSS which determines these groups is given in Diagram 1.3. This is a usual sort of Adams spectral sequence type of chart, with dots representing $\mathbf{Z}/2$'s, positively sloping lines the action of the Hopf map η (or the element h_1 in the spectral sequence), and negatively sloping lines differentials in the spectral sequence, which implies that the elements which they connect do not survive to give nonzero homotopy classes. The dotted lines that look like h_1 means that h_1 is usually present, but perhaps not always.

The reader should observe that this chart, and the homotopy groups which it depicts, has a form very similar to the charts for $v_1^{-1}\pi_*(G_2)$ and $v_1^{-1}\pi_*(F_4)$ depicted in [4, 4.9]. The only difference is the orders of the groups on the 1- and 2-lines, and the lack of some exotic extensions.

Diagram 1.3.

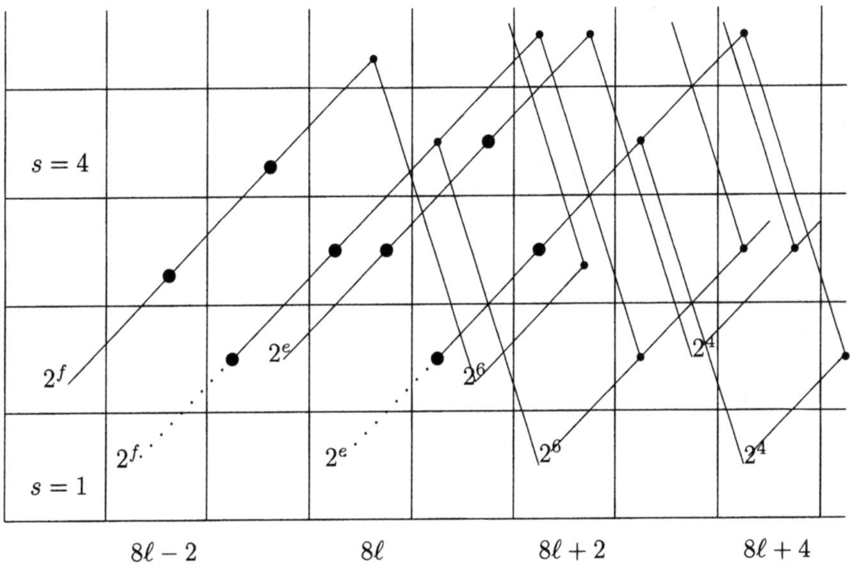

The determination of the E_2-term of the spectral sequence is rather straightforward, given the general results of [4] and calculations of the Adams operations in $K^*(E_6)$ performed in [10]. This is performed in Section 2. The d_3-differentials are determined by showing that there are maps of spaces which relate the classes in question to classes in spaces where d_3 is known. This is performed in Sections 3 and 4. In Section 5, we expand our discussion of whether E_6 satisfies the CTP.

The author would like to thank Martin Bendersky, Pete Bousfield, and Fred Cohen for helpful comments on this work.

2. The E_2-term

In this section we compute the E_2-term of the BTSS converging to $v_1^{-1}\pi_*(\widehat{E}_6)$. We are always using the v_1-periodic BTSS localized at 2, and all v_1-periodic homotopy groups are 2-primary.

In [4, 1.1], it was proved that for spaces X for which $K_*(X)$ and $K^*(X)$ form nice exterior algebras,

$$(2.1)\qquad E_2^{s,t}(X) \approx \begin{cases} \mathrm{Ext}_A^s(QK^1(X;\mathbf{Z}_2^\wedge)/\operatorname{im}(\psi^2), QK^1(S^t;\mathbf{Z}_2^\wedge)) & \text{if } t \text{ is odd} \\ 0 & \text{if } t \text{ is even.} \end{cases}$$

Here $Q(-)$ denote the indecomposable quotient, and A is the category of stable 2-adic Adams modules.

It was proved in [4, 5.5] that simply-connected mod p finite H-spaces whose rational homology is associative satisfy the required hypothesis, and hence (2.1) applies to $X = E_6$. We remark again that although the spectral sequence is denoted as $E_2(X)$, it is only known to converge to $v_1^{-1}\pi_*(\widehat{X})$, and indeed even this is not known to be always true. It was proved in [7, 1.4] that, localized at any prime, if X is KASR, then the BTSS of X converges to $v_1^{-1}\pi_*(\widehat{X})$.

As input for the BTSS, we need the Adams module $K^*(X)$. We use $\mathbf{Z}/2$-graded K-theory. First is the general result of Hodgkin ([14]) which states that, for a simply-connected compact Lie group G, $K^*(G)$ is an exterior algebra on generators in $K^1(G)$ obtained from the fundamental representations of G. For E_6, there are six such generators. Proposition 2.2 follows from [10, 3.9], which gives ψ^k for all integers k. It can also be deduced from [18] after a lot of manipulation.

Proposition 2.2. *There is a basis for $QK^1(E_6)$ on which the matrices of ψ^{-1}, ψ^2, and ψ^3 are given by*

$$(\psi^{-1}) = \begin{pmatrix} -1 & 0 & 0 & 0 & 0 & 0 \\ 1 & 1 & 0 & 0 & 0 & 0 \\ 0 & 0 & -1 & 0 & 0 & 0 \\ 0 & 0 & 0 & -1 & 0 & 0 \\ 0 & 0 & 0 & 1 & 1 & 0 \\ 0 & 0 & 0 & 0 & 0 & -1 \end{pmatrix}$$

$$(\psi^{2}) = \begin{pmatrix} 2 & 0 & 0 & 0 & 0 & 0 \\ 7 & 16 & 0 & 0 & 0 & 0 \\ -33 & 0 & 32 & 0 & 0 & 0 \\ 7 & 0 & -8 & 128 & 0 & 0 \\ -4 & -1 & 4 & 64 & 256 & 0 \\ 1 & 0 & -1 & -24 & 0 & 2048 \end{pmatrix}$$

$$(\psi^{3}) = \begin{pmatrix} 3 & 0 & 0 & 0 & 0 & 0 \\ 39 & 81 & 0 & 0 & 0 & 0 \\ -264 & 0 & 3^5 & 0 & 0 & 0 \\ 147 & 0 & -162 & 3^7 & 0 & 0 \\ -87 & -27 & 81 & 3^7 & 3^8 & 0 \\ 82 & 0 & -81 & -3^7 & 0 & 3^{11} \end{pmatrix}$$

Here each column expresses ψ^k of a basis element in terms of the basis elements.

By [4, 3.10],

$$\text{(2.3)} \qquad \operatorname{Ext}_A^{1,2n+1}(M)^\# \approx M/\operatorname{im}(\psi^{-1} - (-1)^n, \psi^3 - 3^n).$$

Thus the Pontryagin dual of $E_2^{1,2n+1}(E_6)$ is the abelian group presented by the 18×6 matrix

$$\begin{pmatrix} (\psi^2)^T \\ (\psi^{-1})^T - (-1)^n I \\ (\psi^3)^T - 3^n I \end{pmatrix},$$

where $(\psi^k)^T$ refers to the transpose of the matrices of 2.2. Using `Maple`, we simplify the relations, removing generators one-at-a-time. For example, we begin by pivoting on the 1 in position (1,6), and then removing the top row, which expresses the sixth generator in terms of the others, and the last column, since that generator is no longer required. It is convenient to do separate reductions for the two parities of n.

If $n = 2k$ is even, it is most efficient to express entries in terms of $R := 3^{2k} - 3^{72}$. Thus 3^n is replaced by $R + 3^{72}$ at the outset. Of course, it is only in hindsight that one can think to do this. We have $\nu(R) = \nu(k - 36) + 3$. We can eliminate all but the first column, and five nontrivial relations remain on this single generator. Omitting odd coefficients, these are

$$\begin{aligned}
&2^{12} + 2^6 R \\
&2^{25} + 2^{14} R \\
&2^{12} + 2^1 R \\
&2^{16} + 2^1 R + R^2 \\
&2^{13} + 2^5 R.
\end{aligned}$$

If $\nu(R) < 11$, then the relation with the smallest 2-exponent is the third (or fourth), giving $2^{\nu(R)+1}$ as the single relation, the order of the cyclic group. If $\nu(R) \geq 11$, then the smallest relation is the first, giving 2^{12} as the order of the cyclic group. These combine to give $\min(12, \nu(k-36)+4)$ as the exponent of this cyclic 2-group, and this gives the 1-line groups in Diagram 1.3 when $t - s = 8\ell$ or $8\ell + 4$.

Similarly, if $n = 2k + 1$ is odd, it turns out to be most efficient to write the entries of the matrix in terms of $Q := 3^{2k+1} - 3^{11}$. Then $\nu(Q) = \nu(k-5)+3$. The matrix again reduces to one column, with relations (omitting odd coefficients)

$$\begin{aligned}
&2^{14} + 2^6 Q \\
&2^{18} + 2^{14} Q \\
&2^{12} + 2^8 Q + Q^2 \\
&2^{12} + 2^6 Q + Q^2 \\
&2^7 Q + 2^3 Q^2.
\end{aligned}$$

One easily checks that this gives $\min(12, 2\nu(k-5)+6)$ as the minimal 2-exponent, and this yields the 1-line groups in $t - s = 8\ell \pm 2$. We summarize with

Proposition 2.4. *The nonzero groups on the* 1*-line of the BTSS of* E_6 *are given by*

$$E_2^{1,2n+1}(E_6) \approx \begin{cases} \mathbf{Z}/2^{\min(12,\nu(k-36)+4)} & n = 2k \\ \mathbf{Z}/2^{\min(12,2\nu(k-5)+6)} & n = 2k+1. \end{cases}$$

Next, we deal with the 2-line groups. By [4, 3.10], if M is a finite stable 2-adic Adams module and $\theta_n = \psi^3 - 3^n$, then there is a short exact sequence

$$
\begin{aligned}
(2.5) \qquad 0 \;\to\; & \operatorname{coker}(\theta_n|Q_n(M)) \to \operatorname{Ext}_A^{2,2n+1}(M)^{\#} \\
\to\; & \ker(\theta_n|\operatorname{coker}(\psi^{-1} - (-1)^n))|_M \to 0,
\end{aligned}
$$

where the functor Q_n is defined by

$$
(2.6) \qquad Q_n = \frac{\ker(\psi^{-1} - (-1)^n)}{\operatorname{im}(\psi^{-1} + (-1)^n)}.
$$

We evaluate the right-hand part of (2.5). A mild surprise occurred here in that the group is not always cyclic, as had been the case in most other examples.

Proposition 2.7. *For $M = QK^1(E_6)/\operatorname{im}(\psi^2)$,*

$$
\ker(\theta_n|\operatorname{coker}(\psi^{-1} - (-1)^n)) \approx \begin{cases} \mathbf{Z}/2 \oplus \mathbf{Z}/2^{\min(11,\nu(k-36)+3)} & n = 2k \\ \mathbf{Z}/2^{\min(12,2\nu(k-5)+6)} & n = 2k+1. \end{cases}
$$

Proof. Since the kernel and cokernel of an endomorphism of a finite abelian group have the same order, the group has the same order as the corresponding 1-line group given in Proposition 2.4.

We begin with the case $n = 2k+1$. We argue similarly to [4, 4.4]. Let $\overline{M} := M/\operatorname{im}(\psi^{-1}+1)$, $\overline{M}_2 = \ker(\cdot 2|\overline{M})$, and $K = \ker((\psi^3 - 3^n)|\overline{M})$. We will show that $\dim(\overline{M}_2 \cap K) = 1$, which implies that K is cyclic. Using 2.2, the abelian group $\overline{M}$ can be presented by the matrix

$$
\begin{pmatrix}
0 & 1 & 0 & 0 & 0 & 0 \\
0 & 0 & 0 & 0 & 1 & 0 \\
2 & 7 & -33 & 7 & -4 & 1 \\
0 & 16 & 0 & 0 & -1 & 0 \\
0 & 0 & 32 & -8 & 4 & -1 \\
0 & 0 & 0 & 128 & 64 & -24 \\
0 & 0 & 0 & 0 & 0 & 2048
\end{pmatrix}.
$$

If $\{v_1, \ldots, v_6\}$ denotes the basis on which this matrix presents relations, then a basis for $\overline{M}_2$ is given by $\psi^2(v_4)/2$ and $\psi^2(v_6)/2$, the last two rows divided by 2. Using 2.2, we have in $\overline{M}$

$$
(\psi^3 - 1)(\psi^2(v_6)/2) = \psi^2(\psi^3 - 1)(v_6)/2 = \psi^2(\tfrac{3^{11}-1}{2} v_6) \equiv 0
$$

and similarly

$$
(\psi^3 - 1)(\psi^2(v_4)/2) = \psi^2(\tfrac{3^7-1}{2} v_4 + \tfrac{3^7}{2} v_5 - \tfrac{3^7}{2} v_6) \equiv \psi^2(v_6)/2 \not\equiv 0.
$$

Thus $\overline{M}_2 \cap K = \langle \frac{\psi^2 v_6}{2} \rangle$.

The case $n = 2k$ is slightly more delicate. Let $\widetilde{M} = M/\operatorname{im}(\psi^{-1} - 1)$; its presentation matrix can be reduced to

$$(2.8)\qquad \begin{pmatrix} -2 & 1 & 0 & 0 & 0 & 0 \\ 0 & 0 & 2 & 0 & 0 & 0 \\ 16 & 0 & -1 & -1 & 0 & 0 \\ 32 & 0 & 0 & 0 & -1 & 0 \\ 0 & 0 & 0 & 0 & 0 & -1 \\ 2^{12} & 0 & 0 & 0 & 0 & 0 \end{pmatrix}.$$

Thus $\widetilde{M}$ is spanned by v_1 and v_3 of order 2^{12} and 2, respectively. Using 2.2 and the relations in $\widetilde{M}$, we have

$$(\psi^3 - 3^{2k})v_3 \equiv (3^5 - 3^{2k})v_3 - 162(16v_1 - v_3) + 81 \cdot 32v_1 \equiv 0,$$

since $2v_3 = 0$. Thus K has a $\mathbf{Z}/2$-summand generated by v_3. Similarly

$$\begin{aligned}(\psi^3 - 3^{2k})v_1 &= (3 - 3^{2k})v_1 + 39 \cdot 2v_1 + 147(16v_1 - v_3) - 87 \cdot 32v_1 \\ &\equiv v_3 - (3^{2k} + 351)v_1.\end{aligned}$$

Thus v_1 is not in K, but $2v_1$ has a chance. Since $351 = 2^{11}u - 3^{72}$ with u odd,

$$(\psi^3 - 3^{2k})(2v_1) = 2(3^{72} - 3^{2k} - 2^{11}u)v_1 = 2^{3+\nu(2k-72)}u'v_1,$$

with u' odd, where $2^{12}v_1 = 0$. Thus $2^{\max(0,12-(3+\nu(2k-72)))}2v_1$ generates the other summand, which will have 2-exponent $\min(12-1, 3+\nu(2k-72)-1)$, as claimed. □

In Section 4, we will prove that a failure of (2.5) to split compensates for the unexpected splitting in Proposition 2.7, yielding

Proposition 2.9. *The nonzero groups on the 2-line of the BTSS of E_6 are given by*

$$E_2^{2,2n+1}(E_6) \approx \mathbf{Z}/2 \oplus \begin{cases} \mathbf{Z}/2^{\min(12,\nu(k-36)+4)} & n = 2k \\ \mathbf{Z}/2^{\min(12,2\nu(k-5)+6)} & n = 2k+1. \end{cases}$$

Finally, we determine the eta-towers. This refers to elements in filtration > 2, all of which occur in families related by h_1, also known as η. ([4, 3.6]) With Q_n as defined in (2.6), we have, from [4, 3.10], for $s > 2$, a short exact sequence

$$(2.10)\quad 0 \to \operatorname{coker}(\theta_n | Q_{s+n}(M)) \to \operatorname{Ext}_A^{s,2n+1}(M)^{\#} \to \ker(\theta_n | Q_{s+n-1}(M)) \to 0.$$

We establish

Proposition 2.11. *For $M = QK^1(E_6)/\operatorname{im}(\psi^2)$, $Q_n(M) \approx \mathbf{Z}/2$, generated by v_3 if n is odd, and by $2^{17}v_1$ if n is even.*

Proof. If n is odd, $Q_n(M)$ equals $\ker(\psi^{-1} + 1 : \widetilde{M} \to M)$, where $\widetilde{M}$ is presented by (2.8). Then $\widetilde{M} \approx \mathbf{Z}/2 \oplus \mathbf{Z}/2^{12}$, with generators v_3 and v_1. From 2.2, we see that $v_3 \in \ker(\psi^{-1} + 1)$; however, $(\psi^{-1} + 1)(2^{11}v_1) = 2^{11}v_2$, which is nonzero in M.

One way to see that $2^{11}v_2 \neq 0$ is to use pivoting to obtain the following alternate presentation for M, with columns still $v_1, \ldots, v_6$.

$$(2.12)\qquad \begin{pmatrix} 16 & -8 & -40 & 0 & 0 & 1 \\ 0 & 16 & 0 & 0 & -1 & 0 \\ 2 & 7 & -1 & -1 & 0 & 0 \\ 2^7 5 & 2^6 27 & -2^6 17 & 0 & 0 & 0 \\ 0 & 2^{12} & 0 & 0 & 0 & 0 \\ 2^{18} & 0 & 0 & 0 & 0 & 0 \end{pmatrix}.$$

If n is even, we want $\ker(\psi^{-1} - 1 : \overline{M} \to M)$, where $\overline{M}$ is presented by (2.12) with the second and fifth columns omitted. We check $\psi^{-1} - 1$ on the elements of order 2. We have $2^{17}v_1 \mapsto -2^{18}v_1 + 2^{17}v_2 = 0$, while

$$2^6 5v_1 - 2^5 17v_3 \mapsto -2^7 5v_1 + 2^6 5v_2 + 2^6 17v_3 = 2^6(5+27)v_2 = 2^{11}v_2,$$

which has order 2 in M. □

If X is a topological space, let $Q_n(X) = Q_n(QK^1(X)/\operatorname{im}(\psi^2))$.

Corollary 2.13. *For $s > 2$, $E_2^{s,2n+1}(E_6)^\# \approx \mathbf{Z}/2 \oplus \mathbf{Z}/2$, with generators $2^{17}v_1$ and v_3.*

Proof. We use (2.10) and Proposition 2.11, and 2.2 to show that $\theta_n(v_3) = 0$ in $Q_{\mathrm{od}}(E_6)$ and $\theta_n(2^{17}v_1) = 0$ in $Q_{\mathrm{ev}}(E_6)$. Indeed, $\theta_n(v_3) = 2Av_3 - 162v_4 + 81v_5 - 81v_6$, which is 0 in a group presented by (2.8), while $\theta_n(2^{17}v_1) = 0$ in a group presented by (2.12) since there $2^{18}v_1 = 0$ and $2^{17}v_i = 0$ for $2 \le i \le 6$. □

This, with 2.4 and 2.9, completes the determination of the E_2-term, which is as suggested by Diagram 1.3. The h_1-extensions from the 1-line will be discussed in Section 4.

3. d_3-differentials

In this section, we determine the d_3-differentials on the eta towers in the BTSS of E_6. An eta tower consists of elements in $E_2^{s,2s+i}$ for $s \ge s_0$ connected by h_1. Here $s_0 = 1$, 2, or 3. We denote by $\eta_i(X)$ the $\mathbf{Z}_2$-vector space of eta towers passing through $E_2^{s,2s+i}(X)$ for $s > 2$. Note that d_3 is a homomorphism from $\eta_i(X)$ to $\eta_{i-4}(X)$. With $E_2^{s,t}$ depicted as usual in position $(x, y) = (t - s, s)$, then $\eta_i(X)$ is a tower of elements whose position satisfies $x - y = i$. Since $\eta^4 = 0$ in homotopy, d_3-differentials must annihilate all eta towers, except for a few elements at the bottom of the target tower. By (2.10), if $QK^1(X)$ consists of elements whose dimensions are all of the same parity, then so does $\eta_*(X)$, and $\eta_i(X)$ depends only on i mod 4. What must be determined for each family of eta towers is the mod 8 value of i for which $d_3 : \eta_i(X) \to \eta_{i-4}(X)$ is nonzero.

We make heavy use of the fibration

$$(3.1)\qquad F_4 \to E_6 \to EIV,$$

where EIV is the group quotient E_6/F_4. This fibration was studied quite thoroughly in [18]. It was observed there that

$$H^*(EIV;\mathbf{Z}) \approx \Lambda(x_9, x_{17}),$$

and the Serre spectral sequence of (3.1) collapses. From this, the spectral sequence

$$H^*(EIV; K^*(F_4)) \Rightarrow K^*(E_6)$$

implies that there is a short exact sequence

$$0 \to QK^1(EIV) \to QK^1(E_6) \to QK^1(F_4) \to 0, \tag{3.2}$$

with each of the three algebras $K^*(X)$ being exterior. The following key proposition implies that the Adams operations in $K(EIV)$ are as they would be if (1.1) existed.

Proposition 3.3. *There is a basis $\{y_1, y_2\}$ for $QK^1(EIV)$ on which for all k*

$$\psi^k(y_1) = k^4 y_1 + u\tfrac{k^4-k^8}{16} y_2, \tag{3.4}$$

with u a unit in $\mathbf{Z}_{(2)}$, and $\psi^k(y_2) = k^8 y_2$.

Proof. This can be deduced from [18, Thm 2], which computes the Chern character for EIV. We present a proof closer to our methods.

From (3.2), we deduce $QK^1(EIV) = \ker(QK^1(E_6) \to QK^1(F_4))$. This must be the subspace of $QK^1(E_6)$ spanned by v_2 and v_5 (in the basis used in Section 2). Perhaps the easiest way to see this is to use (ψ^{-1}) of 2.2, which shows that $\psi^{-1} = 1$ on this subspace, while it equals -1 on $QK^1(F_4)$.

From 2.2, we find that $240v_2 + v_5$ is an eigenvector for ψ^2 with eigenvalue 16. Thus, considering the rational splitting of E_6 as a product of spheres as in [10], we deduce that $\psi^k(240v_2 + v_5) = k^4(240v_2 + v_5)$ and $\psi^k(v_5) = k^8 v_5$ for all integers k. Thus $\psi^k v_2 = k^4 v_2 + \frac{k^4-k^8}{240} v_5$, from which follows the result with $y_1 = v_2$ and $y_2 = v_5$. □

In [4, 4.2], it was shown that there is a basis $\{x_1, x_2, x_3, x_4\}$ of $QK^1(F_4)$ on which $\psi^{-1} = -1$ and the matrix of ψ^2 is

$$\begin{pmatrix} 2 & 0 & 0 & 0 \\ 3 & 32 & 0 & 0 \\ 1 & -8 & 128 & 0 \\ 0 & -1 & -24 & 2048 \end{pmatrix}$$

We easily verify the following result.

Proposition 3.5. *In terms of the bases introduced, the exact sequence (3.2) is given by $y_1 \mapsto v_2$, $y_2 \mapsto v_5$, $v_1 \mapsto -x_1 - x_2$, $v_3 \mapsto x_2$, $v_4 \mapsto x_3$, and $v_6 \mapsto x_4$.*

Let $\overline{K}(X) := QK^1(X)/\operatorname{im}(\psi^2)$. Since ψ^2 acts injectively in our modules, the Snake Lemma applied to (3.2) yields a short exact sequence

$$0 \to \overline{K}(EIV) \to \overline{K}(E_6) \to \overline{K}(F_4) \to 0 \tag{3.6}$$

and hence by (2.1) a long exact sequence

$$(3.7)\qquad \begin{aligned} 0 &\to E_2^{1,t}F_4 \to E_2^{1,t}E_6 \to E_2^{1,t}EIV \to E_2^{2,t}F_4 \\ &\to E_2^{2,t}E_6 \to E_2^{2,t}EIV \to E_2^{3,t}F_4 \to . \end{aligned}$$

In [4, 4.6], it was shown that

$$E_2^{s,t}(F_4) \approx \begin{cases} \mathbf{Z}/2 & s=1,\ t\equiv 1(4) \\ \mathbf{Z}/2^f & s=1,\ t=4k+3 \\ \mathbf{Z}/2^f \oplus \mathbf{Z}/2 & s=2,\ t=4k+3 \\ \mathbf{Z}/2 \oplus \mathbf{Z}/2 & s=2,\ t\equiv 1(4) \\ \mathbf{Z}/2 \oplus \mathbf{Z}/2 & s>2,\ t \text{ odd} \\ 0 & t \text{ even} \end{cases}$$

with $f = \min(12, 2\nu(k-5)+6)$. Using 3.3 and the method of [4, 4.11–4.12], we obtain

$$E_2^{s,t}(EIV) \approx \begin{cases} \mathbf{Z}/2 & s=1,\ t\equiv 3(4) \\ \mathbf{Z}/2^e & s=1,\ t=4k+1 \\ \mathbf{Z}/2^e \oplus \mathbf{Z}/2 & s=2,\ t=4k+1 \\ \mathbf{Z}/2 \oplus \mathbf{Z}/2 & s=2,\ t\equiv 3(4) \\ \mathbf{Z}/2 \oplus \mathbf{Z}/2 & s>2,\ t \text{ odd} \\ 0 & t \text{ even} \end{cases}$$

with $e = \min(12, \nu(k-36-2^8)+3)$. By just substituting the known E_2-groups into (3.7), we can deduce quite a bit about the morphisms; in particular, the rank of every morphism beginning with $E_2^{2,t}EIV \to E_2^{3,t}F_4$ is 1. This follows by counting the alternating sum of the exponents of the orders of the groups. Thus of the two eta-towers in $E_2^{s,t}(E_6)$ for any odd value of $t-2s$, one comes from F_4 and the other maps nontrivially to EIV.

We need to know more specifically which classes map across in (3.7). For this, we use the following commutative diagram of exact sequences, where $Q_n(X)$ is the functor of (2.6) applied to $M = \overline{K}X$.

$$(3.8)\qquad \begin{array}{ccccccc} Q_{s+n}(EIV) & \xrightarrow{\alpha_1} & \operatorname{Ext}_A^{s,2n+1}(\overline{K}EIV)^\# & \xrightarrow{\beta_1} & Q_{s+n-1}(EIV) & \xrightarrow{\theta} & \\ \downarrow{\scriptstyle f_1} & & \downarrow{\scriptstyle f_2} & & \downarrow{\scriptstyle f_3} & & \\ Q_{s+n}(E_6) & \xrightarrow{\alpha_2} & \operatorname{Ext}_A^{s,2n+1}(\overline{K}E_6)^\# & \xrightarrow{\beta_2} & Q_{s+n-1}(E_6) & \xrightarrow{\theta} & . \\ \downarrow{\scriptstyle g_1} & & \downarrow{\scriptstyle g_2} & & \downarrow{\scriptstyle g_3} & & \\ Q_{s+n}(F_4) & \xrightarrow{\alpha_3} & \operatorname{Ext}_A^{s,2n+1}(\overline{K}F_4)^\# & \xrightarrow{\beta_3} & Q_{s+n-1}(F_4) & \xrightarrow{\theta} & \end{array}$$

Proposition 3.9. *In (3.8) with $s+n$ even, $g_3(v_3) = x_2$ pulls back nontrivially to g_2. If $s+n$ is odd, then $g_3(2^{17}v_1) = 2^{17}x_1$ pulls back nontrivially to g_2. If $s+n$ is even* *(resp. odd), there is an element* $z \in \operatorname{Ext}_A^{s,2n+1}(\overline{K}EIV)^{\#}$ *such that* $\beta_1(z) = 2^7 y_2$ *(resp.* y_1*) and* $f_2(z) = \alpha_2(2^{17}v_1)$ *(resp.* $\alpha_2(v_3)$*).*

Proof. The exact sequence for F_4 is given in [4, 4.5]. Actually, it is more convenient here to use the presentation

$$\begin{pmatrix} 0 & -32 & 8 & 1 \\ 2 & 3 & 1 & 0 \\ 2^7 5 & 2^6 27 & 0 & 0 \\ 2^{18} & 0 & 0 & 0 \end{pmatrix} \tag{3.10}$$

for $\overline{K}(F_4)$ instead of that of [4, 4.2]. This one can be obtained from that one by pivoting. Dividing the third and fourth rows of (3.10) by 2 gives a basis of $Q_{\mathrm{ev}}(F_4)$, and θ sends the first element to the second. This is a computation best done in `Maple`, using the rows of (3.10) to reduce $(\psi^3 - 1)(2^6 5x_1 + 2^6 27x_2)$, obtained from [4, 4.2], to $2^{17}x_1$. As in [4, 4.5], $Q_{\mathrm{od}}(F_4) = \langle x_1, x_2 \sim x_3 \rangle$ with θ sending the first to the second. Now the g_2-part of the proposition in straightforward, using Proposition 3.5.

We have $Q_{\mathrm{od}}(EIV) \approx \mathbf{Z}/2$ with generator $2^7 y_2$, and $Q_{\mathrm{ev}}(EIV) \approx \mathbf{Z}/2$ with generator y_1. One readily checks, using 3.5 and 2.11, that $f_1 = 0$ and $f_3 = 0$ in (3.8). Thus the morphism f_2 must be as claimed because of our previous observation that it has rank 1. □

We can deduce d_3 on half of the eta towers in E_6 from its behavior in F_4. By Corollary 2.13, each $\eta_{\mathrm{od}}(E_6)$ has basis $\{2^{17}v_1, v_3\}$.

Proposition 3.11. *The differential* $d_3 : \eta_{8\ell+1}(E_6) \to \eta_{8\ell-3}(E_6)$ *sends the* v_3*-tower to the* v_3*-tower. The differential* $d_3 : \eta_{8\ell+7}(E_6) \to \eta_{8\ell+3}(E_6)$ *sends the* $2^{17}v_1$*-tower to the* $2^{17}v_1$*-tower.*

Proof. It was shown in [4, 4.13] that in the BTSS of F_4, $d_3 : \eta_{8\ell+1}(F_4) \to \eta_{8\ell-3}(F_4)$ is the "identity map," while $d_3 : \eta_{8\ell+3}(F_4) \to \eta_{8\ell-1}(F_4)$ sends the x_1-tower to the x_1-tower, and $d_3 : \eta_{8\ell+7}(F_4) \to \eta_{8\ell+3}(F_4)$ sends the $2^{17}x_1$-tower to the $2^{17}x_1$-tower. We dualize (3.8) and use 3.9 to see that the v_3-tower is in the image of $\eta_{4*+1}(F_4) \to \eta_{4*+1}(E_6)$, and that $\eta_{4*+3}(F_4) \to \eta_{4*+3}(E_6)$ sends the $2^{17}x_1$-tower to the $2^{17}v_1$-tower. Naturality of d_3 implies the result. □

By using the inclusion map $S^9 \to EIV$, we deduce the third family of d_3-differentials.

Proposition 3.12. *The differential* $d_3 : \eta_{8\ell+3}(E_6) \to \eta_{8\ell-1}(E_6)$ *sends the* v_3*-tower to the* v_3*-tower.*

Proof. Using the dual of f_2 in Proposition 3.9, it suffices to show that $d_3 : \eta_{8\ell+3}(EIV) \to \eta_{8\ell-1}(EIV)$ sends the y_1-tower to the y_1-tower. The map $j : S^9 \to EIV$ induces

$$QK^1 EIV \xrightarrow{j^*} QK^1 S^9$$

satisfying $j^*(y_1) = g$ and $j^*(y_2) = 0$. The generator $g \in QK^1S^9$ gives rise to the stable eta towers in BTSS(S^9), and by [1, p. 58] or [3, p. 488] $d_3 : \eta_{8\ell+3}(S^9) \to \eta_{8\ell-1}(S^9)$ sends the stable eta tower to the stable eta tower. Our conclusion follows by naturality using the map j. □

The final d_3 on eta towers requires a more elaborate argument. The conclusion is that d_3 is as it would be if the fibration (1.1) existed. Our desired result is

Proposition 3.13. *The differential $d_3 : \eta_{8\ell+1}(E_6) \to \eta_{8\ell-3}(E_6)$ sends the $2^{17}v_1$-tower to the $2^{17}v_1$-tower.*

Proof. By Proposition 3.9, it suffices to show $\eta_{8\ell+1}(EIV) \to \eta_{8\ell-3}(EIV)$ sends the 2^7y_2-tower to the 2^7y_2-tower. This is a consequence of the following two propositions. □

Proposition 3.14. *There is an isomorphism of BTSSs*

$$E_r^{s,t}(\Omega EIV) \approx E_r^{s,t+1}(EIV)$$

for $r \geq 2$.

Proposition 3.15. *The differential $d_3 : \eta_{8\ell}(\Omega EIV) \to \eta_{8\ell-4}(\Omega EIV)$ sends the 2^7y_2-tower to the 2^7y_2-tower.*

Proof of Proposition 3.14. We need the following result.

Proposition 3.16. *The algebra $K_*(\Omega EIV)$ is a polynomial algebra on two classes in $K_0(-)$.*

Proof. We first note that $H_*(\Omega EIV)$ is a polynomial algebra on classes in $H_8(-)$ and $H_{16}(-)$. To see this, we first use the computations of the rings $H_*(\Omega F_4; R)$ and $H_*(\Omega E_6; R)$ for $R = \mathbf{Z}/2$ in [15, 2.3] and for $R = \mathbf{Z}/3$ in [13, Thm.1], which imply that $H_*(\Omega EIV; R)$ is polynomial on 8- and 16-dimensional generators. With coefficients $\mathbf{Q}$ or $\mathbf{Z}/p$ for $p > 3$, this is also true because the localizations of F_4 and E_6 are products of spheres whose dimensions are well known. The result with integral coefficients follows by standard methods.

Now the Atiyah-Hirzebruch spectral sequence $H_*(\Omega EIV; K_*) \Rightarrow K_*(\Omega EIV)$ implies the result for $K_*(-)$. □

Returning to the proof of 3.14, we adapt an argument first used in [2, 6.1] to show that for the BP-based unstable Novikov spectral sequence (UNSS) $E_2^{s,t}(\Omega S^{2n+1}) \approx E_2^{s,t+1}(S^{2n+1})$. It was shown in [5, 3.3] that the same argument works if S^{2n+1} is replaced by a space Y for which $BP_*(\Omega Y)$ is the polynomial algebra on classes $\{y_{2i}\}$, and BP_*Y is isomorphic to an exterior algebra on the primitive classes σy_{2i}. It was noted in [6, 4.12] that these arguments can be adapted to the K-based BTSS. This, with Proposition 3.16, yields the proposition when $r = 2$.

To prove the result for all r, we note that the isomorphism of E_2-terms is induced by a map of towers. To see this, we use the following natural map of augmented cosimplicial spaces, where $K(X) = \Omega^\infty(K \wedge \Sigma^\infty X)$.

$$\begin{array}{ccccccccc}
\Omega X & \to & K\Omega X & \rightrightarrows & K(K(\Omega X)) & \substack{\to\\ \to\\ \to} & K^3\Omega X & \substack{\to\\ \to\\ \to\\ \to} & \cdots \\
\downarrow & & \downarrow & & \downarrow & & \downarrow & & \\
\Omega X & \to & \Omega K X & \rightrightarrows & \Omega K(KX) & \substack{\to\\ \to\\ \to} & \Omega K^3 X & \substack{\to\\ \to\\ \to\\ \to} & \cdots
\end{array}$$

Applying $\pi_*(-)$ and taking homology of the alternating sum to the first row yields $E_2^{*,*}(\Omega X)$, and doing this to the second yields $E_2^{*,*-1}(X)$. The induced morphism in homology is the E_2 isomorphism observed above. But these cosimplicial spaces give rise, by the Tot construction, to the towers that define the entire spectral sequence, and so the morphism induces a morphism of spectral sequences. □

Proof of Proposition 3.15. Let F denote the fiber of the inclusion map $S^9 \to EIV$. The Serre spectral sequence of this fibration shows that H^*F, like $H^*(\Omega S^{17})$, is a divided polynomial algebra on a 16-dimensional class. Thus $H_*F = \langle x_{16i} \rangle$ with

$$\psi(x_{16i}) = \sum \binom{i}{j} x_{16j} \otimes x_{16(i-j)}.$$

It follows that the $\mathbf{Z}/2$-graded K_*F has the same coalgebra structure, although the grading is lost. Thus there is an abstract isomorphism of coalgebras $K_*F \approx K_*(\Omega S^{17})$.

This implies that the fibration

$$\Omega S^9 \to \Omega EIV \to F$$

induces a relatively injective extension sequence

$$0 \to K_*(\Omega S^9) \to K_*(\Omega EIV) \xrightarrow{p_*} K_*(F) \to 0,$$

and hence, by [2, 4.3], a long exact sequence of BTSS, commuting with d_3

$$\to E_2^s(\Omega S^9) \to E_2^s(\Omega EIV) \xrightarrow{p_*} E_2^s(F) \to E_2^{s+1}(\Omega S^9) \to .$$

We remark here that the notion of relatively injective extension sequence was defined in [4, 5.14]. It is the notion which was intended in [2, 4.3].

The Hurewicz Theorem gives a map $f : S^{16} \to F$, and we have the map $p : \Omega EIV \to F$ from the above fibration. Both of these maps induce morphisms of BTSS, commuting with d_3. The image under p_* of the eta towers $2^7 y_2$ equal the image under f_* of eta towers in S^{16} which map, in the EHP sequence, to the unstable eta towers in $\eta_{4*+1}(S^{17})$. By [1, p.58] or [3, p.488], the d_3-differential on the unstable eta towers in S^{17} is nonzero from $\eta_{8\ell+1}(S^{17})$ to $\eta_{8\ell-3}(S^{17})$. Hence it is nonzero from $\eta_{8\ell}(S^{16})$ to $\eta_{8\ell-4}(S^{16})$, hence also in F, and thence in ΩEIV. □

4. Fine tuning

In this section we determine the d_3-differentials from the 1-line, prove Proposition 2.9, and show that there are no exotic extensions $(\cdot 2)$ in the BTSS.

In general, the h_1-action on the 1-line of the BTSS can be delicate. See, e.g., [4, 4.15]. It is important, because one way to determine d_3 on the 1-line is to determine d_3 on the eta towers and then use the h_1-action from the 1-line to the eta towers to deduce d_3 on the 1-line. A different method was used for $Sp(n)$ in [5] and illustrated in [5, 4.3] the subtle things that can happen. In that case, the relevant 2-line group contained some classes supporting d_3-differentials and some which did not, and we argued indirectly to see whether d_3x was nonzero, which was equivalent to determining whether h_1x contained any classes of the first type.

In our situation here, the 1-line classes in E_6 map to or from classes in EIV or F_4 on which h_1 and d_3 are known, from which we deduce the following result, which will be proved along with Proposition 2.9.

Proposition 4.1. *The h_1-extensions and d_3-differentials from the* 1- *and* 2*-lines are as depicted in Diagram* 1.3.

Proof of Propositions 2.9 and 4.1. We compare the exact sequences in $(\mathrm{Ext}_A^{*,2n+1})^\#$ and $(\mathrm{Ext}_{\mathrm{GInv}}^{*,2n+1})^\#$ induced by (3.6). Here GInv denotes the category of profinite abelian groups with involution. We will also use Inv, the category of abelian groups with involution.

The analysis is much more delicate when $n = 2k$ is even, and so we focus on this case. Some details of the proof (but not final conclusions) are slightly different when $\nu(k-36) = 8$ due to the $\nu(k - 36 - 2^8)$ which occurs in $E_2(EIV)$, so we assume $\nu(k - 36) \neq 8$ to simplify the exposition.

The exact sequences (2.5) and (2.10) are obtained in [4, §3] from a long exact sequence

$$\begin{aligned}\leftarrow\ & \mathrm{Ext}_A^s(M,N)^\# \leftarrow \mathrm{Ext}_{\mathrm{GInv}}^s(M,N)^\# \overset{\psi_M^3-\psi_N^3}{\longleftarrow} \mathrm{Ext}_{\mathrm{GInv}}^s(M,N)^\# \\ \leftarrow\ & \mathrm{Ext}_A^{s+1}(M,N)^\# \leftarrow\end{aligned}$$

and isomorphisms

$$\begin{aligned} & \mathrm{Ext}_{\mathrm{GInv}}^{s+1,2n+1}(M)^\# = \mathrm{Ext}_{\mathrm{GInv}}^{s+1}(M, QK^1S^{2n+1})^\# \\ \approx\ & \mathrm{Ext}_{\mathrm{Inv}}^s(\mathbf{Z}_{(2)}^{((-1)^n)}, M^\#)^\# \approx \begin{cases} Q_{s+n+1}(M) & s>0 \\ M/(\psi^{-1}-(-1)^n) & s=0. \end{cases}\end{aligned}$$

Then

$$\begin{aligned} 0\ \leftarrow\ & \mathrm{Ext}_{\mathrm{GInv}}^{1,4k+1}(\overline{K}F_4)^\# \leftarrow \mathrm{Ext}_{\mathrm{GInv}}^{1,4k+1}(\overline{K}E_6)^\# \leftarrow \mathrm{Ext}_{\mathrm{GInv}}^{1,4k+1}(\overline{K}EIV)^\# \\ \leftarrow\ & \mathrm{Ext}_{\mathrm{GInv}}^{2,4k+1}(\overline{K}F_4)^\# \leftarrow \mathrm{Ext}_{\mathrm{GInv}}^{2,4k+1}(\overline{K}E_6)^\# \leftarrow \mathrm{Ext}_{\mathrm{GInv}}^{2,4k+1}(\overline{K}EIV)^\# \\ \leftarrow\ & \mathrm{Ext}_{\mathrm{GInv}}^{3,4k+1}(\overline{K}F_4)^\# \end{aligned}$$

is

$$\begin{aligned} 0 &\leftarrow \overline{K}F_4/(\psi^{-1}-1) \leftarrow \overline{K}E_6/(\psi^{-1}-1) \leftarrow \overline{K}EIV/(\psi^{-1}-1) \\ &\leftarrow Q_{\mathrm{ev}}(\overline{K}F_4) \leftarrow Q_{\mathrm{ev}}(\overline{K}E_6) \leftarrow Q_{\mathrm{ev}}(\overline{K}EIV) \leftarrow Q_{\mathrm{od}}(\overline{K}F_4). \end{aligned}$$

Using results and notation from Sections 2 and 3, this sequence becomes, with $x_c = 2^6 5x_1 + 2^5 27x_2$,

$$0 \leftarrow \langle x_2\rangle \leftarrow \langle v_3, \mathbf{v}_1\rangle \leftarrow \langle \mathbf{y}_1\rangle \leftarrow \langle x_c, 2^{17}x_1\rangle \leftarrow \langle 2^{17}v_1\rangle \leftarrow \langle y_1\rangle \leftarrow \langle x_1, x_2\rangle,$$

with all elements having order 2 except $\mathbf{v}_1$ and $\mathbf{y}_1$, which have order 2^{12}. Using 3.5, the explicit morphisms, yielding an exact sequence, are

$$(4.2)\qquad \begin{array}{ccccccccc} x_2 & \hookleftarrow & v_3 & & 0 & \hookleftarrow & 2^{17}x_1 & \hookleftarrow & 2^{17}v_1 \\ x_1+x_2 & \hookleftarrow & \mathbf{v}_1 & & 2^{11}\mathbf{y}_1 & \hookleftarrow & x_c & & 0 & \hookleftarrow y_1 \hookleftarrow x_1 \\ & & 2\mathbf{v}_1 & \hookleftarrow & \mathbf{y}_1. & & & & \end{array}$$

With $\nu = \min(12, \nu(k-36)+3)$, the nonzero occurrences of $\psi^3 - 3^{2k}$ on the groups in this sequence are $x_1 \mapsto x_2$, $\mathbf{v}_1 \mapsto 2^\nu \mathbf{v}_1 + v_3$, $\mathbf{y}_1 \mapsto 2^\nu \mathbf{y}_1$, $x_c \mapsto 2^{17}x_1$, and $x_1 \mapsto x_2$. The reader can verify that these commute with the morphisms of (4.2).

Now the short exact sequences of (2.5) and (2.10), and the isomorphism (2.3), written vertically, become as follows. We write $E_2^s(X)$ for $\mathrm{Ext}_A^{s,4k+1}(\overline{K}X)$, and denote by $x(2^e)$ an element x of order 2^e. Elements not followed by parentheses have order 2.

$$\begin{array}{ccccccc} x_1 & \leftarrow & \mathbf{v}_1(2^{\nu+1}) & \xleftarrow{\cdot 2} & \mathbf{y}_1(2^\nu) & & \\ \downarrow & & \downarrow & & \downarrow & & \\ E_2^1(F_4)^\# & \leftarrow & E_2^1(E_6)^\# & \leftarrow & E_2^1(EIV)^\# & \leftarrow & \end{array}$$

$$\begin{array}{ccccccc} x_c & & 2^{17}v_1 & & y_1 & \leftarrow & x_1 \\ \downarrow & & \downarrow & & \downarrow & & \downarrow \\ E_2^2(F_4)^\# & \leftarrow & E_2^2(E_6)^\# & \leftarrow & E_2^2(EIV)^\# & \leftarrow & E_2^3(F_4)^\# \\ \downarrow & & \downarrow & & \downarrow & & \\ x_2 & \leftarrow & v_3, 2^{12-\nu}\mathbf{v}_1(2^\nu) & \xleftarrow{2} & 2^{12-\nu}\mathbf{y}_1(2^\nu). & & \end{array}$$

In order for the $E_2^s(-)^\#$ sequence to be exact, $2^{12-\nu}\mathbf{v}_1$ must hit x_c in $E_2^2(F_4)^\#$, and $2^{11}\mathbf{y}_1$ must hit $2^{17}v_1$ in $E_2^2(E_6)^\#$. The latter implies the nontrivial extension in $E_2^2(E_6)^\#$, that it contains a $\mathbf{Z}/2^{\nu+1}$ with $2^{17}v_1$ the element of order 2 in this summand. This concludes the proof of 2.9.

The above analysis showed $E_2^{1,4k+1}(E_6)^\# \leftarrow E_2^{1,4k+1}(EIV)^\#$ injective; hence

$$E_2^{1,4k+1}(E_6) \to E_2^{1,4k+1}(EIV)$$

is surjective. In 3.9 and the proof of 3.12, we showed that the stable eta tower in $E_2^{*,4k+2*-1}(E_6)$ maps to the one in EIV which comes from S^9. If $\nu(k) \le 1$, $E_2^{1,4k+1}(S^9) \to E_2^{1,4k+1}(EIV)$ is an isomorphism. So the h_1-extension and d_3-differential from $E_2^{1,8\ell+5}(S^9)$ implies the same in E_6. A nonzero h_1-extension from $E_2^{1,8\ell+1}(E_6)$ is implied when ℓ is odd. It is not so important because d_3 must be

0 on $E_2^{1,8\ell+1}(E_6)$ since h_1^2 times the $\mathbf{Z}/2$ in $(8\ell+1,2)$ is in the image of d_3. It is likely that an adaptation of [4, 3.7] to modules which do not satisfy $\psi^{-1}=-1$ would allow the determination of h_1 on $E_2^{1,8\ell+1}(E_6)$.

Since $E_2^{1,4k+3}(F_4)\to E_2^{1,4k+3}(E_6)$ is an isomorphism, the h_1-extensions and d_3-differentials from $E_2^{1,4k+3}(E_6)$ follow from those in F_4. In [4, 4.14], it was shown that h_1 on $E_2^{1,4k+3}(F_4)$ hits the stable eta tower iff $\nu(k-5)\neq 3$ and hits the unstable eta tower iff $\nu(k-5)=3$. This implies the nonzero h_1 and d_3 from $E_2^{1,8\ell+3}(E_6)$ pictured in Diagram 1.3, and that h_1 is nonzero from $E_2^{1,8\ell-1}(E_6)$ unless $\ell\equiv 7 \bmod 8$, in which case it is 0. The latter is because the unstable eta tower in F_4 in this dimension maps to 0 in E_6. That h_1x is sometimes 0 here is mildly interesting because in [4, 3.8] it was shown that if $\psi^{-1}=-1$ then h_1 acts injectively on $E_2^1(X)/2$.

The analysis of exact sequences earlier in this proof implied that

$$E_2^{2,4k+1}(E_6)\to E_2^{2,4k+1}(EIV)$$

sends the $\mathbf{Z}/2^e$ summand surjectively, and that the generator is dual to $2^{17}v_1$, which is the name of the unstable eta tower whose d_3-differential is described in 3.13. The h_1-extension and d_3-differential on it follow immediately.

The exact sequence in $\mathrm{Ext}_A^{1,4k+3}(-)$ induced by (3.6) easily implies that $E_2^{2,4k+3}(F_4)\to E_2^{2,4k+3}(E_6)$ sends the $\mathbf{Z}/2^f$ bijectively. The h_1-extension (for all k) and d_3-differential when k is odd follow from those in F_4 established in [4]. □

Finally, we have the following result about extensions $(\cdot 2)$ in the spectral sequence.

Proposition 4.3. *The extensions from $E_2^{s,t}(E_6)$ to $E_2^{s+2,t+2}(E_6)$ are all trivial.*

Proof. For $s=1$ or 2, the group $E_2^{s,8\ell+3}(E_6)\approx\mathbf{Z}/2^6$ is mapped isomorphically onto from $E_2^{s,8\ell+3}(F_4)$. In [4, 4.14], it is shown that when $s=1$ (resp. $s=2$) there is a nonzero extension into the element of $E_2^{s+2,8\ell+5}(F_4)$ which is part of the stable (resp. unstable) eta tower. In Proposition 3.9, it is proved that these eta towers map trivially to $E_2^{s+2,8\ell+5}(E_6)$, hence this extension is 0 in E_6.

In the proof of Propositions 2.9 and 4.1, it was shown that for $s=1$ and 2, the element of order 2 in $E_2^{s,8\ell+1}(E_6)$ is in the image from $E_2^{s,8\ell+1}(F_4)$, which does not support an extension in [4, 4.9]. □

5. The CTP for E_6

In this brief section, we mention several issues related to the CTP for E_6. In other words, how might we prove that the groups $v_1^{-1}\pi_*(\widehat{E}_6)$, which we have computed, are in fact isomorphic to $v_1^{-1}\pi_*(E_6)$, which is what we really want?

Not all spaces satisfy the CTP. One example which does not, pointed out to the author by Bousfield, is the Eilenberg-MacLane space $K(\Sigma_\infty,1)$, where Σ_∞ is the infinite symmetric group. On the other hand, the only spaces which are known to satisfy it are certain spaces which can, in some sense, be built from spheres by fibrations.

Localized at an odd prime, the result of [7] which states that a space which is KASR and K-durable satisfies the CTP is very useful. If the 2-primary analogue were known to be true, then we would know that E_6 satisfies the CTP. However, the proof of the result of [7] relies on much delicate machinery developed at the odd primes by Bousfield, especially in [8]. One crucial difference between the odd primes and the prime 2 is that $\mathrm{Ext}^s_A(M,N)$ vanishes for $s > 2$ when p is odd. This is important for existence and uniqueness of realizations of certain morphisms of Adams modules. An adaptation to the prime 2 would be far from straightforward.

A main worry in considering convergence of spectral sequences related to $v_1^{-1}\pi_*(X)$ is the possibility that there could be elements in $v_1^{-1}\pi_*(X)$ not seen by the spectral sequence because they correspond to a family of elements in $\pi_{n_i}(X)$ for $n_i \to \infty$, related to one another by $\cdot v_1^e$ and having increasing filtrations. The algebraic $\cdot v_1$ operation in the UNSS or BTSS preserves filtration, but the homotopy-theoretic operation can increase filtration. For S^{2n+1}, the E_∞-term of the v_1-periodic BTSS yields exactly the v_1-periodic homotopy classes, which we know because they were calculated by another method in [16] ($p = 2$) and [17] (p odd). Spaces built from odd spheres, or perhaps their loop spaces, by fibrations can then guarantee nonexistence of undetected v_1-periodic families by exactness properties. The proof in [7] that at the odd primes X KASR and K-durable implies the CTP for X ultimately boils down to building the localization X_K from various S^{2n+1}_K.

The isomorphism $E_2^{s,t}(S^{2n+1}) \approx E_2^{s,t-1}(\Omega S^{2n+1})$ implies that if a space is nicely fibered by ΩS^{2n+1}'s, then it satisfies the CTP. For spaces fibered by $\Omega^2 S^{2n+1}$, it is not so clear.

If it could be proved that F, the fiber of $S^9 \to EIV$ considered in the proof of 3.15, has the same homotopy type as ΩS^{17}, then the CTP for E_6 could be deduced. In [9, 2.1], a map $\Omega^2 S^{17} \xrightarrow{f} \Omega S^9$ was constructed. If it could be shown that the composite

$$\Omega^2 S^{17} \xrightarrow{f} \Omega S^9 \xrightarrow{\Omega i} \Omega EIV$$

is null-homotopic, then one could deduce existence of a fibration

$$\Omega^2 S^9 \to \Omega^2 EIV \to \Omega^2 S^{17}.$$

It is likely that such a fibration would imply the CTP for E_6, although a detailed argument has not been produced.

References

[1] M. Bendersky, *The v_1-periodic unstable Novikov spectral sequence*, Topology **31** (1992) 47–64.

[2] M. Bendersky, E.B. Curtis, and D.C. Ravenel, *EHP sequences in BP theory*, Topology **21** (1982) 373–391.

[3] M. Bendersky and D.M. Davis, *2-primary v_1-periodic homotopy groups of $SU(n)$*, Amer. Jour. Math. **114** (1991) 529–544.

[4] ——, *A stable approach to an unstable homotopy spectral sequence*, Topology **42** (2003) 1261–1287.

[5] M. Bendersky, D.M. Davis, and M. Mahowald, *v_1-periodic homotopy groups of $Sp(n)$*, Pac. Jour. Math. **170** (1995) 319–378.

[6] M. Bendersky and R.D. Thompson, *The Bousfield-Kan spectral sequence for periodic homology theories*, Amer. Jour. Math. **122** (2000) 599–635.

[7] ——, *Some properties of the K-completion*, Contemp. Math. Amer. Math. Soc. **293** (2002) 35–44.

[8] A.K. Bousfield, *The K-theory localization and v_1-periodic homotopy groups of H-spaces*, Topology **38** (1999) 1239–1264.

[9] F.R. Cohen and P.S. Selick, *Splittings of two function spaces*, Quar. Jour. Math. Oxford **41** (1990) 145–153.

[10] D.M. Davis, *From representation theory to homotopy groups*, Mem. Amer. Math. Soc. **759** (2002).

[11] D.M. Davis and M. Mahowald, *Some remarks on v_1-periodic homotopy groups*, London Math. Society Lecture Notes **176** (1992) 293–327.

[12] ——, *Three contributions to the homotopy theory of the exceptional Lie groups G_2 and F_4*, Jour. Math. Soc. Japan **43** (1991) 661–671.

[13] H. Hamanaka and S. Hara, *The mod 3 homology of the space of loops on the exceptional Lie groups and the adjoint action*, Jour. Math. Kyoto Univ. **37** (1997) 441–453.

[14] L. Hodgkin, *On the K-theory of Lie groups*, Topology **6** (1967) 1–36.

[15] A. Kono and K. Kozima, *The mod 2 homology of the space of loops on the exceptional Lie group*, Proc. Royal Soc. Edin. **112A** (1989) 187–202.

[16] M. Mahowald, *The image of J in the EHP sequence*, Annals of Math. **116** (1982) 65–112.

[17] R.D. Thompson, *The v_1-periodic homotopy groups of an unstable sphere at odd primes*, Trans. Amer. Math. Soc. **319** (1990) 535–559.

[18] T. Watanabe, *The Chern character homomorphism of the compact simply-connected exceptional Lie group E_6*, Osaka Jour. Math. **28** (1991) 663–681.

Donald M. Davis
Lehigh University
Bethlehem, PA 18015, USA
e-mail: `dmd1@@lehigh.edu`

Progress in Mathematics, Vol. 215
Algebraic Topology: Categorical Decomposition Techniques, 125–151

The Homotopy of $L_2V(1)$ for the Prime 3

Paul Goerss, Hans-Werner Henn, and Mark Mahowald

Abstract. Let $V(1)$ be the Toda-Smith complex for the prime 3. We give a complete calculation of the homotopy groups of the L_2-localization of $V(1)$ by making use of the higher real K-theory EO_2 of Hopkins and Miller and related homotopy fixed point spectra. In particular we resolve an ambiguity which was left in an earlier approach of Shimomura whose computation was almost complete but left an unspecified parameter still to be determined.

0. Introduction

The chromatic approach offers at present the most attractive perspective on the stable homotopy category of finite complexes. For any natural prime p there is a tower of localization functors L_n with natural transformations $L_n \to L_{n-1}$ where L_n is Bousfield localization with respect to a certain multiplicative homology theory $E(n)_*$. For a finite complex X the homotopy inverse limit of these localizations gives the p-localization of X. The study of the localization functors L_n is sometimes referred to as the study of the chromatic primes in stable homotopy theory. For more details the reader may consult [Ra3].

The solution of the Adams conjecture led to a good conceptual and calculational understanding of the localization functor L_1 if p is any prime. The case of L_2 is reasonably well understood for primes $p > 3$ at least from a computational point of view [SY]. The case of L_2 at the primes $p = 3$ and $p = 2$ is harder. The standard approach to understand the L_2-localization L_2S^0 (at any prime) is to study L_2X for a "suitable" finite complex X and to work one's way back to L_2S^0 through appropriate Bockstein spectral sequences arising from the skeletal filtration of X. At odd primes the Toda-Smith complexes $V(1)$ (which are defined as cofibre of a self map A of the mod-p Moore spectrum $V(0)$ such that A induces multiplication by v_1 in Brown-Peterson theory BP_*) are suitable in this sense. For primes $p > 3$ the homotopy of $L_2V(1)$ is relatively easy to understand; the Adams-Novikov spectral sequence (ANSS for short) converging to $\pi_*L_2V(1)$ collapses at

The first author was partially supported by the National Science Foundation (USA). The authors would like to thank (in chronological order) the University of Heidelberg, MPI at Bonn, Université Louis Pasteur at Strasbourg and Northwestern University for providing them with the opportunity to work together.

E_2 and its E_2-term is known [Ra1]. Starting from this information Shimomura and Yabe were able to compute the homotopy of L_2S^0 for all primes $p > 3$.

At the prime 3 it is natural to try the same strategy and start with studying $\pi_*L_2V(1)$. In fact, the E_2-term of the ANSS converging to $\pi_*L_2V(1)$ has been computed in [H] (see also [GSS] and [Sh1]) but this time the ANSS for $V(1)$ does not collapse. Using various information about homotopy groups of spheres and related complexes in low dimensions Shimomura studied this spectral sequence and arrived at a calculation modulo some ambiguity; there was an unspecified parameter $k \in \{0,1,2\}$ and several families of homotopy elements which lived in degrees which were only determined up to adding $24k$.

The E_2-term of the ANSS for $L_2V(1)$ can be identified by Morava's change of ring isomorphism with the continuous cohomology of a certain 3-adic Lie group $\mathbb{G}_2$, (the "extended Morava stabilizer group") with coefficients in $\mathbb{F}_9[u^{\pm 1}]$. Devinatz and Hopkins [DH2] have given a homotopy theoretic interpretation of Morava's change of rings isomorphism: the localization $L_{K(2)}S^0$ of the sphere with respect to the second Morava K-theory $K(2)$ has the homotopy type of the homotopy fixed point spectrum $E_2^{h\mathbb{G}_2}$ which is defined by making use of the Hopkins-Miller rigidification of the action of $\mathbb{G}_2$ on the Lubin-Tate spectrum E_2 [Re].

In this paper we analyze the ANSS for $L_2V(1)$ by making serious use of group theoretic and cohomological properties of $\mathbb{G}_2$. In fact, $L_2V(1)$ can be identified with $L_{K(2)}V(1) \simeq L_{K(2)}S^0 \wedge V(1) \simeq E_2^{h\mathbb{G}_2} \wedge V(1)$ and we will study $L_2V(1)$ by comparing it with $E_2^{hN} \wedge V(1)$ where E_2^{hN} is the homotopy fixed point spectrum with respect to the normalizer of an element of order 3 in $\mathbb{S}_2$ (the classical stabilizer group). The use of the spectrum E_2^{hN} was suggested by the calculation of the E_2-term of the ANSS in [H] which made heavy use of centralizers of elements of order 3 in $\mathbb{S}_2$. The other good property of E_2^{hN} on which our method relies is that it can be analyzed in terms of the Hopkins-Miller higher real K-theory spectrum EO_2 at the prime 3. Both properties together allow us to give an independent calculation of $\pi_*L_2V(1)$ which is complete and identifies Shimomura's parameter as $k = 1$.

To state our main result we need some notation. First of all, from now on all spaces or spectra will be localized at 3.

The homotopy of $L_2V(1)$ is annihilated by 3 and is a module over the homotopy of $L_{K(2)}S^0$. Therefore it can be regarded as a module over the algebra $\mathbb{F}_3[\beta] \otimes \Lambda(\zeta)$ where β is the image of the generator $\beta_1 \in \pi_{10}S^0$ and ζ is in $\pi_{-1}L_{K(2)}S^0$. The homotopy groups turn out to be periodic of period 144 and on the E_2-level this periodicity corresponds to multiplication by v_2^9 where v_2 is the polynomial generator in $\pi_{16}BP$. We do not prove that this periodicity arises geometrically but it is convenient to describe $\pi_*L_2V(1)$ nevertheless as a module over $P := \mathbb{F}_3[v_2^{\pm 9}, \beta] \otimes \Lambda(\zeta)$. We will use notation like $P/(\beta^5)\{v_2^l\}_{l=0,1,5}$ to denote the direct sum of P-modules each of which is killed by β^5 (more precisely its annihilator ideal is the ideal generated by β^5) and which have generators named $1 = v_2^0$, v_2 and v_2^5.

Finally we note that the E_2-term has a product structure and it contains elements which deserve to be named v_2 (which is closely related to the generator $v_2 \in \pi_{16}(BP)$), α (which detects the image of the generator $\alpha_1 \in \pi_3(S^0)$), $v_2^{1/2}\beta$, $v_2^{1/2}\alpha$, βa_{35}, αa_{35}, $v_2^{1/2}\beta a_{35}$ and $v_2^{1/2}\beta\alpha a_{35}$ and which live in total degree 16, 3, 18, 11, 45, 38, 53 and 56 respectively. The reason for these names will become clear once we have discussed the spectrum E_2^{hN} and in particular $E_2^{hN} \wedge V(1)$ (see Section 2.3). With these conventions our main result reads as follows (see Theorem 20).

Theorem. *As a module over* $P = \mathbb{F}_3[v_2^{\pm 9}, \beta] \otimes \Lambda(\zeta)$ *there is an isomorphism*

$$\begin{aligned}\pi_* L_2V(1) \cong\ & P/(\beta^5)\{v_2^l\}_{l=0,1,5} \oplus P/(\beta^3)\{v_2^l\alpha\}_{l=0,1,2,5,6,7} \\ \oplus\ & P/(\beta^4)\{v_2^{l+1/2}\beta\}_{l=0,4,5} \oplus\ P/(\beta^2)\{v_2^{l+1/2}\alpha\}_{l=0,1,2,4,5,6} \\ \oplus\ & P/(\beta^4)\{v_2^l\beta a_{35}\}_{l=0,1,5} \oplus\ P/(\beta^3)\{v_2^l\alpha a_{35}\}_{l=0,1,2,5,6,7} \\ \oplus\ & P/(\beta^5)\{v_2^{l+1/2}\beta a_{35}\}_{l=0,4,5} \oplus\ P/(\beta^2)\{v_2^{l+1/2}\beta\alpha a_{35}\}_{l=0,1,2,4,5,6} .\end{aligned}$$

We note that this result has already been used to carry out the program to calculate $\pi_*(L_2S^0)$ that we mentioned above [Sh2], [SW].

The paper is organized as follows. In the first section we discuss the spectrum EO_2 and $EO_2 \wedge V(1)$. In particular we give a detailed discussion of the ANSS converging towards $\pi_*(EO_2)$. In the second section we introduce the homotopy fixed point spectrum E_2^{hN} and we relate E_2^{hN} to EO_2, and consequently $E_2^{hN} \wedge V(1)$ to $EO_2 \wedge V(1)$. In the final section we compare $L_2V(1)$ with $E_2^{hN} \wedge V(1)$ and prove the main theorem.

1. The homotopy of EO_2 and $EO_2 \wedge V(1)$

1.1. The spectrum EO_2

We begin by recalling the construction of the spectrum EO_2 due to Hopkins and Miller. We refer to [Re] for more details.

The point of departure is the Lubin-Tate deformation theory of formal group laws (cf. [LT]), in particular the universal deformation of the formal group law Γ of height 2 over the field $\mathbb{F}_9$ with $[p]$-series $[p](x) = x^9$. The universal deformation is a lift of Γ to a formal group law $\widetilde{\Gamma}$ over $\mathbb{W}_{\mathbb{F}_9}[[u_1]]$ (where $\mathbb{W}_{\mathbb{F}_9}$ are the Witt vectors of $\mathbb{F}_9$ and u_1 is a formal power series variable). Over the graded ring $\mathbb{W}_{\mathbb{F}_9}[[u_1]][u^{\pm 1}]$ (where u is of degree -2 and u_1 of degree 0) this formal group law is isomorphic to the one induced from the universal p-typical formal group law over $BP_* \cong \mathbb{Z}_{(p)}[v_1, v_2, \ldots]$ via the map of algebras which sends v_1 to u_1u^{1-p}, v_2 to u^{1-p^2} and v_n to 0 for $i > 2$. The Landweber exact functor theorem implies that there is a homology theory $(E_2)_*$ represented by a ring spectrum E_2 with coefficients $\pi_*(E_2) = \mathbb{W}_{\mathbb{F}_9}[[u_1]][u^{\pm 1}]$ such that the cohomology theory $(E_2)^*$ is complex oriented with associated formal group law $\widetilde{\Gamma}$. To simplify notation we will abbreviate $\mathbb{W}_{\mathbb{F}_9}$ by $\mathbb{W}$ and E_2 by E throughout.

The group $\mathbb{S}_2$ of automorphisms of Γ (also known as the Morava stabilizer group) acts on the ring spectrum E, up to homotopy, and this action extends in a canonical way to an action of the extended stabilizer group, given as the semidirect product $\mathbb{G}_2 := \mathbb{S}_2 \rtimes C_2$ where the cyclic group C_2 of order 2 acts on $\mathbb{S}_2$ via Galois automorphisms of Γ. (Note that Γ is defined over $\mathbb{F}_3$ and thus we get an action of the Galois group of the extension $\mathbb{F}_3 \subset \mathbb{F}_9$ on $\mathbb{S}_2$.)

Hopkins and Miller have shown how to rigidify this action to a genuine action via A_∞-maps [Re] and subsequently Devinatz and Hopkins [DH2] have shown how to construct homotopy fixed point spectra E^{hH} with respect to closed subgroups H of $\mathbb{G}_2$. Their construction has the following properties: it agrees in the case of finite subgroups with the naive construction of the homotopy fixed point spectrum, if $H = \mathbb{G}_2$ then $E^{hH} \simeq L_{K(2)}S^0$, and for any closed subgroup H and any finite spectrum X there is a spectral sequence

$$E_2^{s,t}(X) = H^s_{cts}(H, E_t(X)) \Longrightarrow \pi_{t-s}(E^{hH} \wedge X)$$

where H^*_{cts} denotes continuous cohomology of the p-adic group H.

The group $\mathbb{S}_2$ can also be identified with the group of units of the maximal order $\mathcal{O}_2$ of the division algebra $\mathbb{D}_2$ over the 3-adic rationals $\mathbb{Q}_3$. The maximal order is a free $\mathbb{W}$-module of rank 2 with basis 1 and S; as a ring it is determined by the relations $S^2 = 3$ and $Sa = \phi(a)S$ if ϕ notes the lift of Frobenius from $\mathbb{F}_9$ to $\mathbb{W}$ [Ra2, Appendix 2]. From this point of view the extended group $\mathbb{G}_2$ is the split extension $\mathbb{S}_2 \rtimes C_2$ where the action of C_2 is given by conjugation with S in $\mathbb{D}_2$.

Let ω be a fixed 8th root of unity in $\mathbb{W}$. The element $s := -\frac{1}{2}(1 + \omega S)$ is easily checked to be of order 3. Furthermore $\omega^2 s\omega^{-2} = s^2$ so that s and $t := \omega^2$ generate a subgroup G_{12} of $\mathbb{S}_2$ which is isomorphic to $C_3 \rtimes C_4$ with C_4 acting non-trivially on C_3. The spectrum EO_2 is defined as the homotopy fixed point spectrum $E^{hG_{12}}$. We add that G_{12} is a maximal finite subgroup of $\mathbb{S}_2$ and every other maximal finite subgroup is conjugate to G_{12}.

1.2. The E_2-term of the ANSS converging to $\pi_*(EO_2)$

We do not claim any originality for the results in this and the following subsection. The ANSS for EO_2 was first investigated by Hopkins and Miller but unfortunately their work remains unpublished. There is a rather brief discussion of this spectral sequence in the still unpublished paper [N]. A discussion of its E_2-term from a different point of view can be found in [GS]. Neither of these sources suits well our needs and therefore we have decided to give a self-contained treatment here.

In order to describe the E_2-term $E_2^{s,t} \cong H^s(G_{12}, E_t)$ we start by analyzing E_* as a G_{12}-algebra. The first step is to locate an appropriate subrepresentation in E_{-2}.

Let χ be the representation of G_{12} on $\mathbb{W}$ which is trivial on s and on which t acts by multiplication by ω^2. Define a G_{12}-module ρ by the short exact sequence

$$0 \to \chi \to \mathbb{W}[G_{12}] \otimes_{\mathbb{W}[C_4]} \chi \to \rho \to 0$$

in which in the middle term χ is considered as a representation of the subgroup C_4 generated by t and where the first map takes a generator e of χ to

$$(1+s+s^2)e := (1+s+s^2) \otimes e \in \mathbb{W}[G_{12}] \otimes_{\mathbb{W}[C_4]} \chi \,.$$

Lemma 1. *There is a morphism of G_{12}-modules*

$$\rho \longrightarrow E_{-2}$$

so that the induced map

$$\rho \otimes_{\mathbb{W}} \mathbb{F}_9 \to E_{-2} \otimes_{E_0} E_0/(3, u_1^2)$$

is an isomorphism.

Proof. We need to know something about the action of $\mathbb{G}_2$ on E_*. Let $\mathfrak{m} = (3, u_1) \subseteq E_0$ be the maximal ideal. Then Proposition 3.3 and Lemma 4.9 of [DH1] together imply that, modulo $\mathfrak{m}^2 E_{-2}$

$$\begin{aligned} s_*(u) &\equiv -\frac{1}{2}(u + \omega^3 u u_1) \\ s_*(uu_1) &\equiv -\frac{1}{2}(3\omega u + uu_1) \\ t_*(u) &\equiv \omega^2 u \\ t_*(uu_1) &\equiv -\omega^2 uu_1 \,. \end{aligned}$$

(In fact, the formulae for t_* are true on the nose.) In particular, we see that $E_{-2} \otimes_{E_0} E_0/(3, u_1^2)$ is isomorphic to $\rho \otimes_{\mathbb{W}} \mathbb{F}_9$ as a G_{12}-module and that the residue class of u is a G_{12}-module generator. Thus we would like to find a class $x \in E_{-2}$ with the same reduction as u so that $t_*(x) = \omega^2 x$ and $x + s_*(x) + s_*^2(x) = 0$. In fact, because the action of t on $E_{-2} \otimes_{E_0} E_0/(3, u_1^2)$ is diagonalizable with distinct eigenvectors it suffices to find x such that $x \equiv u \bmod (3, u_1)$, up to a scalar in $\mathbb{F}_9^\times$.

Such an x can be obtained as follows: we start with the element $u^{-2}u_1$ which is the image of $v_1 \in BP_*$ with respect to the map $BP_* \to E_*$ which classifies $\widetilde{\Gamma}$ (we will denote this element simply by v_1 in the sequel); v_1 is invariant modulo 3 with respect to the action of all of $\mathbb{S}_2$. More precisely, the structure formulae in BP_*BP [Ra2, Appendix 2] yield

$$g_*(v_1) = v_1 + (3 - 3^3)t_1(g) \equiv v_1 + 3t_1(g) \bmod (3^2)$$

for every $g \in \mathbb{S}_2$. Here we use the identification of E_*E with the continuous functions from the profinite group $\mathbb{S}_2$ with values in E_* equipped with the profinite topology (see [St, Thm. 12] for a convenient reference) and $t_1 \in E_4E$ is the image of the element with the same name in BP_*BP.

By definition of t_1 we have $t_1(-\frac{1}{2}(1+\omega S)) \equiv \omega u^{-2} \bmod (3, u_1)$. Hence, if $z = \frac{1}{3}(s_*(v_1) - v_1)$ then

$$z \equiv \omega u^{-2} \mod (3, u_1) ,$$

in particular z is non-zero. Clearly we have $z + s_*(z) + s_*^2(z) = 0$ but z does not yet have the right degree.

Therefore we consider the class

$$y = us_*(u)s_*^2(u)z .$$

Then

$$y \equiv \omega u \mod (3, u_1)$$

and we still have

$$y + s_*(y) + s_*^2(y) = 0 .$$

However, y might not yet have the correct invariance property with respect to the element t of order 4. Therefore we average and set

$$x = \frac{1}{4}(y + \omega^{-2}t_*(y) + \omega^{-4}t_*^2(y) + \omega^{-6}t_*^3(y)) .$$

Then we get $x \equiv \omega u \bmod(3, u_1^2)$ and we are done. □

The morphism of G_{12}-modules constructed in the lemma defines a morphism of $\mathbb{W}[G_{12}]$-algebras

$$S(\rho) \longrightarrow E_*$$

where $S(\rho)$ denotes the symmetric algebra on ρ. We note that as an algebra $S(\rho)$ is polynomial over $\mathbb{W}$ on two generators e and $s_*(e)$, and we can choose e such that its image is the element x of the proof of the lemma above and is therefore invertible in E_*. Let

$$N = \prod_{g \in G_{12}} g_* e \in S(\rho) ;$$

then we have a morphism of $\mathbb{W}[G_{12}]$-algebras

$$S(\rho)[N^{-1}] \longrightarrow E_* .$$

Note that inverting N inverts e as well, but in an invariant manner. Let $I \subset S(\rho)[N^{-1}]$ be the preimage of the maximal ideal $\mathfrak{m} = (3, u_1) \subset E_*$ (now considered as a homogeneous graded ideal).

Proposition 2. *The induced map of complete algebras*

$$S(\rho)[N^{-1}]_I^\wedge \longrightarrow E_*$$

is an isomorphism.

Proof. It is enough to show that the induced maps

$$\frac{I^k}{I^{k+1}} \to \frac{\mathfrak{m}^k}{\mathfrak{m}^{k+1}}$$

are isomorphisms for each k.

If we identify $S(\rho)$ with $\mathbb{W}[e, s_*(e)]$ then we get $N = -\big(es_*(e)(e+s_*(e))\big)^4$. Furthermore it is straightforward to check that the homogeneous graded ideal I is generated by 3 and $e - s_*(e)$ and that the maps in question are isomorphisms. □

Our next step is to study $H^*(G_{12}, S(\rho))$. We will see in Theorem 6 below how the calculation of the E_2-term can be reduced to that of $H^*(G_{12}, S(\rho))$.

If $e \in \rho$ is the generator, let d be the multiplicative norm of e with respect to the subgroup generated by s, i.e., $d = es_*(e){s^2}_*(e)$. We note that d is of degree -6, it is invariant with respect to s and furthermore $d^4 = -N$.

For a finite group G and any G module M, let

$$\mathrm{tr}_G = \mathrm{tr} : M \longrightarrow M^G = H^0(G, M)$$

be the transfer: $\mathrm{tr}(x) = \sum_{g\in G} gx$. The following calculates $H^*(G_{12}, S(\rho))$ completely if $* > 0$ and gives partial information if $* = 0$; an element listed as being in bidegree (s,t) is in $H^s(G, S(\rho)_t)$. (Note that for this the t-degree of ρ is -2!)

Lemma 3. *Let $C_3 \subseteq G_{12}$ be the normal subgroup generated by s. Then there is an exact sequence*

$$S(\rho) \xrightarrow{tr} H^*(C_3, S(\rho)) \to \mathbb{F}_9[b, d] \otimes \Lambda(a) \to 0$$

where a has bidegree $(1, -2)$, b has bidegree $(2, 0)$ and d has bidegree $(0, -6)$. Furthermore the action of t is described by

$$t_*(a) = -\omega^2 a \qquad t_*(b) = -b \qquad t_*(d) = \omega^6 d\ .$$

(By abuse of notation we have denoted the image of the invariant class d in the quotient $H^0(C_3, S_{-6}(\rho))/Im\ tr$ still by d.)

Proof. Let F be the G_{12}-module $\mathbb{W}[G_{12}] \otimes_{\mathbb{W}[C_4]} \chi$. We can choose a $\mathbb{W}$-basis x_1, x_2, x_3 of F such that $x_2 := s_*(x_1)$ and $x_3 := s_*(x_2)$, and then we get an identification

$$S(F) = \mathbb{W}[x_1, x_2, x_3]$$

with all x_i in degree -2. The kernel of the canonical C_3-linear algebra map which sends F to ρ is the principal ideal generated by $\sigma_1 = x_1 + x_2 + x_3$, i.e., we have a short exact sequence of graded C_3-modules

$$(*) \qquad 0 \to S(F) \otimes \chi \to S(F) \to S(\rho) \to 0\ .$$

(In the tensor product χ has to be treated as a representation in degree -2 in order to make the maps degree preserving.)

As a C_3-module $S(F)$ splits into a direct sum of free modules and trivial modules where the trivial modules are generated by the powers of the monomial $\sigma_3 := x_1x_2x_3$. Therefore we obtain a short exact sequence

$$S(F) \xrightarrow{\mathrm{tr}} H^*(C_3, S(F)) \to \mathbb{F}_9[b, \sigma_3] \to 0$$

where b has bidegree $(2,0)$ and σ_3 has bidegree $(0,-6)$. Here b is a generator of $H^2(C_3,\mathbb{W}) \cong \mathbb{W}/3$ and $\mathbb{W} \subseteq S(F)$ is the submodule generated by the unit of the algebra $S(F)$. The action of t is given by the following formula

$$t_*(\sigma_3) = \omega^6\sigma_3 = -\omega^2\sigma_3 \qquad \text{and} \qquad t_*(b) = -b.$$

The short exact sequence (*) and the fact that $H^1(C_3,S(F)) = 0$ now imply that there is an exact sequence

$$S(\rho) \xrightarrow{\text{tr}} H^*(C_3,S(\rho)) \to \mathbb{F}_9[a,b,d]/(a^2) \to 0.$$

where d is the image of σ_3 and a is the preimage of

$$b \in H^2(C_3,\mathbb{W}) = H^2(C_3,S_0(F)\otimes\chi)$$

with respect to the isomorphism

$$H^1(C_3,\rho) = H^1(C_3,S_1(\rho)) \to H^2(C_3,S_0(F)\otimes\chi)$$

given by the obvious connecting homomorphism. Thus a has bidegree $(1,-2)$ and the action of t is twisted by χ:

$$t_*(a) = -\omega^2 a = \omega^6 a\ .$$

□

The next step is to compute the invariants $S(\rho)^{C_3}$ together with the action of t. For this we start with the invariants of $S(F)$ and then we use the exact sequence (*). The action of the cyclic group C_3 on $S(F) = \mathbb{W}[x_1,x_2,x_3]$ extends in an obvious way to an action of the symmetric group Σ_3 on three letters; thus we have an inclusion of algebras

$$\mathbb{W}[\sigma_1,\sigma_2,\sigma_3] = \mathbb{W}[x_1,x_2,x_3]^{\Sigma_3} \subseteq S(F)^{C_3}.$$

It is clear that the following element

$$\epsilon = x_1^2x_2 + x_2^2x_3 + x_3^2x_1 - x_2^2x_1 - x_1^2x_3 - x_3^2x_2$$

(the "anti-symmetrization" of $x_1^2x_2$) is also invariant under the action of C_3. We use the same notation for the images of these elements in $S(\rho)$ and we note that σ_1 becomes 0 in $S(\rho)$.

Lemma 4.

a) *There is an isomorphism of* $\mathbb{W}$*-algebras*

$$\mathbb{W}[\sigma_1,\sigma_2,\sigma_3,\epsilon]/(\epsilon^2 - f) \cong S(F)^{C_3}$$

where f is the following polynomial in σ_1, σ_2, σ_3

$$f = -27\sigma_3^2 - 4\sigma_2^3 - 4\sigma_3\sigma_1^3 + 18\sigma_1\sigma_2\sigma_3 + \sigma_1^2\sigma_2^2\ .$$

Furthermore, the action of $t \in G_{12}$ is given by

$$t_*(\sigma_1) = \omega^2\sigma_1 \qquad t_*(\sigma_2) = -\sigma_2 \qquad t_*(\sigma_3) = \omega^6\sigma^3 \qquad t_*(\epsilon) = \omega^2\epsilon\ .$$

b) *There is an isomorphism of* $\mathbb{W}$*-algebras*

$$\mathbb{W}[\sigma_2,\sigma_3,\epsilon]/(\epsilon^2 - g) \cong S(\rho)^{C_3}$$

where g is the following polynomial in σ_2, σ_3

$$g = -27\sigma_3^2 - 4\sigma_2^3 \ .$$

Furthermore, the action of $t \in G_{12}$ is given by

$$t_*(\sigma_2) = -\sigma_2 \qquad t_*(\sigma_3) = \omega^6\sigma^3 \qquad t_*(\epsilon) = \omega^2\epsilon \ .$$

Proof. a) It is clear that ϵ^2 is Σ_3 invariant and therefore it can be expressed as a polynomial in σ_1, σ_2 and σ_3. To find the precise relation is an elementary exercise. We also leave it to the reader to verify that the action of t is as claimed. Thus it remains to determine the algebra structure of $S(F)^{C_3}$.

As a graded C_3-module $S(F)$ decomposes into a direct sum of free modules of rank 3 and of trivial modules of rank 1, and each of these summands contributes a summand $\mathbb{W}$ to $S(F)^{C_3}$. From this it is easy to calculate the Poincaré series of the invariants and we find

$$\chi_{S(F)^{C_3}}(t) = \frac{1+t^6}{(1-t^2)(1-t^4)(1-t^6)} \ .$$

(In this calculation we regrade $S(F)$ such that F is homogeneous of degree +2.)

On the other hand there is still an action of C_2 on $S(F)^{C_3}$, and $S(F)$ splits as direct sum of eigenspaces

$$S(F) \cong S(F)^+ \oplus S(F)^- \ .$$

Furthermore $S(F)^+ \cong S(F)^{\Sigma_3}$ and $S(F)^-$ is a module over $S(F)^+$. By the Poincaré series calculation it is therefore enough to verify that $S(F)^-$ is free as an $S(F)^+$-module with generator ϵ.

In fact, it is clear that ϵ is in $S(F)^-$ and because $\mathbb{W}[x_1, x_2, x_3]$ is without zero divisors it is also clear that ϵ generates a free $S(F)^+$ module with the correct Poincaré series. Now suppose $p \in S(F)^-$. We can choose a constant $c \in \mathbb{W}$ of minimal valuation, say r, such that $cp = \epsilon q$ for a unique polynomial $q \in \mathbb{W}[\sigma_1, \sigma_2, \sigma_3]$. Then

$$\epsilon(cp) = \epsilon^2 q = fq \ .$$

If c is divisible by 3 then the formula for f shows that q must be divisible by 3 and then r was not minimal. Hence $r = 0$, $c \in \mathbb{W}^\times$ and p is in the submodule generated by ϵ.

b) This is an immediate consequence of (a) and the vanishing of $H^1(C_3, S(F))$. □

The next step is to invert the element N. This element is the image of σ_3^4; thus, we are effectively inverting the element $\sigma_3 \in S(\rho)^{C_3}$. We begin with the observation that if we divide

$$\epsilon^2 = -27\sigma_3^2 - 4\sigma_2^3$$

by $4\sigma_3^6$ we obtain

$$\left(\frac{\epsilon}{2\sigma_3^3}\right)^2 + \left(\frac{\sigma_2}{\sigma_3^2}\right)^3 = -\frac{27}{4\sigma_3^4} .$$

Thus if we set

$$c_4 = -\frac{\sigma_2}{\sigma_3^2}, \quad c_6 = \frac{\epsilon}{2\sigma_3^3}, \quad \Delta = -\frac{1}{4\sigma_3^4}$$

then we get the following relation

$$c_6^2 - c_4^3 = 27\Delta$$

which corresponds to the famous relation from the theory of modular forms [D] (except that in our case 2 is invertible and hence the usual factor 1728 can be simplified to 27). The reader is referred to [GS] for an explanation of this coincidence.

Furthermore, c_4, c_6, and Δ are all invariant under the action of the entire group G_{12}. The elements

$$\alpha := ad^{-1} \in H^1(C_3, (S(\rho)[N^{-1}])_4)$$

and

$$\beta := bd^{-2} \in H^2(C_3, (S(\rho)[N^{-1}])_{12})$$

are clearly fixed by the action of t and by degree reasons they are acted on trivially by c_4 and c_6. The following result is now straightforward to verify.

Proposition 5.

a) *The inclusion*

$$\mathbb{W}[c_4, c_6, \Delta^{\pm 1}]/(c_6^2 - c_4^3 = 27\Delta) \to S(\rho)[N^{-1}]^{G_{12}}$$

is an isomorphism.

b) *There is an exact sequence*

$$S(\rho)[N^{-1}] \xrightarrow{tr} H^*(G_{12}, S(\rho)[N^{-1}]) \to \mathbb{F}_9[\alpha, \beta, \Delta^{\pm 1}]/(\alpha^2) \to 0$$

and c_4 and c_6 act trivially on α and β.

The final step is now to investigate what happens under completion. We continue to use c_4, c_6 etc. for the images of these elements in $H^*(G_{12}, E_*)$ with respect to the map $S(\rho)[N^{-1}] \to E_*$ studied in Proposition 2.

Theorem 6.

a) *There is an isomorphism*

$$(E_*)^{G_{12}} \cong \mathbb{W}[[c_4^3\Delta^{-1}]][c_4, c_6, \Delta^{\pm 1}]/(c_6^2 - c_4^3 = 27\Delta)$$

b) *There is an exact sequence*

$$E_* \xrightarrow{tr} H^*(G_{12}, E_*) \to \mathbb{F}_9[\alpha, \beta, \Delta^{\pm 1}]/(\alpha^2) \to 0$$

and c_4 and c_6 act trivially on α and β.

Proof. a) We use Proposition 2 and we use that completion commutes with taking invariants. We abbreviate $S(\rho)[N^{-1}]$ by A and we recall that the ideal $I \subset A$ is generated by 3 and $e - s_*(e)$. With this it is straightforward to check that both c_4 and c_6 belong to I. The relation $c_6^2 - c_4^3 = 27\Delta$ implies then that $I \cap A_0^{C_3}$ is generated by 3 and $c_4^3\Delta^{-1}$. This implies (a).

b) It is also straightforward to verify that $\sigma_2 \equiv -(e - s_*(e))^2 \bmod (3)$ and this implies that the ideals I and $(3, c_4^3\Delta^{-1})$ define the same completion. Abbreviate $c_4^3\Delta^{-1}$ by z. We have an isomorphism

$$E_* \cong \lim_k A/(z^k) \ .$$

Now we consider the short exact sequence

$$0 \to A \xrightarrow{z^k} A \to A/(z^k) \to 0 \ .$$

Because z acts trivially on $H^q(G_{12}, A)$ for $q > 0$ (by Proposition 5b) we obtain for each $q > 0$ a tower (indexed by k) of short exact sequences

$$0 \to H^q(G_{12}, A) \to H^q(G_{12}, A/(z^k)) \to H^{q+1}(G_{12}, A) \to 0.$$

The maps on the right-hand side of this tower are also induced by multiplication with z, hence they are trivial and therefore we obtain an isomorphism

$$H^q(G_{12}, A) \cong \lim_k H^q(G_{12}, A/(z^k)) \ .$$

On the other hand the graded quotients $A/(z^k)$ are of finite type for each k and this implies that the usual short exact sequences

$$0 \to \lim_k^1 H^{q-1}(G_{12}, A/(z^k)) \to H^q(G_{12}, \lim_k A/(z^k)) \to$$
$$\to \lim_k H^q(G_{12}, A/(z^k)) \to 0$$

degenerate into isomorphisms

$$H^q(G_{12}, \lim_k A/(z^k)) \cong \lim_k H^q(G_{12}, A/(z^k))$$

and the proof is complete. □

Remark. With the same reasoning we can also compute the E_2-terms for the homotopy fixed point spectra $E_2^{hC_3}$ and $E_2^{hC_6}$ where C_3 is, as before, the subgroup generated by s, while C_6 is that generated by s and t^2. In fact, the G_{12}-invariant $\Delta = -1/4\sigma_3^4$ has a C_3-invariant 4-th root $\Delta^{1/4}$ in $S(\rho)[N^{-1}]$ and we get

$$(E_*)^{C_3} \cong \mathbb{W}[[c_4^3\Delta^{-1}]][c_4, c_6, \Delta^{\pm 1/4}]/(c_6^2 - c_4^3 = 27\Delta) \ .$$

Furthermore there is an exact sequence

$$E_* \xrightarrow{\text{tr}} H^*(C_3, E_*) \to \mathbb{F}_9[\alpha, \beta, \Delta^{\pm 1/4}] \to 0$$

and c_4 and c_6 act trivially on α and β.

Similarly,

$$(E_*)^{C_6} \cong \mathbb{W}[[c_4^3\Delta^{-1}]][c_4, c_6, \Delta^{\pm 1/2}]/(c_6^2 - c_4^3 = 27\Delta) \ ,$$

there is an exact sequence

$$E_* \xrightarrow{\text{tr}} H^*(C_6, E_*) \to \mathbb{F}_9[\alpha, \beta, \Delta^{\pm 1/2}] \to 0$$

and c_4 and c_6 act trivially on α and β.

1.3. The homotopy of EO_2

Before we turn to the discussion of the differentials of the spectral sequence we relate the elements Δ, c_4, c_6, α and β to well-known quantities in homotopy theory.

We start by recalling that the elements $v_1^k \in Ext^{0,4k}_{BP_*BP}(BP_*, BP_*/(3))$ define permanent cycles in the classical ANSS of the mod-3 Moore space $V(0)$. Similarly, the element $v_2 \in Ext^{0,16}_{BP_*BP}(BP_*, BP_*/(3,v_1))$ defines a permanent cycle in the classical ANSS for the cofibre $V(1)$ of the Adams self map $\Sigma^4 V(0) \to V(0)$. Furthermore v_1 and v_2 give rise via the Greek letter construction to generators $\alpha_1 \in \pi_3(S^0) \cong \mathbb{Z}/3$ resp. $\beta_1 \in \pi_{10}(S^0) \cong \mathbb{Z}/3$ which are detected in the classical ANSS by elements with the same name in $Ext^{1,4}_{BP_*BP}(BP_*, BP_*)$ resp. $Ext^{2,12}_{BP_*BP}(BP_*, BP_*)$.

Finally we note that the localization map from a finite spectrum X to $L_{K(2)}X$ together with the Morava change of rings isomorphism and the obvious restriction homomorphism in group cohomology induce a natural homomorphism

$$\lambda_X : Ext^{s,t}_{BP_*BP}(BP_*, BP_*X) \to H^s(\mathbb{G}_2; E_tX) \to H^s(G_{12}; E_tX) .$$

We will denote the images of the elements v_1, v_2 with respect to $\lambda_{V(0)}$ resp. $\lambda_{V(1)}$ still by v_1 resp. v_2.

Proposition 7.

a) *Reduction modulo $(3, u_1)$ sends Δ^2 to the image of v_2^3 in $H^0(G_{12}, E_{32}/(3, u_1))$.*

b) *The mod-3 reduction map*

$$H^0(G_{12}, E_t) \to H^0(G_{12}, E_t/(3))$$

sends c_4 resp. c_6 to the image of v_1^2 resp. v_1^3, up to multiplication by a unit in $H^0(G_{12}, E_0/(3)) \cong \mathbb{F}_9[[c_4^3\Delta^{-1}]]$. Furthermore there is an element

$$\widetilde{\alpha} \in H^1(G_{12}, E_{12}/(3))$$

and an isomorphism (of modules over $\mathbb{F}_9[[v_1^6\Delta^{-1}]][v_1, \Delta^{\pm 1}, \beta] \otimes \Lambda(\alpha)$)

$$H^*(G_{12}, E_*/3) \cong \mathbb{F}_9[[v_1^6\Delta^{-1}]][v_1, \Delta^{\pm 1}, \beta] \otimes \Lambda(\alpha)\{1, \widetilde{\alpha}\}/(v_1\alpha, v_1\widetilde{\alpha}, \alpha\widetilde{\alpha} + v_1\beta)$$

c) *The map λ_{S^0} sends α_1 to α and β_1 to β up to a nontrivial constant in $\mathbb{W}/3$.*

Proof. a) The definition of Δ implies immediately that its reduction modulo $(3, u_1)$ is equal to that of u^{-12}. On the other hand the reduction of u^{-24} is equal to v_2^3.
b) It is clear from our calculation of $H^*(G_{12}, E_*)$ and the short exact sequence of G_{12}-modules

$$0 \to E_* \to E_* \to E_*/(3) \to 0 \qquad (*)$$

that v_1^2 and v_1^3 are in the image of mod-3 reduction. Furthermore they generate the invariants in degree 8 resp. 12 as module over $H^0(G_{12}, E_0/(3))$. On the other hand the G_{12}-invariants in degree 8 resp. 12 of E_* are freely generated (as modules over $H^0(G_{12}, E_0)$) by c_4 resp. c_6. This proves the statement on c_4 and c_6 and also gives the result for $H^0(G_{12}, E_*/(3))$. We define $\widetilde{\alpha}$ such that $\delta^0(\widetilde{\alpha}) = \beta$ where δ^0 denotes the boundary homomorphism associated to the exact sequence (*). Then everything else except perhaps the relation $v_1\beta = \alpha\widetilde{\alpha}$ is straightforward to check. This relation is obtained by calculating

$$\delta^0(v_1\beta + \alpha\widetilde{\alpha}) = \delta^0(v_1)\beta - \alpha\delta^0(\widetilde{\alpha}) = \alpha\beta - \alpha\beta = 0$$

and by noting that δ^0 is a monomorphism in the relevant bidegree.

c) This is a consequence of the compatibility (with respect to the maps λ_X) of the Greek letter construction for $Ext_{BP_*BP}(BP_*, -)$ and an analogous Greek letter construction for $H^*(G_{12}, -)$.

In fact, the image of $v_1 \in Ext^0_{BP_*BP_*}(BP_*, BP_*/(3))$ (which is the class u_1u^{-2}) defines an element in $H^0(G_{12}, E_*/(3))$. The short exact sequence

$$0 \to E_* \to E_* \to E_*/(3) \to 0$$

shows that this class has a nontrivial image $\delta^0(v_1) \in H^1(G_{12}, E_4) \cong \mathbb{W}/3$ and this latter group is generated by α.

Similarly, for β_1 we just need to check that the result of the Greek letter construction on $u_2^{-8} \in H^0(G_{12}, E_{16}/(3, u_1))$ yields a nontrivial element in $H^2(G_{12}, E_{12})$. First we note that the boundary map δ^1 associated to the exact sequence

$$0 \to \Sigma^4 E_*/(3) \xrightarrow{v_1} E_*/(3) \to E_*/(3, u_1) \to 0$$

maps u^{-8} nontrivially and hence to $\widetilde{\alpha}$, up to a nonconstant multiple. In the proof of (b) we have seen that $\delta^0(\widetilde{\alpha}) = \beta$ and hence we are done. □

In the sequel we redefine α resp. β such that $\alpha = \lambda_{S^0}(\alpha_1)$ and $\beta = \lambda_{S^0}(\beta_1)$. We are now ready to describe the differentials in our SS.

Theorem 8. *In the spectral sequence*

$$H^s(G_{12}, E_t) \Longrightarrow \pi_{t-s}(EO_2)$$

we have an inclusion of subrings

$$E_\infty^{0,*} \cong \mathbb{W}[[c_4^3\Delta^{-1}]][c_4, c_6, c_4\Delta^{\pm 1}, c_6\Delta^{\pm 1}, 3\Delta^{+1}, \Delta^{\pm 3}]/(c_4^3 - c_6^2 = 27\Delta) .$$

In positive filtration $E_\infty^{s,t}$ is additively generated by the elements α, $\alpha\beta$, $\Delta\alpha$, $\Delta\alpha\beta$, β^j, $j = 1, 2, 3, 4$ and their multiples by powers of $\Delta^{\pm 3}$. All these elements are of order 3 and c_4 and c_6 act trivially on elements in positive filtration.

Proof. First we observe that every element in the image of the transfer is a permanent cycle. The last proposition implies furthermore that the elements α and β are permanent cycles detecting homotopy classes with the same name. Next we use Toda's relation $\alpha_1\beta_1^3 = 0$ in $\pi_*(S^0)$. This implies that $\alpha\beta^3 = 0$ in $\pi_*(EO_2)$ and this can only happen if $d_5(\Delta) = a_1\alpha\beta^2$ for some $a_1 \in \mathbb{F}_9^\times$.

Then we use the Toda bracket relation $\beta_1 \in \pm\langle\alpha_1, \alpha_1, \alpha_1\rangle$ in $\pi_*(S^0)$. Consequently we have $\beta \in \pm\langle\alpha, \alpha, \alpha\rangle$ in $\pi_*(EO_2)$. This and $\alpha\beta^2 = 0$ imply that β^3 is in the indeterminacy of the bracket $\langle\alpha\beta^2, \alpha, \alpha\rangle$. This is only possible if $\alpha\Delta$ is a permanent cycle and $\alpha(\alpha\Delta) = \beta^3$, up to a nontrivial constant.

The next possible differential is d_9. Up to nontrivial constants we have $\beta^5 = \beta^2\beta^3 = \beta^2\alpha(\alpha\Delta) = 0$ in $\pi_*(EO_2)$ and this forces $d_9(\Delta^2\alpha) = a_2\beta^5$ for some $a_2 \in \mathbb{F}_9^\times$. Then there is no more room for further differentials and $E_\infty \cong E_{10}$ is as stated in the theorem. □

Remark. With the same reasoning we can also compute the homotopy of $E_2^{hC_3}$ and $E_2^{hC_6}$. In the case of C_3 we obtain an inclusion of subrings

$$E_\infty^{0,*} \cong \mathbb{W}[[c_4^3\Delta^{-1}]][c_4, c_6, c_4\Delta^{\pm 1/4}, c_6\Delta^{\pm 1/4}, 3\Delta^{\pm 1/4}, \Delta^{\pm 3/4}]/(c_4^3 - c_6^2 = 27\Delta)$$

and in positive filtration $E_\infty^{s,t}$ is additively generated by the elements α, $\alpha\beta$, $\Delta\alpha$, $\Delta\alpha\beta$, β^j, $j = 1,2,3,4$ and their multiples by powers of $\Delta^{\pm 3/4}$. These elements are of order 3 and c_4 and c_6 act trivially on elements in positive filtration.

In the case of C_6 we obtain an inclusion of subrings

$$E_\infty^{0,*} \cong \mathbb{W}[[c_4^3\Delta^{-1}]][c_4, c_6, c_4\Delta^{\pm 1/2}, c_6\Delta^{\pm 1/2}, 3\Delta^{\pm 1/2}, \Delta^{\pm 3/2}]/(c_4^3 - c_6^2 = 27\Delta)$$

and in positive filtration $E_\infty^{s,t}$ is additively generated by the elements α, $\alpha\beta$, $\Delta\alpha$, $\Delta\alpha\beta$, β^j, $j = 1,2,3,4$ and their multiples by powers of $\Delta^{\pm 3/2}$. Again these elements are of order 3 and c_4 and c_6 act trivially on elements in positive filtration.

In particular we see that EO_2 is 72 periodic with periodicity generator Δ^3, $E_2^{hC_3}$ is 18 periodic with periodicity generator $\Delta^{3/4}$ and $E_2^{hC_6}$ is 36 periodic with periodicity generator $\Delta^{3/2}$.

1.4. The ANSS converging to $\pi_*(EO_2 \wedge V(1))$

In this section we calculate $\pi_*(EO_2 \wedge V(1))$. We can do this by using Theorem 8 and the long exact sequences associated to the defining cofibre sequences of $V(0)$ and $V(1)$. However, later on we will make use of the structure of the ANSS for $L_2V(1)$ and so we choose to give a presentation in terms of the ANSS

$$E_2^{s,t}(V(1)) = H^s(G_{12}, E_*V(1)) \Longrightarrow \pi_{t-s}(EO_2 \wedge V(1))\ .$$

First we note that $E_*(V(1))$ is given as $\mathbb{F}_9[u^{\pm 1}]$. The element $s \in G_{12}$ acts necessarily trivially on this ring while t acts via $t_*(u) = \omega^2 u$ (cf. the proof of Lemma 1). This gives us the following E_2-term

$$E_2^{*,*} \cong \left(\mathbb{F}_9[u^{\pm 1}, y] \otimes \Lambda(x)\right)^{C_4}$$

in which the (s,t)-bidegree of u is $(0,-2)$, that of y is $(2,0)$ and that of x is $(1,0)$. The invariants can then be identified with $\mathbb{F}_9[u^{\pm 4}, \beta] \otimes \Lambda(\alpha)$ where $\beta := u^{-6}y$ is a permanent cycle detecting $\beta \in \pi_{10}(V(1))$ and $\alpha := u^{-2}x$ is a permanent cycle detecting $\alpha \in \pi_3(V(1))$ and where β and α are the images of the classical elements β_1 and α_1 in $\pi_*(S^0)$. (This can easily be checked via the long exact sequences of homotopy groups mentioned above.) The element u^{-8} is the image of v_2 in the E_2-term of the ANSS for $\pi_*V(1)$ with respect to the localization map (cf. the proof of Proposition 7 above).

If k is an integer than we will write from now on $v_2^{k/2}$ instead of u^{-4k}. If x is an element of E_2 we will denote $u^{-4k}x$ by $v_2^{k/2}x$. We note that the periodicity generator Δ^3 of $\pi_{72}EO_2$ projects to $v_2^{9/2}$.

Theorem 9. *There are elements $v_2^{k/2} \in \pi_{8k}(EO_2 \wedge V(1))$, $k = 0, 1, 2$ and $v_2^{k/2}\alpha \in \pi_{8k+3}(EO_2 \wedge V(1))$, $k = 0, 1, 2, 3, 4, 5$, such that as a module over $\mathbb{F}_9[v_2^{\pm 9/2}, \beta]$ there is an isomorphism*

$$\pi_*(EO_2 \wedge V(1))$$
$$\cong \mathbb{F}_9[v_2^{\pm 9/2}] \otimes \left(\mathbb{F}_9[\beta]/(\beta^5)\{1, v_2^{1/2}, v_2\} \oplus \mathbb{F}_9[\beta]/(\beta^2)\{\alpha, \ldots, v_2^{5/2}\alpha\}\right).$$

Remark. We remark that additively $\pi_k(EO_2)$ is nontrivial and of dimension 1 over $\mathbb{F}_9$ if $k \equiv 10m + 8k \bmod (72)$ with $0 \le m \le 4$, $0 \le k \le 2$, or if $k = 10m + 8k + 3$ if $0 \le m \le 1$ and $0 \le k \le 5$. For all other k the homotopy group is trivial.

Proof. Because α and β are permanent cycles, the first possible nontrivial differential is d_5 and it is determined by its value on the powers of $v_2^{1/2}$. By using the long exact homotopy sequences associated to the defining cofibre sequences of $V(0)$ and $V(1)$ together with Theorem 8 and Proposition 7b it is easy to verify that the elements 1, $v_2^{1/2}$ and v_2 are permanent cycles.

Now $E_r^{*,*}(V(1))$ is a differential graded module over $E_r^{*,*}(S^0)$. This implies

$$d_5(v_2^{k/2}) = \begin{cases} 0 & \text{if } k = 0, 1, 2 \bmod 9 \\ c_k v_2^{k-3/2}\alpha\beta^2 & \text{if } k = 3, 4, 5, 6, 7, 8 \bmod 9 \end{cases}$$

for suitable nontrivial constants c_k and therefore

$$E_6 \cong \mathbb{F}_9[v_2^{\pm 9/2}, \beta]\{1, v_{1/2}, v_2\} \oplus \mathbb{F}_9[v_2^{\pm 9/2}, \beta]\{v_2^{6/2}\alpha, v_2^{7/2}\alpha, v_2^{8/2}\alpha\}$$
$$\oplus\, \mathbb{F}_9[v_2^{\pm 9/2}, \beta]/(\beta^2)\{\alpha, v_2^{1/2}\alpha, v_2\alpha, v_2^{3/2}\alpha, v_2^2\alpha, v_2^{5/2}\alpha\}.$$

The next possible differential is d_9 and by using the module structure again we obtain

$$d_9(v_2^{k/2}\alpha) = \begin{cases} 0 & \text{if } k = 0, 1, 2, 3, 4, 5 \bmod 9 \\ c'_k v_2^{k/2-3}\beta^5 & \text{if } k = 6, 7, 8 \bmod 9 \end{cases}$$

for nontrivial constants c'_k. The resulting E_{10}-term is isomorphic to the stated result, and in fact, there is no room for further differentials. □

2. The homotopy fixed point spectrum E^{hN}

2.1. The subgroups N and N^1

Next we introduce certain infinite closed subgroups of $\mathbb{S}_2$ which are closely related to the subgroup G_{12} which is used to define EO_2. We refer to [H, Section 3] for more details on the following discussion. The centralizer $C := C_{\mathbb{S}_2}(C_3)$ of the subgroup $C_3 \subset G_{12}$ generated by s can be identified with the maximal order of the units in the cyclotomic extension $\mathbb{Q}_3(\zeta_3)$ of $\mathbb{Q}_3$ generated by a third root of unity ζ_3, and is hence isomorphic to $C_3 \times C_2 \times \mathbb{Z}_3^2$. Furthermore C is of index 2 in its normalizer $N := N_{\mathbb{S}_2}(C_3)$. The action of the element n of order 2 in N/C on C is via the Galois automorphism of the cyclotomic extension. The action can be diagonalized, i.e., the splitting of $C_3 \times C_2 \times \mathbb{Z}_3^2$ can be chosen to be invariant with respect to the action of n, and n acts trivially on C_2 and on one copy of $\mathbb{Z}_3$ while it acts by multiplication by -1 on the other copy and on C_3.

Furthermore, there is a homomorphism $\mathbb{S}_2 \to \mathbb{Z}_3^\times \to (\mathbb{Z}_3^\times)/\{\pm 1\}$ given as the composition of the reduced norm and the canonical projection. Its kernel is denoted by $\mathbb{S}_2^1$, and $\mathbb{S}_2$ decomposes as $\mathbb{S}_2^1 \times \mathbb{Z}_3$ where the complementary factor $\mathbb{Z}_3$ comes from the center of the division algebra. There is a corresponding splitting $N \cong N^1 \times \mathbb{Z}_3$ and this splitting is preserved by the conjugation action of n (where n acts trivially on the complementary factor $\mathbb{Z}_3$). We observe that the subgroup G_{12} is contained in N^1 and that N^1 is a (nonsplit) extension of C_2 by $C^1 := C_3 \times C_2 \times \mathbb{Z}_3$ where the C_2-action preserves the splitting of C^1 and is nontrivial on the factors C_3 and $\mathbb{Z}_3$.

In the sequel we will make use of the homotopy fixed point spectra E^{hN} and E^{hN^1}. These spectra are closer to $L_{K(2)}S^0$ but we will see that they are also closely related to EO_2.

2.2. The spectra E^{hN}, E^{hN^1} and EO_2

The splitting $N \cong N^1 \times \mathbb{Z}_3$ implies the following result.

Theorem 10. *There is a cofibration sequence*

$$E^{hN} \to E^{hN^1} \to E^{hN^1}$$

where the map $E^{hN^1} \to E^{hN^1}$ is given by $id - k$ if k denotes a topological generator of the central $\mathbb{Z}_3$.

Proof. There is a canonical map $f : E^{hN} \to E^{hN^1}$ induced by the inclusion $N^1 \subset N$. Furthermore k induces a self map of E^{hN^1} and $(id - k) \circ f$ is null. This gives us a map g from E^{hN} to the fibre F of $id - k$. Then one shows that g, or equivalently $L_{K(2)}(g \wedge id_E)$ is an equivalence. In fact, as in [DH2, Prop. 7.1] we can identify $\pi_* L_{K(2)}(E^{hN} \wedge E)$ with $\mathrm{map}_{cts}(\mathbb{G}_2/N, E_*)$, the continuous maps from the coset space $\mathbb{G}_2/N$ to E_*, and likewise $\pi_* L_{K(2)}(E^{hN^1} \wedge E)$ with $\mathrm{map}_{cts}(\mathbb{G}_2/N_1, E_*)$. Then one sees that the map $id - k$ induces a surjection on $\pi_*(L_{K(2)}(- \wedge E))$ and g induces an isomorphism between $\mathrm{map}_{cts}(\mathbb{G}_2/N, E_*)$ and the kernel. □

The spectrum E^{hN^1} itself can be obtained from EO_2 in a slightly more sophisticated fashion. For this we consider the ring spectrum $E^{hC_6} = E^{h(C_3 \times C_2)}$. The group C_6 is normal in G_{12} and we obtain an induced action of the quotient $G_{12}/C_6 \cong C_2$ on the ring spectrum E^{hC_6}. The spectrum E^{hC_6} splits with respect to this action as $E^+ \vee E^-$ where E^+ and E^- are the "eigenspectra" of E^{hC_6} with respect to the two nontrivial characters of C_2. Furthermore $E^+ \simeq (E^{hC_6})^{hC_2}$ can be identified with EO_2 and thus E^{hC_6} and E^- become EO_2-module spectra.

The following elementary observation introduces a suspension which may seem surprising at first but which becomes very important for the sequel.

Proposition 11. *There is an equivalence of EO_2-module spectra*

$$E^- \simeq \Sigma^{36} EO_2 \ .$$

Proof. We have seen in Section 1.3 above that E^{hC_6} is a periodic ring spectrum with periodicity generator $\Delta^{3/2}$ of degree 36. Furthermore, the periodicity generator is in the -1 eigenspace of the action of C_2 on $\pi_*(E^{hC_6})$ and represents an element in $\pi_{36}(E^-)$. Using the structure of E^- as a module spectrum it defines an equivalence between $\Sigma^{36}EO_2$ and E^-. □

We have other elements of order 2 acting on E^{hC_6}, e.g., all elements of order 2 in the group $N^1/C_6 \cong \mathbb{Z}_3 \rtimes C_2$. In particular, if k_1 is a topological generator of $\mathbb{Z}_3$ and if we choose the image of $t \in G_{12}$ as generator of C_2, then $k_1 t$ is such an element. We refer to the corresponding eigenspectra of any element τ of order 2 as $E^{\pm}_{\tau}$. In particular we have $E^+_t = EO_2$, $E^-_t = \Sigma^{36}EO_2$.

Theorem 12.

a) *There is a cofibration sequence*

$$E^{hN^1} \longrightarrow E^+_t \longrightarrow E^-_{k_1 t}$$

and the map between the eigenspectra is induced by $id - k_1$ (on the level of E^{hC_6}).

b) *There is an equivalence*

$$E^-_{k_1 t} \simeq \Sigma^{36} EO_2 \ .$$

Proof. a) This follows the same strategy as the proof of Theorem 10. The map $(id - k_1)$ induces a self map of E^{hC_6} which we can easily check to induce a map $E^+_t \longrightarrow E^-_{k_1 t}$. Let F be the fibre of this map. As before the canonical map $f : E^{hN^1} \to E^+_t$ becomes null after composing it with $id - k_1$ so that we obtain a map $g : E^{hN^1} \to F$. This time we get an identification

$$\pi_* L_{K(2)}(E^{hC_6} \wedge E) \cong \ \mathrm{map}_{cts}(\mathbb{G}_2/C_6, E_*) \cong \mathrm{Hom}_{cts}(\mathbb{Z}_3[[\mathbb{G}_2/C_6]], E_*)$$

where for a profinite $\mathbb{G}_2$-set $X = \lim_\alpha X_\alpha$ with finite $\mathbb{G}_2$-sets X_α we write $\mathbb{Z}_3[[X]]$ for $\lim_{\alpha,n} \mathbb{Z}_3/(3^n)[X_\alpha]$ and where Hom_{cts} denotes continuous homomorphisms. The elements t and $k_1 t$ act on the coset space and after linearization we can pass to the corresponding $\pm$ eigenspaces which we denote $\mathrm{Hom}^{t,\pm}_{cts}$ etc.

Now $id - k_1$ induces as before a surjective map

$$\mathrm{Hom}^{t,+}_{cts}(\mathbb{Z}_3[[\mathbb{G}_2/C_6]], E_*) \to \mathrm{Hom}^{k_1 t,-}_{cts}(\mathbb{Z}_3[[\mathbb{G}_2/C_6]], E_*)$$

whose kernel gets identified via g with $\mathrm{Hom}_{cts}(\mathbb{Z}_3[[\mathbb{G}_2/N^1]], E_*)$. This finishes the proof of (a).

b) This is an immediate consequence of Proposition 11 and the fact that the elements t and $k_1 t$ are conjugate in N^1 because 2 is a unit in $\mathbb{Z}_3$. Hence they have equivalent eigenspectra. □

Remark. Theorem 10 and Theorem 12 have the following discrete analogues which may, in particular in the case of Theorem 12, help to explain the situation.

It is well known that in the case of an action of a discrete infinite cyclic group C_∞ on a spectrum X one can obtain X^{hC_∞}, up to homotopy, as the fibre of the map $X \to X$ given by $id - k$ where k is a generator of C_∞. In fact, if we take the real line $\mathbb{R}$ as a model for the universal C_∞-space EC_∞ with C_∞ acting via translations then the usual equivariant cell structure gives a presentation of EC_∞ as a pushout and hence of X^{hC_∞} as a pullback, and then it is easy to see that in the case of spectra this pullback is equivalent to the homotopy fibre of $id - k$. Similarly, if we consider $C_\infty \rtimes C_2$ as acting on $\mathbb{R}$ via translations and reflections and if we ignore 2-primary phenomena then $\mathbb{R}$ is still a good substitute for the universal $C_\infty \rtimes C_2$-space $E(C_\infty \rtimes C_2)$. This time the 0-simplices are the integral points on the real line with isotropy isomorphic to C_2, and the isotropy group of a 1-simplex is C_2 with C_2 acting nontrivially on the 1-simplex. From this cell structure we get, as before, a pullback description of $map^{C_\infty \rtimes C_2}(\mathbb{R}, X)$, and if X is a spectrum localized away from 2 then the pullback is on the one hand equivalent to the fibre of a map as in Theorem 12 and on the other hand to $X^{hC_\infty \rtimes C_2}$.

Corollary 13. *There is a cofibration sequence*

$$E^{hN^1} \longrightarrow EO_2 \longrightarrow \Sigma^{36} EO_2 \ .$$ □

2.3. The homotopy groups of $E^{hN} \wedge V(1)$ and of $E^{hN^1} \wedge V(1)$

It is not hard to see that the spectrum E^{hN^1} is no longer periodic. Nevertheless the following lemma shows that the homotopy groups $\pi_k(E^{hN^1} \wedge V(1))$ remain 72-periodic.

Lemma 14. *The map*

$$\pi_*(EO_2 \wedge V(1)) \to \pi_*(\Sigma^{36} EO_2 \wedge V(1))$$

which is induced by $id - k_1$ is trivial.

Proof. Theorem 9 implies that if $\pi_n(EO_2 \wedge V(1))$ and $\pi_n(\Sigma^{36} EO_2 \wedge V(1))$ are both nonzero then

$$n \in \{0, 10, 20, 36, 46, 56\} \bmod 72 \ .$$

Because $id - k_1$ commutes with the action of β we see that the map is trivial if $n \equiv 36, 46, 56$ and that it suffices to concentrate on the case $n \equiv 0$. Now the periodicity generator $v_2^{9/2} \in \pi_{72}(EO_2 \wedge V(1))$ comes from $\Delta^3 \in \pi_{72}(EO_2)$ and therefore it suffices to show that the composition

$$\pi_{72}(EO_2) \to \pi_{72}(\Sigma^{36} EO_2) \to \pi_{72}(\Sigma^{36} EO_2 \wedge V(1)))$$

annihilates Δ^3. In fact, we see from Proposition 7b and Theorem 8 that the image of $\pi_{72}(\Sigma^{36}EO_2)$ in $\pi_{72}(\Sigma^{36}EO_2 \wedge V(0))$ is divisible by v_1 and becomes therefore trivial in $\pi_{72}(\Sigma^{36}EO_2 \wedge V(1)))$. □

The lemma allows us to analyze the ANSS

$$E_2^{s,t} \cong H^s(N^1, E_*V(1)) \Longrightarrow \pi_{t-s}(E^{hN^1} \wedge V(1)) .$$

Its E_2-term is easily calculated to be

$$H^*(N^1, E_*V(1)) \cong \left(\mathbb{F}_9[u^{\pm 1}, y] \otimes \Lambda(x, a')\right)^{C_4} \cong \mathbb{F}_9[v_2^{\pm 1/2}, \beta] \otimes \Lambda(\alpha, u^{-18}a')$$

where as before $\alpha = u^{-2}x$, $\beta = u^{-6}y$, $v_2^{1/2} = u^{-4}$ and the exterior generator a' is the contribution of the factor $\mathbb{Z}_3$ in the centralizer $C^1 \cong C_3 \times C_2 \times \mathbb{Z}_3$. Its bidegree is $(1, 0)$ and the generator t of C_4 acts on it by multiplication by -1. Therefore $a_{35} := u^{-18}a'$ is a new invariant class. We note that the elements x and y are a priori not canonically defined, not even up to a nontrivial constant because the corresponding groups are of rank 2 over $\mathbb{F}_9$. We can and will choose them such that α and β detect the images of α_1 and β_1 in $\pi_*(E_2^{hN^1} \wedge V(1)$ for example by defining them via Greek letter constructions in $H^*(N, -)$.

Proposition 15.

a) *The ANSS*

$$E_2^{s,t} \cong H^s(N^1, E_*V(1)) \cong \mathbb{F}_9[v_2^{\pm 1/2}, \beta] \otimes \Lambda(\alpha)\{1, a_{35}\} \Longrightarrow \pi_{t-s}(E^{hN^1} \wedge V(1))$$

splits as the direct sum of the ANSS for $EO_2 \wedge V(1)$ and that of $\Sigma^{35}EO_2 \wedge V(1)$ (where the summand indexed by 1 corresponds to $EO_2 \wedge V(1)$ and that by a_{35} to $\Sigma^{35}EO_2 \wedge V(1)$).

b) *As modules over $\mathbb{F}_9[v_2^{\pm 9/2}, \beta]$ there is an isomorphism*

$$\pi_*(E^{hN_1} \wedge V(1)) \cong \pi_*(EO_2 \wedge V(1))\{1, a_{35}\} .$$

Remark. We emphasize that the module structure over $v_2^{\pm 9/2}$ is (at least at this point) a purely algebraic accident induced by an algebraic module structure on the level of E_2-terms.

Proof. The fibration sequence

$$E^{hN^1} \wedge V(1) \to EO_2 \wedge V(1) \to \Sigma^{36}EO_2 \wedge V(1)$$

induces an exact sequence

$$0 \to E_*(E^{hN^1} \wedge V(1)) \to E_*(EO_2 \wedge V(1)) \to E_*(\Sigma^{36}EO_2 \wedge V(1)) \to 0$$

where E_*X has to be interpreted as $\pi_*(L_{K(2)}(E \wedge X))$. In fact, as in the proof of Theorem 12 this sequence can be identified with the sequence

$$0 \to \mathrm{Hom}_{cts}(\mathbb{Z}_3[[\mathbb{G}_2/N^1]], E_*V(1)) \to \mathrm{Hom}_{cts}^{t,+}(\mathbb{Z}_3[[\mathbb{G}_2/C_6]], E_*V(1))$$
$$\to \mathrm{Hom}_{cts}^{k_1t,-}(\mathbb{Z}_3[[\mathbb{G}_2/C_6]], E_*V(1)) \to 0 .$$

In cohomology (i.e., on E_2-terms of the relevant ANSS) this sequence induces short exact sequences for all $s \geq 0$

$$0 \to H^{s-1}(G_{12}, E_t(\Sigma^{36}V(1))) \to H^s(N^1, E_tV(1)) \to H^s(G_{12}, E_t(V(1)) \to 0$$

where the monomorphism converges towards the map

$$\pi_{t-s}(\Sigma^{35}EO_2 \wedge V(1)) \to \pi_{t-s}(E^{hN^1} \wedge V(1))$$

by the geometric boundary theorem and the epimorphism converges towards the map

$$\pi_{t-s}(E^{hN^1} \wedge V(1)) \to \pi_{t-s}(EO_2 \wedge V(1)) .$$

by naturality. The proposition follows. □

Now we turn towards $E^{hN} \wedge V(1)$ and consider the ANSS spectral sequence

$$E_2^{s,t} \cong H^s(N, E_tV(1)) \Longrightarrow \pi_{t-s}(E^{hN} \wedge V(1)) .$$

The E_2 -term of the SS is easily calculated to be

$$H^*(N, E_*V(1)) \cong \left(\mathbb{F}_9[u^{\pm 1}, y] \otimes \Lambda(x, a', \zeta)\right)^{C_4} \cong \mathbb{F}_9[v_2^{\pm 1/2}, \beta] \otimes \Lambda(\alpha, a_{35}, \zeta)$$

where the new exterior generator ζ is the contribution of the central factor $\mathbb{Z}_3$ in the centralizer C. Its bidegree is $(1, 0)$ and it is fixed by the action of t.

Proposition 16.

a) *The ANSS*

$$E_2^{s,t} \cong H^s(N, E_tV(1)) \cong \mathbb{F}_9[v_2^{\pm 1/2}, \beta] \otimes \Lambda(\alpha, \zeta)\{1, a_{35}\} \Longrightarrow \pi_{t-s}(E^{hN} \wedge V(1))$$

splits as the direct sum of the ANSS of $E^{hN^1} \wedge V(1)$ and of $\Sigma^{-1}E^{hN^1} \wedge V(1)$ (where the summand indexed by 1 corresponds to $E_2^{hN_1} \wedge V(1)$ and that by ζ to $\Sigma^{-1}E_2^{hN_1} \wedge V(1)$.)

b) *As modules over $\mathbb{F}_9[v_2^{\pm 9/2}, \beta]$ there is an isomorphism*

$$\pi_*(E_2^{hN^1} \wedge V(1)) \cong \pi_*(EO_2 \wedge V(1))\{1, a_{35}, \zeta, \zeta a_{35}\} .$$

□

Remark. We emphasize that as before the module structure over $v_2^{\pm 9/2}$ is (at least at this point) a purely algebraic accident on the level of E_∞-terms.

Proof. As before the proof will be an easy consequence of the following result (which is analogous to Lemma 14) whose proof will, however, make use of the structure of the SS considered in Proposition 16. □

Lemma 17. *The map*

$$(id - k)_* : \pi_*(E^{hN^1} \wedge V(1)) \to \pi_*(E^{hN^1} \wedge V(1))$$

is trivial.

Proof. First note that the action of k on $E_*V(1)) \cong \mathbb{F}_9[u^{\pm 1}]$ is trivial. Indeed, the generator k of the central $\mathbb{Z}_3 \subset \mathbb{D}_2^\times$ is necessarily congruent 1 mod 3 (for example, one can take $k = 4$). Then k acts on $u \in \pi_{-2}E$ by multiplication with $k \equiv 1 \bmod 3$ and therefore the action of $id - k$ is trivial on $E_*V(1)$ and hence on the E_2-term of the ANSS. This shows that the action on $\pi_k(E^{hN^1} \wedge V(1))$ is trivial except possibly in degrees

$$n \equiv 0, 3, 8, 11, 16, 19, 35, 38, 43, 45, 46, 48, 53, 56 \bmod 72$$

where the total degree n of the E_∞-term of the SS is made of two copies of $\mathbb{F}_9$. Degrees 35 and 43 resp. 45 and 53 can be excluded from the list because both copies have the same filtration (= 1 resp. 3). Next the action of α and β and the compatibility of k with the fibration sequence of Theorem 12 resp. Corollary 13 imply that degrees 38, 46, 48 and 56 can also be excluded. Similarly the action of α and β show that it is enough to consider the cases $n \equiv 0, 8, 16 \bmod 72$. If $id - k$ acts nontrivially in one of these dimensions then there exists some integer p and there exists $q \in \{0, 1, 2\}$ such that

$$(id - k)_*(v_2^{9p+q/2}) = cv_2^{(9(p-1)+q+3)/2}\beta\alpha a_{35}$$

for some nontrivial constant c. This implies then that in the ANSS for $E^{hN} \wedge V(1)$ the element $v_2^{(9(p-1)+q+3)/2}\beta\alpha\zeta a_{35}$ (which is a permanent cycle by Proposition 14 and the fact that ζ detects a homotopy class which comes from $L_{K(2)}S^0$) does not survive and hence that it is in the image of a differential. At this point we turn attention towards the analysis of the SS to show that this cannot happen.

In the ANSS converging to $\pi_*(E^{hN} \wedge V(1))$ the elements α, β and ζ come from the sphere (or at least from the $K(2)$-local sphere). Furthermore we know from [Ra2, Table A3.4] that the Greek letter element $\beta_{6/3} \in Ext^{2,84}_{BP_*BP}(BP_*, BP_*)$ is a permanent cycle in the ANSS for S^0. By [Sh1, Lemma 2.4] and Corollary 19 below this element is detected in $E_2^{2,84}$ in our SS and therefore agrees with $v_2^{9/2}\beta$, up to a nontrivial constant.[1] Because E_2 is free over $\mathbb{F}_9[\beta]$ we deduce that the first differential is linear with respect to multiplication by $v_2^{9/2}$.

So for the first differential we need to study the elements

$$v_2^{r/2}, \quad a_{35}v_2^{r/2} \text{ if } r \equiv 0, 1, \ldots, 8 \bmod 9 .$$

[1] **Note added in Proof**: Shimomura shows that $\beta_{6/3}$ is detected by an element which in his notation is given by $-v_2^3 b_{11}$ and in [Sh1, Proposition 5.9] he proves that $b_{11}^2 = -v_2^3\beta^2$. In view of the structure of the E_2-term, given below in Corollary 19, this implies that $-v_2^3 b_{11}$ agrees, up to a constant, with $v_2^{9/2}\beta$.

The following gives an independent and more elementary justification of the $v_2^{9/2}$-linearity of d_5 which is within the framework of our methods: First we note that $\Delta\beta$ in the E_2-term of EO_2 can be pulled back (not necessarily uniquely) to a class in the E_2-term of E_2^{hN}. The third power of this class will be a cycle for d_5 in the ANSS for E^{hN}, and when we project this third power to the E_2-term of $E^{hN} \wedge V(1)$ then we get $v_2^{9/2}\beta^3$. Therefore we know that in the ANSS for $E_2^{hN} \wedge V(1)$ the differential d_5 is linear with respect to $v_2^{9/2}\beta^3$, and then also with respect to $v_2^{9/2}$, because the E_5-term is free with respect to β.

Degree reasons (i.e., calculating modulo total degree 8) shows that the first possibility for a differential is d_5. The possible targets are as follows:

- $d_5(v_2^{r/2})$ is a linear combination of the elements $v_2^{(r-3)/2}\beta^2\alpha$, $v_2^{(r-7)/2}\beta^2 a_{35}$ and $v_2^{(r-6)/2}\beta\alpha\zeta a_{35}$.
- $d_5(a_{35}v_2^{r/2})$ is a multiple of $v_2^{(r-3)/2}\beta^2\alpha a_{35}$.

Now we use that 1, v_2 and v_2^5 are permanent cycles coming from $V(1)$ (see [Sh1, Thm 2.6]). This implies that for $r \equiv 0, 1, 2 \bmod 9$ we have

$$d_5(v_2^{r/2}) = 0 \ ,$$

in particular $v_2^{(9(p-1)+q+3)/2}\beta\alpha\zeta a_{35}$ is not in the image of d_5 for $q = 0, 1, 2$, hence it survives and the proof of the lemma is complete. □

3. The homotopy of $L_2V(1)$

In this section we will calculate the homotopy of $L_2V(1) \simeq E^{h\mathbb{G}_2} \wedge V(1)$ by comparing it to that of $E_2^{hN} \wedge V(1)$. The E_2-term of the ANSS converging to $\pi_* L_2V(1)$ is isomorphic to

$$H^*(\mathbb{G}_2, \mathbb{F}_9[u^{\pm 1}]) \cong (H^*(\mathbb{S}_2, \mathbb{F}_9[u^{\pm 1}])^{C_2}$$

where C_2 acts via conjugation on $\mathbb{S}_2$ and via Frobenius on $\mathbb{F}_9$ (hence the action is free and the spectral sequence of the extension $\mathbb{S}_2 \to \mathbb{G}_2 \to C_2$ degenerates into the stated isomorphism). Furthermore there is a canonical isomorphism

$$H^*(\mathbb{S}_2, \mathbb{F}_9[u^{\pm 1}]) \cong (H^*(S_2, \mathbb{F}_9) \otimes_{\mathbb{F}_9} \mathbb{F}_9[u^{\pm 1}])^{\mathbb{F}_9^\times}$$

where S_2 is the 3-Sylow subgroup of $\mathbb{S}_2$ and acts trivially, and the invariants are taken with respect to the action of the quotient $\mathbb{S}_2/S_2$ which can be naturally identified with $\mathbb{F}_9^\times$ generated by ω. The generator ω of $\mathbb{F}_9^\times$ acts diagonally on this tensor product, via conjugation on S_2 and via multiplication with ω on u, so that taking invariants amounts to taking the eigenspace decomposition of $H^*(S_2, \mathbb{F}_9)$ with respect to the action of ω (determined implicitly by Theorem 18 below) and tensoring the eigenspace E_{ω^i} (with eigenvalue ω^i) with powers u^{-i+8k} to get invariants.

In [H, Prop. 3.4 and Thm. 4.2] $H^*(S_2, \mathbb{F}_3)$ was studied via the restriction map to the centralizers $C_{S_2}(E_i) \cong C_3 \times \mathbb{Z}_3^2$, $i = 1, 2$, where the E_i are representatives of the two different conjugacy classes of C_3's in S_2. We can choose E_1 to be the subgroup generated by $s \in G_{12}$ and then E_2 can be chosen to be $\omega^{-1}E_1\omega$ so that the restriction map

$$H^*(S_2, \mathbb{F}_3) \to \prod_{i=1}^{2} H^*(C_{S_2}(E_i), \mathbb{F}_3) \cong \prod_{i=1}^{2} \mathbb{F}_3[y_i] \otimes \Lambda(x_i, \zeta_i, a_i{}')$$

becomes $\mathbb{F}_9^\times$-equivariant where the $\mathbb{F}_9^\times$-action on the right is induced from the conjugation action of $N_{\mathbb{S}_2}(E_i)/C_{S_2}(E_i) \cong C_4 \subset \mathbb{F}_9^\times$. We note that $t \in G_{12} \subset N_{\mathbb{S}_2}(E_1)$ projects to a generator in C_4. We can choose the cohomology classes such that y_i and x_i correspond to the generators of the cohomology of the cyclic subgroup, ζ_i to the cohomology of the central factor $\mathbb{Z}_3$, and $a_i{}'$ to that of the noncentral factor $\mathbb{Z}_3$ on which t acts by multiplication by -1. This notation differs somewhat from that in [H] but is consistent with our notation in Section 2.

Theorem 18 [H].

a) *The restriction map*

$$H^*(S_2,\mathbb{F}_3)\to\prod_{i=1}^{2}H^*(C_{S_2}(E_i),\mathbb{F}_3)\cong\prod_{i=1}^{2}\mathbb{F}_3[y_i]\otimes\Lambda(x_i,\zeta_i,a_i{}')$$

is an $\mathbb{F}_9^\times$-invariant monomorphism whose image is the $\mathbb{F}_3$-subalgebra generated by x_1, x_2, y_1, y_2, $\zeta_1+\zeta_2$, $x_1a_1{}'-x_2a_2{}'$, $y_1a_1{}'$ and $y_2a_2{}'$.

b) *In particular $H^*(S_2,\mathbb{F}_3)$ is a free module over $\mathbb{F}_3[y_1+y_2]\otimes\Lambda(\zeta_1+\zeta_2)$ generated by 1, x_1, x_2, y_1, $x_1a_1{}'-x_2a_2{}'$, $y_1a_1{}'$, $y_2a_2{}'$ and $y_1x_1a_1{}'$.* □

We note that a priori the elements x_i, y_i and $a_i{}'$ are not canonical (because they depend on the chosen decomposition of $C_{S_2}(E_1)$). The theorem implies, however, that x_i and y_i (as the Bockstein of x_i) are distinguished (at least up to nontrivial constants).

Next we describe the invariants of $H^*(S_2,\mathbb{F}_9[u^{\pm1}])$ with respect to the action of $\mathbb{F}_9^\times$ (which is determined by $\omega_*(u)=\omega u$). By using the $\mathbb{F}_9^\times$-linear monomorphism these invariants can be identified with a subring of

$$\Big(\prod_{i=1}^{2}H^*(C_{S_2}(E_i),\mathbb{F}_9[u^{\pm1}]\Big)^{\mathbb{F}_9^\times}\cong(H^*(C_{S_2}(E_1),\mathbb{F}_9[u^{\pm1}])^{C_4}\cong H^*(N_{\mathbb{S}_2}(E_1),\mathbb{F}_9[u^{\pm1}])$$

where C_4 is as before generated by $t\in G_{12}$. Its action on $H^*(C_{S_2}(E_1),\mathbb{F}_3)$ is given by

$$t_*(y)=-y,\quad t_*(x)=-x,\quad t_*(a')=-a',\quad t_*(\zeta)=\zeta$$

(where we have omitted the indices for simplicity). The following corollary is now straightforward to verify. As in Section 1.4. and Chapter 2 we write $v_2^{k/2}$ for u^{-4k}.

Corollary 19.

a) *The restriction map*

$$H^*(\mathbb{S}_2,\mathbb{F}_9[u^{\pm1}])\to H^*(N,\mathbb{F}_9[u^{\pm1}])\cong\big(\mathbb{F}_9[u^{\pm1},y]\otimes\Lambda(x,\zeta,a')\big)^{C_4}$$

is a monomorphism. Its target is isomorphic to

$$\mathbb{F}_9[v_2^{\pm1/2},\beta]\otimes\Lambda(\alpha,\zeta,a_{35})$$

(with $\beta=u^{-6}y$, $\alpha=u^{-2}x$, ζ and $a_{35}=u^{-18}a'$) and its image is the $\mathbb{F}_9$-subalgebra of $\mathbb{F}_9[v_2^{\pm1/2},\beta]\otimes\Lambda(\alpha,\zeta,a_{35})$ generated by $v_2^{\pm1}$, α, $v_2^{1/2}\alpha$, β, $v_2^{1/2}\beta$, ζ, αa_{35}, βa_{35} and $v_2^{1/2}\beta a_{35}$.

b) *In particular $H^*(\mathbb{S}_2;\mathbb{F}_9[u^{\pm1}])$ is free as a $\mathbb{F}_9[v_2^{\pm1},\beta]\otimes\Lambda(\zeta)$-module and identifies with the free submodule of $\mathbb{F}_9[v_2^{\pm1/2},\beta]\otimes\Lambda(\alpha,\zeta,a_{35})$ generated by 1, α, $v_2^{1/2}\alpha$, $v_2^{1/2}\beta$, αa_{35} βa_{35}, $v_2^{1/2}\beta a_{35}$, and $v_2^{1/2}\beta\alpha a_{35}$.* □

The restriction map above is the comparison map at the E_2-level between the two ANSS converging to $\pi_{t-s}(E^{h\mathbb{S}_2} \wedge V(1))$ resp. $\pi_{t-s}(E^{hN} \wedge V(1))$. We still have to deal with the Galois action of C_2 if we want to get at $L_2V(1)$.

Now the Galois generator $\phi \in C_2$ acts on $\mathbb{S}_2 \subset \mathbb{D}_2^\times$ by conjugation by S, hence it is clear that $\omega\phi$ centralizes $s = -\frac{1}{2}(1+\omega S)$ and thus everything in the commutative subfield of $\mathbb{D}_2$ generated by s. In particular $\omega\phi$ commutes with the units in the maximal order of this subfield, i.e., with $C_{\mathbb{S}_2}(E_1)$. Therefore the group $C_{\mathbb{G}_2}(E_1)$ (which is generated by $C_{\mathbb{S}_2}(E_1)$ and $\omega\phi$) is an abelian group and $\omega\phi$ acts trivially on $H^*(C_{S_2}(E_1), \mathbb{F}_3)$. The action on the coefficient ring $\mathbb{F}_9[u^{\pm 1}]$ is given by $(\omega\phi)_*(cu^k) = \phi(c)\omega^k u^k$ if $c \in \mathbb{F}_9$.

The monomorphism of Theorem 18 (with coefficients extended to $\mathbb{F}_9[u^{\pm 1}]$) is actually linear even with respect to $\mathbb{F}_9^\times \rtimes C_2$ where the Galois generator ϕ acts on the target on the level of groups by conjugation by S while it acts on $\mathbb{F}_9[u^{\pm 1}]$ by Frobenius again. The $\mathbb{F}_9^\times \rtimes C_2$-invariants in the target of this monomorphism can therefore be identified with

$$\Big(\prod_{i=1}^{2} H^*(C_{S_2}(E_i), \mathbb{F}_9[u^{\pm 1}])\Big)^{\mathbb{F}_9^\times \rtimes C_2} \cong \big(H^*(C_{S_2}(E_1), \mathbb{F}_9[u^{\pm 1}])^{C_4}\big)^{\langle\omega\phi\rangle} .$$

The preceding two paragraphs show that we can modify (if necessary) the elements $v_2^{1/2}$, β, α, ζ and a_{35} of Corollary 19 by scalars in $\mathbb{F}_9$ so that they become invariant with respect to the action of $\omega\phi$. After having done this $H^*(\mathbb{G}_2, \mathbb{F}_9[u^{\pm 1}])$ can be identified with the $\mathbb{F}_3$-subalgebra of $\mathbb{F}_9[v_2^{1/2}, \beta] \otimes \Lambda(\alpha, \zeta, a_{35})$ generated by elements with the same name as in Corollary 19a and $H^*(\mathbb{G}_2, \mathbb{F}_9[u^{\pm 1}])$ is a free module over $\mathbb{F}_3[v_2^{\pm 1}, \beta] \otimes \Lambda(\zeta)$ on elements with the same name as in Corollary 19b.

Now we are ready to compare the differentials in the two ANSS converging to $L_2V(1) \simeq E_2^{h\mathbb{G}_2} \wedge V(1)$ resp. $E_2^{hN} \wedge V(1)$. We refer to them as the source SS resp. the target SS.

For the target SS we deduce from Section 2.3 that $v_2^{\pm 9/2}$, β, α, ζ and a_{35} are permanent cycles and that the differentials are linear with respect to the algebra

$$R := \mathbb{F}_9[v_2^{\pm 9/2}, \beta] \otimes \Lambda(\alpha, \zeta, a_{35}) .$$

To better compare with the spectral sequence of the source we should consider this SS as one of modules over $S := \mathbb{F}_9[v_2^{\pm 9}, \beta] \otimes \Lambda(\alpha, \zeta, a_{35})$. In fact, the E_2-term of the target SS is a free module over S on generators $v_2^{k/2}$, $k = 0, \dots, 17$.

We consider the E_2-term of the ANSS of the source as a free module over

$$P := \mathbb{F}_3[v_2^{\pm 9}, \beta] \otimes \Lambda(\zeta)$$

(note that here we have taken the prime field) on the following generators (where $l = 0, 1, \dots, 8$):

- v_2^l, $v_2^{l+1/2}\beta$, $v_2^l\beta a_{35}$, $v_2^{l+1/2}\beta a_{35}$
- $v_2^l\alpha$, $v_2^{l+1/2}\alpha$, $v_2^l\alpha a_{35}$, $v_2^{l+1/2}\beta\alpha a_{35}$

We have seen in Section 1 and 2 that the first differential for the target SS is d_5. It is determined by

$$d_5(v_2^{k/2}) = \begin{cases} 0 & \text{if } k \equiv 0,1,2 \bmod 9 \\ c_k v_2^{(k-3)/2}\alpha\beta^2 & \text{if } k \equiv 3,\dots,8 \bmod 9 . \end{cases}$$

This implies that the first differential of the source SS is also d_5. Furthermore by the discussion above the nontrivial constants c_k have to be in $\mathbb{F}_3$ and d_5 in the source SS is given by

$$d_5(v_2^l) = \begin{cases} 0 & \text{if } l \equiv 0,1,5 \bmod 9 \\ \pm v_2^{l-2+1/2}\alpha\beta^2 & \text{if } l \equiv 2,3,4,6,7,8 \bmod 9 \end{cases}$$

$$d_5(v_2^{l+1/2}\beta) = \begin{cases} 0 & \text{if } l \equiv 0,4,5 \bmod 9 \\ \pm v_2^{l-1}\alpha\beta^3 & \text{if } l \equiv 1,2,3,6,7,8 \bmod 9 \end{cases}$$

$$d_5(v_2^l\beta a_{35}) = \begin{cases} 0 & \text{if } l \equiv 0,1,5 \bmod 9 \\ \pm v_2^{l-2+1/2}\alpha\beta^3 a_{35} & \text{if } l \equiv 2,3,4,6,7,8 \bmod 9 \end{cases}$$

$$d_5(v_2^{l+1/2}\beta a_{35}) = \begin{cases} 0 & \text{if } l \equiv 0,4,5 \bmod 9 \\ \pm v_2^{l-1}\alpha\beta^3 a_{35} & \text{if } l \equiv 1,2,3,6,7,8 \bmod 9 \end{cases}$$

$$d_5(v_2^l\alpha) = d_5(v_2^{l+1/2}\alpha) = 0 \quad \text{if } l = 0,\dots,8 \bmod 9$$

$$d_5(v_2^l\alpha a_{35}) = d_5(v_2^{l+1/2}\beta\alpha a_{35}) = 0 \quad \text{if } l = 0,\dots,8 \bmod 9 .$$

This yields the following E_6-term (which is already presented in a form which is adapted to the discussion of the next differential):

$$E_6 \cong E_{6,1} \oplus E_{6,2} \oplus E_{6,3} \oplus E_{6,4}$$

with

$$\begin{aligned} E_{6,1} &= P\{v_2^l\}_{l=0,1,5} \oplus P\{v_2^l\alpha\}_{l=3,4,8} \oplus P/(\beta^3)\{v_2^l\alpha\}_{l=0,1,2,5,6,7} \\ E_{6,2} &= P\{v_2^{l+1/2}\beta\}_{l=0,4,5} \oplus P\{v_2^{l+1/2}\alpha\}_{l=3,7,8} \oplus P/(\beta^2)\{v_2^{l+1/2}\alpha\}_{l=0,1,2,4,5,6} \\ E_{6,3} &= P\{v_2^l\beta a_{35}\}_{l=0,1,5} \oplus P\{v_2^l\alpha a_{35}\}_{l=3,4,8} \oplus P/(\beta^3)\{v_2^l\alpha a_{35}\}_{l=0,1,2,5,6,7} \\ E_{6,4} &= P\{v_2^{l+1/2}\beta a_{35}\}_{l=0,4,5} \oplus P\{v_2^{l+1/2}\beta\alpha a_{35}\}_{l=3,7,8} \oplus \\ &\oplus P/(\beta^2)\{v_2^{l+1/2}\beta\alpha a_{35}\}_{l=0,1,2,4,5,6} \cdot \end{aligned}$$

We know that the next differential in the target SS is d_9 and is determined by

$$d_9(v_2^{k/2}\alpha) = \begin{cases} 0 & \text{if } k = 0,1,2,3,4,5 \bmod 9 \\ c'_k v_2^{k/2-3}\beta^5 & \text{if } k = 6,7,8 \bmod 9 \end{cases}$$

for suitable nontrivial constants c'_k. Again by comparing with the spectral sequence of the target we deduce that the next differential in the source SS is also d_9, the constants have to be in $\mathbb{F}_3$ and d_9 in the source SS is given by

$$d_9(v_2^l\alpha) = \begin{cases} 0 & \text{if } l \equiv 0,1,2,5,6,7 \bmod 9 \\ \pm v_2^{l-3}\beta^5 & \text{if } l \equiv 3,4,8 \bmod 9 \end{cases}$$

$$d_9(v_2^{l+1/2}\alpha) = \begin{cases} 0 & \text{if } l \equiv 0,1,2,4,5,6 \bmod 9 \\ \pm v_2^{l-3+1/2}\beta^5 & \text{if } l \equiv 3,7,8 \bmod 9 \end{cases}$$

$$d_9(v_2^l\alpha a_{35}) = \begin{cases} 0 & \text{if } l \equiv 0,1,2,5,6,7 \bmod 9 \\ \pm v_2^{l-3}\beta^5 a_{35} & \text{if } l \equiv 3,4,8 \bmod 9 \end{cases}$$

$$d_9(v_2^{l+1/2}\beta\alpha a_{35}) = \begin{cases} 0 & \text{if } l \equiv 0,1,2,4,5,6 \bmod 9 \\ \pm v_2^{l-3+1/2}\beta^6 a_{35} & \text{if } l \equiv 3,7,8 \bmod 9 \end{cases}$$

and we obtain the following E_{10}-term

$$\begin{aligned} E_{10} \cong\ & P/(\beta^5)\{v_2^l\}_{l=0,1,5} \oplus P/(\beta^3)\{v_2^l\alpha\}_{l=0,1,2,5,6,7} \\ & \oplus\ P/(\beta^4)\{v_2^{l+1/2}\beta\}_{l=0,4,5} \oplus\ P/(\beta^2)\{v_2^{l+1/2}\alpha\}_{l=0,1,2,4,5,6} \\ & \oplus\ P/(\beta^4)\{v_2^l\beta a_{35}\}_{l=0,1,5} \oplus\ P/(\beta^3)\{v_2^l\alpha a_{35}\}_{l=0,1,2,5,6,7} \\ & \oplus\ P/(\beta^5)\{v_2^{l+1/2}\beta a_{35}\}_{l=0,4,5} \oplus\ P/(\beta^2)\{v_2^{l+1/2}\beta\alpha a_{35}\}_{l=0,1,2,4,5,6}\,. \end{aligned}$$

At this point there is no more room for further nontrivial differentials and we arrive at the desired result below in which the names of the generators are chosen so as to give representatives in the E_2-term of the ANSS for $\pi_*(E_2^{hN} \wedge V(1))$.

Theorem 20. *As a module over $P = \mathbb{F}_3[v_2^{\pm 9}, \beta] \otimes \Lambda(\zeta)$ there is an isomorphism*

$$\begin{aligned} \pi_* L_2V(1) \cong\ & P/(\beta^5)\{v_2^l\}_{l=0,1,5} \oplus P/(\beta^3)\{v_2^l\alpha\}_{l=0,1,2,5,6,7} \\ & \oplus\ P/(\beta^4)\{v_2^{l+1/2}\beta\}_{l=0,4,5} \oplus\ P/(\beta^2)\{v_2^{l+1/2}\alpha\}_{l=0,1,2,4,5,6} \\ & \oplus\ P/(\beta^4)\{v_2^l\beta a_{35}\}_{l=0,1,5} \oplus\ P/(\beta^3)\{v_2^l\alpha a_{35}\}_{l=0,1,2,5,6,7} \\ & \oplus\ P/(\beta^5)\{v_2^{l+1/2}\beta a_{35}\}_{l=0,4,5} \oplus\ P/(\beta^2)\{v_2^{l+1/2}\beta\alpha a_{35}\}_{l=0,1,2,4,5,6}\,. \quad \square \end{aligned}$$

We finish by observing that this matches with Shimomura's result (if his parameter is taken to be $k = 1$!).

References

[D] P. Deligne, *Courbes elliptiques: Formulaire (d'après J. Tate)*, in: Modular Functions of One Variable IV, Lecture Notes in Math. **476** (1975), Springer Verlag.

[DH1] E. Devinatz and M. Hopkins, *The action of the Morava stabilizer group on the Lubin-Tate moduli space of lifts*, Amer. J. Math. **117** (1995), 669–710.

[DH2] E. Devinatz and M. Hopkins, *Homotopy fixed point spectra for closed subgroups of the Morava stabilizer groups*, to appear in Topology.

[GS] V. Gorbounov and P. Symonds, *Towards the homotopy groups of the higher real K theory EO_2*, Homotopy theory via algebraic geometry and group representations. Contemp. Math. **220** (1998), 103–115.

[GSS] V. Gorbounov, S. Siegel and P. Symonds, *Cohomology of the Morava stabilizer group S_2 at the prime* 3, Proc. Amer. Math. Soc. **126** (1998), 933–941.

[H] H.-W. Henn, *Centralizers of elementary abelian p-subgroups and mod-p cohomology of profinite groups*, Duke Math. J. **91** (1998), 561–585.

[LT] J. Lubin and J. Tate, *Formal moduli for one-parameter formel Lie groups*, Bull. Soc. Math. France **94** (1966), 49–60.

[N] L. Nave, *On the nonexistence of Smith-Toda complexes*, Preprint (available at http:// hopf.math.purdue.edu)

[Ra1] D. Ravenel, *The Cohomology of the Morava Stabilizer Algebras*, Math. Zeit. **152** (1977), 287–297.

[Ra2] D. Ravenel, *Complex cobordism and stable homotopy groups of spheres*, Academic Press, 1986.

[Ra3] D. Ravenel, *Nilpotence and Periodicity in Stable Homotopy Theory*, Ann. of Math. Studies, Princeton University Press, 1992.

[Re] C. Rezk, *Notes on the Hopkins-Miller theorem*, Homotopy theory via algebraic geometry and group representations. Contemp. Math. **220** (1998), 313–366.

[Sh1] K. Shimomura, *The homotopy groups of the L_2-localized Toda-Smith complex $V(1)$ at the prime* 3, Trans. Amer. Math. Soc. **349** (1997), 1821–1850.

[Sh2] K. Shimomura, *The homotopy groups of the L_2-localized mod-3 Moore spectrum at the prime* 3, J. Math. Soc. Japan **52** (2000), 65–90.

[SY] K. Shimomura and A. Yabe, *The homotopy groups $\pi_*(L_2S^0)$*, Topology **34** (1995), 261–289.

[SW] K. Shimomura and X. Wang, *The homotopy groups $\pi_*(L_2S^0)$ at the prime* 3, Topology **41** (2002), 1183–1198.

[St] N. Strickland, *Gross-Hopkins duality*, Topology **39** (2000), 1021–1033.

Paul Goerss
Department of Mathematics
Northwestern University
2033 Sheridan Road
Evanston, IL, 60208-2730, USA

Hans-Werner Henn
Institut de Recherche Mathématique Avancée
Université Louis Pasteur et CNRS
7 rue René Descartes
F-67084 Strasbourg Cedex, France

Mark Mahowald
Department of Mathematics
Northwestern University
2033 Sheridan Road
Evanston, IL, 60208-2730, USA

Progress in Mathematics, Vol. 215
Algebraic Topology: Categorical Decomposition Techniques, 153–165

Spherical Space Forms – Homotopy Types and Self-equivalences

Marek Golasiński and Daciberg Lima Gonçalves

Abstract. Let Q_{4m} be a generalized quaternion group and $X(n)$ an n-dimensional CW-complex with the homotopy type of an n-sphere. We compute the number of distinct homotopy types of spherical space forms with respect to all Q_{4m}-actions on all CW-complexes $X(4n-1)$ and deduce an existence of finite space forms given by some free cellular Q_{4m}-actions on CW-complexes $X(3)$ which do not have the homotopy type of a Clifford-Klein form provided that m is not a power of 2. We show that homotopy types of $(2mn-1)$-space forms with respect to G-actions are exhausted by homotopy types of orbit spaces of joins of CW-complexes $X(2n-1)$ with appropriate G-actions.

Given a free cellular action γ of a finite periodic group G on a CW-complex $X(n)$, we derive an exact sequence relating the group $\mathcal{E}(X(n)/\gamma)$ of homotopy classes of homotopy self-equivalences of the space form $X(n)/\gamma$ with the set of homotopy types of space forms with respect to all free cellular G-actions γ. At the end, the group of homotopy self-equivalences of a lens space and a space form associated with a free cellular Q_{4m}-action are studied.

Introduction

The cohomology of any discrete group that acts freely on a finite-dimensional CW-complex and the homotopy type of a sphere is periodic. Wall [20] has asked if the converse is true. A positive answer is given in the recent result [2] under the assumption that the periodicity isomorphism is a cup product with integral cohomology class. Swan [16] has shown that any finite group with periodic cohomology of period s acts freely and cellularly on an $(s-1)$-dimensional CW-complex of the homotopy type of an $(s-1)$-sphere. Furthermore, also in virtue of [16], there exists a finite $(ds-1)$-dimensional CW-complex with the homotopy type of a $(ds-1)$-sphere and a free cellular action of that periodic group, for some d with $ds > 2$. By means of [10], d cannot be taken equal to one in general. However, due to a Milnor result [11], this CW-complex may not have the homotopy type of any closed manifold for some finite groups with periodic cohomology.

2000 *Mathematics Subject Classification.* Primary 55M35, 55P15; secondary 57S17.
Key words and phrases. automorphism group, CW-complex, free cellular G-action, join, lens space, quaternion group, spherical space form.

The study of free cellular actions of a group on CW-complexes with the homotopy type of a sphere is related to an investigation of their orbit spaces called spherical space forms and a classification of their homotopy types has been studied extensively in [8, 18, 19]. Useful cohomological and geometric aspects associated to group actions are presented in [3] and lists of basic conjectures are provided. An old result of homotopy type theory (see e.g., [12, 17]) says that, up to homotopy, the set of lens spaces exhausts all homotopy types of orbit spaces of the cyclic group $\mathbb{Z}/m$ free cellular actions on $(2n-1)$-dimensional CW-complexes with the homotopy type of a $(2n-1)$-sphere. Thus the classification of all $\mathbb{Z}/m$-actions on these CW-complexes is equivalent to the classification of manifolds with the homotopy type of lens spaces studied by Browder in [4]. The set of homotopy types of lens spaces is evidently finite and their homotopy classification has been presented in [6, 12]. The paper [7] deals with homotopy types of space forms for cyclic groups. An explicit formula for the number of lens spaces has been obtained there and the structure of the group of homotopy classes of homotopy self-equivalences of a lens space described in the case of $\mathbb{Z}/p$ with p an odd prime. By means of [16], it is shown in [17] that the set of homotopy types of spherical space forms of all free cellular G-actions on $(2n-1)$-dimensional CW-complexes with the homotopy type of a $(2n-1)$-sphere is in one-one correspondence with the orbits, which contain a generator, in $H^{2n}(G;\mathbb{Z}) = \mathbb{Z}/|G|$ under the action of $\pm\mathrm{Aut}\, G$ (see also [7] for another approach).

The problem of immersing or embedding in Euclidean space of space forms $\mathbb{S}^{4n-1}/\gamma$ of free orthogonal Q_{2^m}-actions γ with $m \geq 3$ on the sphere $\mathbb{S}^{4n-1}$ has been considered in [14]. The aim of this paper is to calculate the number of homotopy types of spherical space forms for a generalized quaternion group Q_{4m} and describe the structure of the group of homotopy classes of homotopy self-equivalences of these space forms and lens spaces as well. These results are presented in Theorem 2.2, Theorem 3.3 and Theorem 3.4. The paper is divided into three sections. In Section 1 we recall the results presented in [7] on orbits, which contain a generator, in $H^{2n}(\mathbb{Z}/m;\mathbb{Z})$ under the action of $\pm\mathrm{Aut}\,\mathbb{Z}/m$ for the cyclic group $\mathbb{Z}/m$ of order m, describe the automorphism group $\mathrm{Aut}\, Q_{4m}$ and compute the induced maps by elements in $\mathrm{Aut}\, Q_{4m}$ on the cohomology $H^*(Q_{4m};\mathbb{Z})$.

In Section 2, we establish in Theorem 2.2 the number of homotopy types of orbit spaces given by free cellular Q_{4m}-actions on $(4n-1)$-dimensional CW-complexes with the homotopy type of a $(4n-1)$-sphere. In particular (Corollary 2.4), there are only two homotopy types of space forms determined by all free cellular $Q_{2^{m'+2}m}$-actions (with an odd number m and $m' \geq 0$) on 3-dimensional CW-complexes with the homotopy type of a 3-sphere if and only if either m is a positive power of an odd prime and $m' = 0$ or $m = 1$ and $m' > 0$. Otherwise, the number of those homotopy types is a power of 2 with exponent greater than one. Then, we extend the result in [18] and deduce in Theorem 2.5 that there is a finite space form given by a free cellular $Q_{2^{m'+2}m}$-action on a 3-dimensional CW-complex with the homotopy type of a 3-sphere which does not have the homotopy type of a Clifford-Klein form provided that $m' \geq 0$ and $m > 1$ is an odd number, and any finite space form has the homotopy type of a Clifford-Klein one for $m' > 0$ and $m = 1$. We also correct the numbers of generators in $H^4(Q_{2^{m'+2}m}; \mathbb{Z})$ presented in [8] corresponding to 3-dimensional space forms from the finite, infinite and Clifford-Klein form classes, respectively. At the end we generalize (Theorem 2.7) the classical result of homotopy type theory [12, 17] on CW-complexes with the homotopy type of spheres and cellular free $\mathbb{Z}/k$-actions by showing that homotopy types of orbit spaces with respect to G-actions on $(2mn-1)$-dimensional CW-complexes with the homotopy type of $(2mn-1)$-spheres is exhausted by homotopy types of orbit spaces of joins of $(2n-1)$-dimensional CW-complexes with the homotopy type of $(2n-1)$-spheres furnished with appropriate G-actions.

In Section 3, we study the structure of the group of homotopy classes of homotopy self-equivalences for lens spaces and space forms given by free cellular Q_{4m}-actions. We prove that all these groups are finite and abelian with orders described by means of the order of the acting group and the dimension of that space form.

The authors are grateful to the referee for carefully reading the original manuscript and his many valuable suggestions.

1. Arithmetical backgrounds

Throughout this paper X denotes a finite-dimensional CW-complex with the homotopy type of a sphere and of the same dimension as that sphere, and a free cellular action γ of a finite group G of order $|G|$. Write X/γ for the corresponding orbit space called an $(n-1)$-*spherical space form* or a *Swan* $(n-1)$-*complex* (see e.g., [3]) of dimension $(n-1)$. Then, by [5, Chap. XVI, §9], the group G is periodic with period s dividing n and $H^n(G;\mathbb{Z}) = \mathbb{Z}/|G|$, where $\mathbb{Z}/|G|$ is the cyclic group of order $|G|$. By means of [16], it is shown in [17] the homotopy types of space forms are in one-one correspondence with the orbits, which contain a generator in $H^n(G;\mathbb{Z}) = \mathbb{Z}/|G|$ under the action of $\pm\operatorname{Aut} G$ (see also [7] for another approach). But generators of the group $\mathbb{Z}/|G|$ are given by the unit group $(\mathbb{Z}/|G|)^*$ of the ring $\mathbb{Z}/|G|$. Thus, these homotopy types are in one-one correspondence with the quotient $(\mathbb{Z}/|G|)^*/\{\pm\varphi^*;\ \varphi \in \operatorname{Aut} G\}$, where φ^* is the induced automorphism on the cohomology $H^n(G;\mathbb{Z}) = \mathbb{Z}/|G|$ by φ in the automorphism group $\operatorname{Aut} G$. We intend to examine this quotient for G being the generalized quaternion group $Q_{4m} = \{x, y;\ x^m = y^2, xyx = y\}$. Note that in this group $x^{2m} = y^4 = e$ and each of its elements can be uniquely written in the form $x^k y^l$ with $k = 0, \ldots, 2m-1$ and $l = 0, 1$.

Let g_2 be a generator in the cohomology group $H^2(\mathbb{Z}/m;\mathbb{Z}) = \mathbb{Z}/m$. The results in [5, Chap. XII, §11] show that g_2^n generates the group $H^{2n}(\mathbb{Z}/m;\mathbb{Z}) = \mathbb{Z}/m$. Then, for any φ in $\operatorname{Aut}\mathbb{Z}/m = (\mathbb{Z}/m)^*$, the induced automorphism φ^* of the group $H^{2n}(\mathbb{Z}/m;\mathbb{Z}) = \mathbb{Z}/m$ is determined by the power k^n with k corresponding to φ in the group $(\mathbb{Z}/m)^*$.

Let now g_4 be a generator in the group $H^4(Q_{4m};\mathbb{Z}) = \mathbb{Z}/4m$. Then by [16], the element g_4^n generates $H^{4n}(Q_{4m};\mathbb{Z}) = \mathbb{Z}/4m$. The results presented in the second part of the proposition below are developed in [16] for generalized quaternion groups of order being a power of 2 and ≥ 8. However, for any generalized quaternion group Q_{4m} with $m > 2$, they are also probably known to some experts but it seems that they are not in the literature.

Proposition 1.1. *Let $\varphi : Q_{4m} \to Q_{4m}$ be a map with $m \geq 2$.*

(1) *If φ is an automorphism then $\varphi(x) = x^r$, $\varphi(y) = x^i y$ for some i such that $r = 0, \ldots, 2m-1$ with r and $2m$ relatively prime for $m > 2$. Conversely, any map φ defined on the generators x and y as above extends uniquely to an automorphism of Q_{4m}.*

(2) *If $\varphi^* : H^{4n}(Q_{4m};\mathbb{Z}) \to H^{4n}(Q_{4m};\mathbb{Z})$ is the induced map in cohomology by an automorphism φ of the group Q_{4m} with $\varphi(x) = x^r$ then $\varphi^*(g_4^n) = r^{2n} g_4^n$ for $n \geq 0$.*

Proof. (1) Given an automorphism φ of the group $Q_{4m} = \{x, y;\ x^m = y^2, xyx = y\}$, let $\varphi(x) = x^r y$ for some $r = 0, \ldots, 2m-1$. But $xyx = y$, hence $\varphi(x^4) = (x^r y x^r) y (x^r y x^r) y = y^4 = e$ and consequently $x^4 = e$. This is impossible since the generator x is of order $2m$ with $m > 2$. Therefore, $\varphi(x) = x^r$ for some $r = 0, \ldots, 2m-1$ and relatively prime with $2m$. Because φ is surjective, so $\varphi(y) = x^i y$ for some $i = 0, \ldots, 2m-1$.

Conversely, let φ be a map defined on generators x and y as above. Then, the extension of φ on the whole Q_{4m} by taking $\varphi(x^k y^l) = x^{rk}(x^i y)^l$, for any element $x^k y^l$ in Q_{4m} with $k = 0, \ldots, 2m-1$ and $l = 0, 1$ is well defined and preserves the multiplication on Q_{4m}. The cyclic subgroup (x) of Q_{4m} with index two is properly contained in the image of φ. Consequently φ is an automorphism of Q_{4m}.

(2) Let φ be an automorphism of Q_{4m} and $\iota : (x) \to Q_{4m}$ the inclusion homomorphism. Then, by [16, Proposition 8.5] $\iota^*(g_4) = g_2^2$, where g_2 is a generator in the group $H^2((x);\mathbb{Z})$. But, any automorphism φ of the group Q_{4m} restricts to one of the subgroup (x). By the naturality, one gets $\varphi^*(g_4) = r^2 g_4$ and consequently, the results follows. □

Thus we are in a position to state the following conclusion proved in [15] as well.

Corollary 1.2. *The automorphism group $\operatorname{Aut} Q_{4m}$ of the group Q_{4m} with $m > 2$ is metabelian and there exists an isomorphism*

$$\operatorname{Aut} Q_{4m} \cong \mathbb{Z}/2m \rtimes (\mathbb{Z}/2m)^\star,$$

where $\mathbb{Z}/2m \rtimes (\mathbb{Z}/2m)^\star$ denotes the semi-direct product of the groups $\mathbb{Z}/2m$ and $(\mathbb{Z}/2m)^\star$.

Proof. By the first part of Proposition 1.1, we see that any automorphism φ of the group Q_{4m} is uniquely determined by the pair (i, r) of the number satisfying the indicated conditions. Given two automorphisms determined by the pairs (i, r) and (i', r'), their composition yields the pair $(i + i'r, rr')$. Then, the map $\operatorname{Aut} Q_{4m} \to$

$\mathbb{Z}/2m \rtimes (\mathbb{Z}/2m)^*$ assigning to any automorphism φ the associated pair (i, r) is the required isomorphism. □

Remark 1.3. It is well known (see e.g., [1, Lemma 6.9]) that the group $\operatorname{Aut} Q_8$ is isomorphic to the symmetric group S_4.

In the light of the above, to describe homotopy types of spherical space forms, we are led to compute the quotient group $A(m,n) = (\mathbb{Z}/m)^*/\{\pm x^n;\ x \in (\mathbb{Z}/m)^*\}$. If $m = p_1^{l_1} \cdots p_s^{l_s}$ is the prime factorization of an integer $m \geq 1$ with $l_i \geq 1$ for $i = 1, \dots, s$, then it is well known (see e.g., [21, Chapter IV]) that $(\mathbb{Z}/m)^* = (\mathbb{Z}/p_1^{l_1})^* \times \cdots \times (\mathbb{Z}/p_s^{l_s})^*$. Moreover, $(\mathbb{Z}/p^l)^* = \mathbb{Z}/(p-1) \times \mathbb{Z}/p^{l-1}$ for p an odd prime or $l < 3$ and $(\mathbb{Z}/2^l)^* = \mathbb{Z}/2 \times \mathbb{Z}/2^{l-2}$ for $l \geq 3$ with generators, multiplication by -1 giving the element of order two, and multiplication by 5 or -3 giving the element of order 2^{l-2}.

Now we recall some results presented in [7] on the quotient group $A(m,n)$ which are useful in the next section.

Proposition 1.4. *If $m = p_1^{l_1} \cdots p_s^{l_s}$ is the prime factorization of an integer $m > 1$, then for any integer $n \geq 1$ there is an extension*

$$0 \to (\mathbb{Z}/2)^t \to A(m,n) \to \prod_{i=1}^{s} A(p_i^{l_i}, n) \to 0,$$

where t is determined as follows:

(1) *$t = 0$, if $-1 \in \{x^n;\ x \in (\mathbb{Z}/p_i^{l_i})^*\}$ for any $i = 1, \dots, s$;*
(2) *$t = s - 1 - \#\{i;\ -1 \in \{x^n;\ x \in (\mathbb{Z}/p_i^{l_i})^*\}\}$, otherwise.*

Of course, we can assume that $n = q_1^{t_1} \cdots q_l^{t_l}$ and $p - 1 = q_1^{u_1} \cdots q_l^{u_l}$, where $q_1, \dots, q_l$ are different primes, $t_1, \dots, t_l, u_1, \dots, u_l$ are non-negative integers and t_i or u_i is positive for all $i = 1, \dots, l$. Then, in [7] we have stated

Proposition 1.5. *Let $O(p^m, n)$ denote the order of the group $A(p^m, n)$ for positive integers m, n and a prime p.*

(1) *If p is an odd prime and $(p-1)/2$ is divisible by n in the group $\mathbb{Z}/(p-1)$, or $p = 2$ and n is an odd integer, then $O(p^m, n)$ is equal to:*
   (i) *$q_1^{\min(t_1,u_1)} \cdots q_l^{\min(t_l,u_l)}$, for $p \neq 2$ and $p \neq q_1, \dots, q_l$;*
   (ii) *$q_1^{\min(t_1,u_1)} \cdots q_{i_0-1}^{\min(t_{i_0-1},u_{i_0-1})} p^{\min(m-1,t_{i_0})} q_{i_0+1}^{\min(t_{i_0+1},u_{i_0+1})} \cdots q_l^{\min(t_l,u_l)}$, if $p \neq 2$ and $p = q_{i_0}$ for some $1 \leq i_0 \leq l$;*
   (iii) *1, if $p = 2$.*
(2) *Otherwise $O(p^m, n)$ is equal to:*
   (i) *$2^{-1} q_1^{\min(t_1,u_1)} \cdots q_l^{\min(t_l,u_l)}$, if $p \neq 2$ and $p \neq q_1, \dots, q_l$;*
   (ii) *$2^{-1} q_1^{\min(t_1,u_1)} \cdots q_{i_0-1}^{\min(t_{i_0-1},u_{i_0-1})} p^{\min(m-1,t_{i_0})} q_{i_0+1}^{\min(t_{i_0+1},u_{i_0+1})} \cdots q_l^{\min(t_l,u_l)}$, if $p \neq 2$ and $p = q_{i_0}$ for some $1 \leq i_0 \leq l$;*
   (iii) *$2^{\min(t_{i_0},m-2)}$, for $p = 2$, $m \geq 3$ and $q_{i_0} = 2$;*
   (iv) *1, for $p = 2$, $m \leq 2$ and $q_{i_0} = 2$.*

In particular, one gets

Corollary 1.6. *Let p be an odd prime, $2^u \mid p-1$ and $2^{u+1} \nmid p-1$ and $m,n \geq 1$. Then $O(p^m, 2^n) = 2^{u-1}$ for $u \leq n$ and $O(p^m, 2^n) = 2^n$ otherwise. Furthermore, $O(2^m, 2^n) = 2^{\min(n,m-2)}$ for $m \geq 3$ and $O(2^m, 2^n) = 1$ for $m \leq 2$.*

2. Homotopy types of space forms

In this section we give formulas for the number of homotopy types of orbit spaces under actions of a generalized quaternion group Q_{4m} on finite-dimensional CW-complexes with the homotopy type of a sphere. For some finite groups G with periodic cohomology with the period four, the numbers of generators in $H^4(G;\mathbb{Z})$ corresponding to infinite, finite and Clifford-Klein forms, respectively are studied and listed in [8, 18]. We make use of these and [7] to present some new results on infinite, finite and Clifford-Klein forms for the group Q_{4m}.

Observe that by identifying elements in the group $(\mathbb{Z}/m)^\star$ with all positive integers less than m and relatively prime with m one gets the inclusion $(\mathbb{Z}/2m)^\star \subseteq (\mathbb{Z}/4m)^\star$.

Lemma 2.1. *For any $m \geq 1$, the canonical homomorphism*

$$(\mathbb{Z}/4m)^\star/\{\pm x^{2n};\ x \in (\mathbb{Z}/2m)^\star\} \to A(4m, 2n)$$

is an isomorphism.

Proof. It is enough to show that $\{\pm x^{2n};\ x \in (\mathbb{Z}/2m)^\star\} = \{\pm x^{2n};\ x \in (\mathbb{Z}/4m)^\star\}$. Any element in $(\mathbb{Z}/4m)^\star$ which does not belong to $(\mathbb{Z}/2m)^\star$ is of the form $x+2m$ for some x in $(\mathbb{Z}/2m)^\star$. By means of the congruence $(x+2m)^{2n} = (x^2+4xm+4m^2)^n \equiv x^{2n} \pmod{4m}$ the proof is complete. □

Now consider the family of orbit spaces X/γ with respect to all free cellular G-actions γ on $(2n-1)$-dimensional CW-complexes X with the homotopy type of a $(2n-1)$-sphere. Two orbit spaces X/γ and X'/γ' are called equivalent if they are homeomorphic. Let $\mathcal{K}_G^{2n-1}$ denote the set of all such classes. We say that two classes $[X/\gamma]$ and $[X'/\gamma']$ are homotopic if the orbit spaces X/γ and X'/γ' are homotopy equivalent. Write $\mathcal{K}_G^{2n-1}/_\simeq$ for the associated quotient set of $\mathcal{K}_G^{2n-1}$ and $\operatorname{card}\mathcal{K}_G^{2n-1}/_\simeq$ for its cardinality, respectively. Then, we can state

Theorem 2.2. *Let $m \geq 2$, $n \geq 1$ and $4m = 2^l p_1^{l_1} \cdots p_s^{l_s}$ be the prime factorization of $4m$. Then, $\operatorname{card}\mathcal{K}_{Q_{4m}}^{4n-1}/_\simeq = 2^t\, O(2^l, 2n)\, O(p_1^{l_1}, 2n) \cdots\ O(p_s^{l_s}, 2n)$, where the number t is given by Proposition 1.4.*

In particular, $\operatorname{card}\mathcal{K}_{Q_{2^{m'+2}}}^{2^{n+1}n'-1}/_\simeq = 2^{\min(n,m')}$ for any $m', n \geq 1$, provided that n' is an odd positive integer.

Proof. By means of [16], it was shown in [17] that elements of the set $\mathcal{K}_G^{2n-1}/_\simeq$ are in one-one correspondence with the orbits, which contain a generator in the group $H^{2n}(G;\mathbb{Z}) = \mathbb{Z}/|G|$ under the action of $\pm\operatorname{Aut} G$. But generators of the group $\mathbb{Z}/|G|$ are given by the unit group $(\mathbb{Z}/|G|)^\star$ of the ring $\mathbb{Z}/|G|$. Thus, the homotopy types of space forms are in one-one correspondence with the quotient $(\mathbb{Z}/|G|)^\star/\{\pm\varphi^*;\ \varphi \in \operatorname{Aut} G\}$, where φ^* is the induced automorphism on the cohomology $H^{2n}(G;\mathbb{Z}) = \mathbb{Z}/|G|$ by φ in the automorphism group $\operatorname{Aut} G$.

Now examine this quotient for G being the generalized quaternion group $Q_{4m} = \{x, y;\ x^m = y^2, xyx = y\}$. Then the first claim follows straightforward from Proposition 1.4 and Lemma 2.1. The second one is a direct consequence of Proposition 1.5. □

In the group $\mathbb{Z}/(p-1) \times \mathbb{Z}/p^{m-1}$, if p is an odd prime, $(p-1)/2$ is the only element of order 2, which corresponds to -1 in the group $(\mathbb{Z}/p^m)^\star$ by the isomorphism above. Therefore, the equation $-1 = x^n$ has a solution in the multiplicative group $(\mathbb{Z}/p^m)^\star$ if and only if $(p-1)/2$ is divisible by the integer n in the additive group $\mathbb{Z}/(p-1)$. Then, in the light of Proposition 1.4 and Corollary 1.6, we may state

Corollary 2.3. *Let $m' \geq 0$, $n \geq 1$ and $m = p_1^{l_1} \cdots p_s^{l_s}$ be the prime factorization of an odd positive integer $m > 1$, $2^{u_i} \mid p_i - 1$ and $2^{u_i+1} \nmid p_i - 1$ for $i = 1, \ldots, s$ and $p_{k_1}, \ldots, p_{k_t}$ be all such primes from this factorization that $u_j \leq n$ for $j = 1, \ldots, t$. Then $\operatorname{card} \mathcal{K}^{2^{n+1}-1}_{Q_{2^{m'+2}m}}/\!\simeq\, = 2^{\min(n,m')+u_{k_1}+\cdots+u_{k_t}+(s-t)n}$.*

In particular, for m a positive power of an odd prime p with $2^u \mid p-1$ and $2^{u+1} \nmid p-1$ $\operatorname{card} \mathcal{K}^{2^{n+1}-1}_{Q_{2^{m'+2}m}}/\!\simeq\, = 2^{\min(n,m')+u}$, if $u \leq n$ and $2^{\min(n,m')+n}$, otherwise.

Consequently, for 3-dimensional space forms, one can state

Corollary 2.4. *Let $m' \geq 0$ and $m = p_1^{l_1} \cdots p_s^{l_s}$ be the prime factorization of an odd positive integer $m > 1$. Then $\operatorname{card} \mathcal{K}^3_{Q_{2^{m'+2}m}}/\!\simeq\, = 2^{\min(1,m')+s}$.*

In particular, $\operatorname{card} \mathcal{K}^3_{Q_{2^{m'+2}m}}/\!\simeq\, = 2$ if and only if either m is a positive power of an odd prime number and $m' = 0$ or $m = 1$ and $m' > 0$.

Let $m' \geq 0$ and m be an odd positive integer. The number of generators in the cohomology group $H^4(Q_{2^{m'+2}m};\mathbb{Z}) = \mathbb{Z}/2^{m'+2}m$ is given by $\phi(2^{m'+2}m) = 2^{m'+1}\varphi(m)$, where ϕ denotes Euler's function. By [8, Theorem 1], all $2\phi(m)$ those generators correspond to finite space forms, for $m' = 0$ and only $2^{m'}\phi(m)$ of them correspond to such space forms, otherwise. Furthermore, by means of [18], to each generator of the form $\pm x^2$ in $\mathbb{Z}/2^{m'+2}m$ corresponds a Clifford-Klein form. In the light of our investigations above, all orbits under the action (on generators of $\mathbb{Z}/2^{m'+2}m$) of $\pm\operatorname{Aut} Q_{2^{m'+2}m}$ have the same cardinality. On the other hand, by [13, Chapter 6], Clifford-Klein forms (other than lens spaces) are determined up to homeomorphism by their fundamental groups. Then, the following result (stated partially in [18]) based on Corollary 2.4 summarizes our discussion.

Theorem 2.5. *Let $m' \geq 0$ and m be an odd positive integer.*

(1) *If $m' \geq 0$ and $m > 1$ then there is a finite space form given by a free cellular $Q_{2^{m'+2}m}$-action on a 3-dimensional CW-complex with the homotopy type of a 3-sphere which does not have the homotopy type of a Clifford-Klein form.*

(2) *If $m' > 0$ and $m = 1$ then any finite space form given by a free cellular $Q_{2^{m'+2}}$-action on a 3-dimensional CW-complex with the homotopy type of a 3-sphere has the homotopy type of a Clifford-Klein form.*

More precisely, let $m' \geq 0$ and $m = p_1^{l_1} \cdots p_s^{l_s}$ be the prime factorization of an odd positive integer $m > 1$. Following [8], divide all 3-dimensional space forms X corresponding to all $Q_{2^{m'+2}m}$-free cellular actions into three classes:

I, X is infinite, hence not equivalent to a manifold;

F, X is finite;

CK, X is equivalent to a Clifford-Klein form.

Basing on results presented in [13, Chapter 6] and our calculations above, numbers of generators in $H^4(Q_{2^{m'+2}m};\mathbb{Z})$ listed in [8] and corresponding to 3-dimensional space forms from each of these three classes should be stated as follows:

m'	m	I	CK	F but not CK
0	>1	—	$2^{1-s}\phi(m)$	$2\phi(m) - 2^{1-s}\phi(m)$
>0	>1	$2^{m'}\phi(m)$	$2^{m'-s}\phi(m)$	$2^{m'}\phi(m) - 2^{m'-s}\phi(m)$
>0	1	$2^{m'}$	$2^{m'}$	—

Remark 2.6. In virtue of [9], there are two finite homotopy types of space forms given by free cellular Q_{4m}-actions on $(4n-1)$-dimensional CW-complexes with the homotopy type of a $(4n-1)$-sphere provided that m is an odd prime and $n \geq 1$. One is realized by an orthogonal Q_{4m}-action, the other is not.

Write X_γ for a $(2n-1)$-dimensional CW-complex X with the homotopy type of a $(2n-1)$-sphere and a free cellular G-action $\gamma : G \times X \to X$ and let X/γ be the associated orbit space. In virtue of [7], there is a G-map $f_\gamma : X_\gamma \to X^0_{\gamma_0}$ with degree $d(f_\gamma)$ to an arbitrary $(2n-1)$-dimensional CW-complex X^0 with the homotopy type of a $(2n-1)$-sphere and a free cellular G-action γ_0. Then the map

$$\mathcal{D}_G^{2n-1} : \mathcal{K}_G^{2n-1} \longrightarrow (\mathbb{Z}/|G|)^\star/\{\pm\varphi^*;\ \varphi \in \operatorname{Aut} G\}$$

given by $\mathcal{D}_G^{2n-1}([X/\gamma]) = [d(f_\gamma)]$, where $[d(f_\gamma)]$ is the class of the automorphism in $(\mathbb{Z}/|G|)^\star$ determined by the integer $d(f_\gamma)$, is well defined. In the light of [7] the induced map

$$\widetilde{\mathcal{D}_G^{2n-1}} : \mathcal{K}_G^{2n-1}/\simeq \longrightarrow (\mathbb{Z}/|G|)^\star/\{\pm\varphi^*;\ \varphi \in \operatorname{Aut} G\}$$

is injective and by means of [17] it is surjective as well. On the other hand, by [7, Theorem 1.1] given a G-map $f : X^1 \to X^2$ of $(2n-1)$-dimensional CW-complexes with the homotopy type of $(2n-1)$-spheres and $d \equiv d(f) \pmod{|G|}$ there exists a G-map $h : X^1 \to X^2$ with $d(h) = d$. Consequently, the surjection $\mathcal{D}_G^{2n-1} : \mathcal{K}_G^{2n-1} \longrightarrow (\mathbb{Z}/|G|)^\star/\{\pm\varphi^*;\ \varphi \in \operatorname{Aut} G\}$ lifts to a surjection

$$\widehat{\mathcal{D}}_G^{2n-1} : \mathcal{K}_G^{2n-1} \longrightarrow (\mathbb{Z}/|G|)^\star.$$

Given $(2n-1)$-dimensional CW-complexes $X^1, \ldots, X^m$ with the homotopy type of a $(2n-1)$-sphere the join $X^1 * \cdots * X^m$ is a $(2mn-1)$-dimensional

CW-complex with the homotopy type of a $(2mn-1)$-sphere. Free cellular actions $\gamma_1,\dots,\gamma_m$ of G on $X^1,\dots,X^m$, respectively determine the free cellular action

$$\gamma_1*\cdots*\gamma_m : G\times(X^1*\cdots*X^m)\to X^1*\cdots*X^m$$

given by $(\gamma_1*\cdots*\gamma_m)(g,x_1*\cdots*x_m)=\gamma_1(g,x_1)*\cdots*\gamma_m(g,x_m)$. Moreover, cellular G-maps $f_k : X^k_{\gamma_k}\to X'^k_{\gamma'_k}$, for $k=1,\dots,m$ yield a cellular G-map

$$f_1*\cdots*f_n : (X^1*\cdots*X^m)_{(\gamma_1*\cdots*\gamma_m)}\to(X'^1*\cdots*X'^m)_{(\gamma'_1*\cdots*\gamma'_m)}.$$

It is easy to check that for a cellular G-map $f : X\gamma\to X^0_{\gamma_0}$ we get the associated cellular G-map

$$\mathrm{id}_{(X^0)^{*(m-1)}}*f : ((X^0)^{*^{(m-1)}}*X)_{(\gamma_0^{*(m-1)}*\gamma)}\longrightarrow((X^0)^{*^m})_{(\gamma_0^{*m})}$$

with $d(\mathrm{id}_{(X^0)^{*(m-1)}}*f)=d(f)$. Then we are ready for

Theorem 2.7. *Let G be a finite group with the bijection*

$$\widetilde{\mathcal{D}_G^{2n-1}} : \mathcal{K}_G^{2n-1}/\simeq\longrightarrow(\mathbb{Z}/|G|)^*/\{\pm\varphi^*;\ \varphi\in\operatorname{Aut}G\}.$$

Given a $(2mn-1)$-dimensional CW-complex X with the homotopy type of a $(2mn-1)$-sphere and a cellular G-action γ there are $(2n-1)$-dimensional CW-complexes $X^1,\dots,X^m$ with the homotopy type of a $(2n-1)$-sphere and cellular G-actions $\gamma_1,\dots,\gamma_m$, respectively such that the orbit spaces X/γ and $(X^1*\cdots*X^m)/(\gamma_1*\cdots*\gamma_m)$ have the same homotopy types for any $m\geq 1$.

If G is the cyclic group $\mathbb{Z}/k$ of order k then the surjection $\widehat{\mathcal{D}}^1_{\mathbb{Z}/k} : \mathcal{K}^1_{\mathbb{Z}/k}\longrightarrow(\mathbb{Z}/k)^*$ restricts to a surjective on the homeomorphism classes $[\mathbb{S}^1/\gamma]$ for all free cellular $\mathbb{Z}/k$-actions γ on the circle $\mathbb{S}^1$. Thus we deduce one of the classical result of homotopy theory [12, 17].

Corollary 2.8. *Given a free cellular action γ of the cyclic group $\mathbb{Z}/k$ on a $(2n-1)$-dimensional CW-complex X, the orbit space X/γ is homotopy equivalent to a lens space.*

On the other hand, by [10, 11] there are finite groups acting freely and cellularly on the sphere $\mathbb{S}^{2mn-1}$ which do not admit a free cellular action on the sphere $\mathbb{S}^{2n-1}$ for some $m,n>1$.

3. The group of homotopy self-equivalences

We make use of the previous sections and some results from [7] to study the structure of the group $\mathcal{E}(X/\gamma)$ of homotopy classes of homotopy self-equivalences for lens spaces and space forms X/γ given by free cellular Q_{4m}-actions γ on a finite-dimensional CW-complex X with the homotopy type of a sphere and compute the order of this group.

Given a $(2n-1)$-dimensional CW-complex with the homotopy type of a $(2n-1)$-sphere and a free action of a finite group G, for a fixed base point in X, the fundamental group functor π_1 yields a homomorphism $\mathcal{E}(X/\gamma) \to \operatorname{Aut} G$. Take the surjection $(\mathbb{Z}/|G|)^\star/\{\pm 1\} \longrightarrow \mathcal{K}_G^{2n-1}/_\simeq$ as the composition of the quotient map $(\mathbb{Z}/|G|)^\star/\{\pm 1\} \to (\mathbb{Z}/|G|)^\star/\{\pm\varphi^*;\ \varphi \in \operatorname{Aut} G\}$ with the inverse map $(\widetilde{\mathcal{D}_G^{2n-1}})^{-1}$. Finally, let the map $\operatorname{Aut} G \to (\mathbb{Z}/|G|)^\star/\{\pm 1\}$ be given as the composition of $\operatorname{Aut} G \to (\mathbb{Z}/|G|)^\star$ (induced by the cohomology functor H^{2n}) with the quotient map $(\mathbb{Z}/|G|)^\star \to (\mathbb{Z}/|G|)^\star/\{\pm 1\}$. Fix a free cellular G-action γ on X and take the homotopy class of the homeomorphism class $[X/\gamma]$ as a base point in the set $\mathcal{K}_G^{2n-1}/_\simeq$. Then, one may state

Proposition 3.1. *Let X be a $(2n-1)$-dimensional CW-complex with the homotopy type of a $(2n-1)$-sphere and a free cellular action γ of a finite group G with $|G| > 2$. Then there is an exact sequence*

$$1 \to \mathcal{E}(X/\gamma) \to \operatorname{Aut} G \to (\mathbb{Z}/|G|)^\star/\{\pm 1\} \to \mathcal{K}_G^{2n-1}/_\simeq \to *$$

with the maps defined above.

Proof. The exactness at $\mathcal{E}(X/\gamma)$ and $\operatorname{Aut} G$ is a consequence of [7, Proposition 1.5] and at $(\mathbb{Z}/|G|)^\star/\{\pm 1\}$ follows straightforward from the definition of the indicated maps. □

From the exact sequence above it follows

Corollary 3.2. *If X is a $(2n-1)$-dimensional CW-complex with the homotopy type of a $(2n-1)$-sphere and a free cellular G-action γ then the cardinality of the set $\mathcal{K}_G^{2n-1}/_\simeq$ is equal to*

$$2^{-1}\phi(|G|)|\mathcal{E}(X/\gamma)|(|\operatorname{Aut}(G)|)^{-1},$$

where ϕ denotes Euler's function.

At the end, the group $\mathcal{E}(X/\gamma)$ of homotopy classes of homotopy self-equivalences is studied, where γ is given either by a free cellular action of an arbitrary finite cyclic group or a generalized quaternion group on a finite-dimensional CW-complex with the homotopy type of a sphere. For a cyclic group with an odd prime order, the group $\mathcal{E}(X/\gamma)$ has been established in [7, Proposition 2.6].

In view of Proposition 3.1, the group $\mathcal{E}(X/\gamma)$ is the kernel of the map $\operatorname{Aut} G \to (\mathbb{Z}/|G|)^\star/\{\pm 1\}$. In particular, by Section 1 the group $\mathcal{E}(L)$ is isomorphic to the subgroup of $(\mathbb{Z}/p^l)^\star$ consisting of all elements k with $k^n \equiv \pm 1 (\operatorname{mod} p^l)$ for a lens space L given by a free cellular $\mathbb{Z}/p^l$-action on a homotopy $(2n-1)$-sphere with p an odd prime and any $l \geq 1$ or $p = 2$ and $l \geq 2$.

For an odd prime p and $l \geq 1$, one has that $(\mathbb{Z}/p^l)^\star = \mathbb{Z}/(p-1) \times \mathbb{Z}/p^{l-1}$, so the corresponding additive subgroup of all solutions of that equation consists of all elements (m_1, m_2) with $n(m_1, m_2) = (0, 0)$ or $n(m_1, m_2) = ((p-1)/2, 0)$, because $((p-1)/2, 0)$ is the only element of order two in the group $\mathbb{Z}/(p-1) \times \mathbb{Z}/p^{l-1}$. If d_1 (resp. d_2) is the greatest common divisor of the integers n and $p-1$ (resp. n and p^{l-1}), then the subgroup $G_{d_1} \times G_{d_2}$ of the group $\mathbb{Z}/(p-1) \times \mathbb{Z}/p^{l-1}$ generated by the element $((p-1)/d_1, p^{l-1}/d_2)$ consists of all elements (m_1, m_2) in the group $\mathbb{Z}/(p-1) \times \mathbb{Z}/p^{l-1}$ with $n(m_1, m_2) = (0,0)$. Moreover, if there is a solution m_0 in the group $\mathbb{Z}/(p-1)$ of the equation $nm_1 = (p-1)/2$, then the set of all solutions of the equation $n(m_1, m_2) = ((p-1)/2, 0)$ in the group $\mathbb{Z}/(p-1) \times \mathbb{Z}/p^{l-1}$ is equal to the coset $(m_0, 0) + G_{d_1} \times G_{d_2}$. Denote by $G(p^l, n)$ either the subgroup $G_{d_1} \times G_{d_2}$, if such m_0 does not exist or $G_{d_1} \times G_{d_2} \cup ((m_0, 0) + G_{d_1} \times G_{d_2})$ otherwise.

For $p = 2$ and $l \geq 2$, one has that $(\mathbb{Z}/2^l)^\star = \mathbb{Z}/2 \times \mathbb{Z}/2^{l-2}$ and multiplication by -1 yields the only element of order two (in the group $\mathbb{Z}/2$). Take the greatest common divisor d of the integers n and 2^{l-2} and consider the subgroup G_d of the group $\mathbb{Z}/2^{l-2}$ generated by the element $2^{l-2}/d$. It is easy to check that the corresponding additive subgroup $G(2^l, n)$ of the group $\mathbb{Z}/2 \times \mathbb{Z}/2^{l-2}$ (formed by appropriate solutions) is equal to $\mathbb{Z}/2 \times G_d$. Consequently, $\mathcal{E}(L) = G(p^l, n)$ for a lens space given by a free cellular $\mathbb{Z}/p^l$ on a homotopy $(2n-1)$-sphere. Let $G(2, n)$ denote the trivial group. Then, in view of [7, Proposition 1.5] the following result holds.

Theorem 3.3. *Let L be a lens space determined by a free cellular $\mathbb{Z}/m$-action on a homotopy $(2n-1)$-sphere and $m = p_1^{k_1} \cdots p_s^{k_s}$ the prime factorization of an integer $m > 2$ with $k_i \geq 1$ for $i = 1, \ldots, s$. Then $\mathcal{E}(L) = G(p_1^{k_1}, n) \times \cdots \times G(p_s^{k_s}, n)$.*

Now let X be a $(4n-1)$-dimensional CW-complex with homotopy type of a $(4n-1)$-sphere and a free cellular Q_{4m}-action. For the greatest common divisor d of the integers $2n$ and 2^{l-1}, consider the subgroup $\widetilde{G}_d$ of the group $\mathbb{Z}/2^{l-2}$ generated by the element $2^{l-1}/d$. Then, it is easy to check that the corresponding additive subgroup $\widetilde{G}(2^l, 2n)$ (studied above) is equal to $\mathbb{Z}/2 \times \widetilde{G}_d$.

Finally, we can conclude the paper with the following result.

Theorem 3.4. *Let Q_{4m} be a generalized quaternion group, γ a free cellular action on a $(4n-1)$-dimensional CW-complex X with the homotopy type of a $(4n-1)$-sphere and $2m = 2^{k_0+1} p_1^{k_1} \cdots p_s^{k_s}$ the prime factorization of $2m$ with $m \geq 2$. Then $\mathcal{E}(X/\gamma)$ is isomorphic to:*

(1) *the semi-direct product*

$$\mathbb{Z}/2m \rtimes (\widetilde{G}(2^{k_0+1}, 2n) \times G(p_1^{k_1}, 2n) \times \cdots \times G(p_s^{k_s}, 2n))$$

for $m \geq 3$;

(2) *the symmetric group S_4 for $m = 2$.*

Proof. Part one is a consequence of Corollary 1.2 and the discussion before Theorem 3.3. By [16, Proposition 8.3] any automorphism of Q_8 induces the identity automorphism on the cohomology $H^{4n}(Q_8; \mathbb{Z})$. Then, in the light of the exact sequence given in Proposition 3.1, one gets $\mathcal{E}(X/\gamma) = \operatorname{Aut} Q_8 = S_4$ and the second part follows. □

References

[1] A. Adem, R.J. Milgram, *Cohomology of Finite Groups*, Grundl. der Math. Wissen. **309** (1994), Springer-Verlag, New York (1994).

[2] A. Adem, J.H. Smith, *Periodic complexes and group actions*, Ann. of Math. **154** (2001), no. 2, 407–435.

[3] A. Adem, J.F. Davis, *Topics in Transformation Groups*, Handbook of geometric topology, 1–54, North-Holland, Amsterdam (2002).

[4] W. Browder, *Free $\mathbb{Z}_p$-actions on homotopy spheres*, Topology of Manifolds, Proceedings of the University of Georgia, Topology of Manifolds Institute (1969), 217–226.

[5] H. Cartan, S. Eilenberg, *Homological Algebra*, Princeton, New Jersey (1956).

[6] M.M. Cohen, *A Course in Simple-Homotopy Theory*, Springer-Verlag, New York-Heidelberg-Berlin (1973).

[7] M. Golasiński, D.L. Gonçalves, *Homotopy spherical space forms – a numerical bound for homotopy types*, Hiroshima Math. J. **31**, no. 1 (2001), 107–116.

[8] R. Lee, C.B. Thomas, *Free finite group actions on $\mathbb{S}^3$*, Bull. Amer. Math. Soc. **79** (1973), 211–215.

[9] I. Madsen, *Reidemeister torsion, surgery invariants and spherical space forms*, Proc. London Math. Soc. (3) **46** (1983), 193–240.

[10] I. Madsen, C.B. Thomas, C.T.C. Wall, *Topological spherical space form problem III: dimensional bounds and smoothing*, Pacific J. Math. **106**, no. 1 (1983), 135–143.

[11] J. Milnor, *Groups which act on $\mathbb{S}^n$ without fixed points*, Amer. J. Math. **79** (1957), 623–630.

[12] P. Olum, *Mappings of manifolds and the notion of degree*, Ann. of Math. **58**, no. 3 (1953), 458–480.

[13] P. Orlik, *Seifert Manifolds*, Lecture Notes in Math. **291**, Springer-Verlag, New York (1972).

[14] D. Pitt, *Free actions of generalized quaternion groups on spheres*, Proc. London Math. Soc. (3) **26** (1973), 1–18.

[15] S. Roß, W. Prax, *Die Automorphismengruppe der Quaternionengruppe Q_n für $n \geq 3$*, Wiss. Z. Pädagog. Hochsch. Erfurt/Mühlhausen Math.-Natur. Reihe **22** (1986), no. 1, 96–101.

[16] R.G. Swan, *Periodic resolutions for finite groups*, Ann. of Math. **72**, no. 2 (1960), 267–291.

[17] C.B. Thomas, C.T.C. Wall, *The topological spherical space form problem I*, Compositio Math. **23** (1971), 101–114.

[18] C.B. Thomas, *Homotopy classification of free actions by finite groups on $\mathbb{S}^3$*, Proc. London Math. Soc. (3) **40** (1980), 284–297.

[19] C.T.C. Wall, *Surgery on Compact Manifolds*, Academic Press, London-New York (1970).

[20] C.T.C. Wall, *Periodic projective resolutions*, Proc. London Math. Soc. **39** (1979), 509–533.

[21] H.J. Zassenhaus, *The Theory of Groups*, Chelsea, New York (1958).

Marek Golasiński
Faculty of Mathematics and Computer Science,
Nicholas Copernicus University,
Chopina 12/18, 87-100 Toruń, Poland
e-mail: marek@mat.uni.torun.pl

Daciberg Lima Gonçalves
Department of Mathematics-IME
University of São Paulo
Caixa Postal 66.281-AG.
Cidade de São Paulo
05311-970 São Paulo, Brasil
e-mail: dlgoncal@ime.usp.br

Progress in Mathematics, Vol. 215
Algebraic Topology: Categorical Decomposition Techniques, 167–178

The Homotopy Type of Two-regular K-theory

Luke Hodgkin and Paul Arne Østvær

Abstract. We identify the 2-adic homotopy type of the algebraic K-theory space for rings of integers in two-regular exceptional number fields. The answer is given in terms of well-known spaces considered in topological K-theory.

1. Introduction

Let E be a number field, $\mathcal{O}_E$ its ring of algebraic integers, and $R_E = \mathcal{O}_E[\frac{1}{2}]$ the corresponding ring of 2-integers. For the definition of the algebraic K-theory space $K(\mathcal{O}_E)$ (resp. $K(R_E)$) we refer to [7]; it is the usual 'plus construction' on the stable classifying space $BGL(\mathcal{O}_E)$ (resp. $BGL(R_E)$). A related, homotopically more accessible space is the étale topological K-theory space $K(R_E)_{ét}$ of R_E, see [3]. The purpose of this paper is to pin down the 2-adic homotopy type of $K(R_E)$ for some special E. Our results rely on the following recent advances.

In [9], Rognes and Weibel determined up to extensions the 2-completed groups $K_n(R_E)\hat{_2}$ $(= K_n(\mathcal{O}_E)\hat{_2}$ for $n \geq 2)$. Their computation is expressed in terms of the étale cohomology groups $H^*(R_E; \mathbb{Z}\hat{_2}(i))$ of R_E with coefficients twisted by the action of the roots of unity. A case where the extension problems disappear is the '2-regular case', (see definition below) for which Rognes and Østvær [11] gave a complete description of $K_n(R_E)\hat{_2}$, for all n. The above results are among the consequences of Voevodsky's solution of the Milnor conjecture [13], as developed in subsequent work [12], [2]. A particular interesting end-product is 'étale descent for K-theory of number fields at 2', i.e., the strong Quillen-Lichtenbaum conjecture is true for number fields at the prime 2. See [10] for the case of totally imaginary number fields, and [14] for the case of real number fields. In particular, we have:

(a) the spaces $K(R_E)$ and $K(R_E)_{ét}$ are the same on zero-connected components at the prime 2, and
(b) in consequence one can try to adopt the étale homotopy methods from [3] to determine the 2-completed homotopy type of $K(R_E)$ $(= K(\mathcal{O}_E)$ on one-connected components).

2000 *Mathematics Subject Classification.* Primary 19D99; Secondary 55N15, 55P15.
Key words and phrases. Homotopy type of K-theory spaces, rings of integers in number fields.
The second author was supported in part by NSF.

Part (b) was solved in the case where E is 2-regular and 'non-exceptional' (see definition below), in [15]; the result is satisfyingly simple, in that the homotopy type is a product of well-known components. In this paper we apply a similar analysis to the more complicated 'exceptional' case. Here the product structure is replaced by a twisting (fibration), but the degree of complication is minimal, and almost all of the factors are untwisted, and again of a simple type. As a check, we verify that our result gives the homotopy groups calculated in [11].

To state the results, we need some definitions and notation. Let r_1 respectively r_2 denote the number of real embeddings respectively pairs of conjugate complex embeddings of E. We say that E is *2-regular* if any of the following equivalent conditions is satisfied:

1) The 'modified tame kernel' of E vanishes. This latter can be identified with the kernel of the natural surjection:

$$\alpha^2 : H^2(R_E; \mathbb{Z}_2(2)) \to \bigoplus^{r_1} H^2(\mathbb{R}; \mathbb{Z}_2(2)) \cong r_1.\mathbb{Z}/2$$

summed over the real embeddings of E.

2) The ideal (2) does not split in E and the narrow Picard group $Pic_+(R_E)$ of R_E has odd order.

3) $$H^2(R_E; \mathbb{Z}_2(i)) = \begin{cases} r_1.\mathbb{Z}/2 & (i \neq 0 \text{ even}) \\ 0 & (i \neq 1 \text{ odd}). \end{cases}$$

For a discussion, and further criteria, see Proposition 2.2 of [11].

Next, let ζ_m denote an mth root of unity and $\overline{\zeta}_m$ its conjugate. For each r, we have the cyclotomic extension $E(\zeta_{2^r})$. E is said to be *exceptional* if Gal $(E(\zeta_{2^r})/E)$ is not cyclic for some r; otherwise it is *non-exceptional.* It is easy to check that E is non-exceptional if and only if $\zeta_4 \in E$ or $\zeta_{2^k} + \overline{\zeta}_{2^k} \in E$ for some $k \geq 3$. Any real number field is exceptional, while any 2-cyclotomic field $\mathbb{Q}(\zeta_{2^r})$, $r \geq 2$, is non-exceptional. In our case (exceptional), the description of $K(\mathcal{O}_E)\hat{_2}$ is complicated by the above-mentioned twisting, and we require some names for the spaces which will be our building blocks.

We consider all spaces completed at the prime 2, where not explicitly stated; we hope that the reader will accept statements such as '$\pi_1(S^1) = \mathbb{Z}\hat{_2}$' which result from this. We also abusively write Spec R for what is properly the étale homotopy type $(\text{Spec}\, R)_{ét}$. As usual U, O are the stable unitary and orthogonal groups, and BU, BO their classifying spaces. The complexification c maps O into U as a subgroup, with quotient U/O. For q odd, let ψ^q be the Adams operation on BU. Note in particular that ψ^{-1} is the conjugation map. By Quillen's fundamental result [8], when q is an odd prime-power the K-theory space of the finite field $\mathbb{F}_q$ is the fibre of $\psi^q - 1 : BU \to BU$; we shall denote this space by $F\psi^q$. Similarly, we write $F\psi^{-1}$ for the fibre of $\psi^{-1} - 1$. We require two variants of $F\psi^q$:

1) In analogy with a construction of Bökstedt, define $JK(q)$ to be the fibre of the composite:
$$BO \xrightarrow{c} BU \xrightarrow{\psi^q - 1} BU.$$
(This is $K(\mathbb{Z})$ if $q \equiv \pm 3 \bmod 8$.)
2) Let $j : F\psi^{-1} \to BU$ be the inclusion of the fibre, and define $J_c(q)$ to be the fibre of the composite $(\psi^q - 1) \circ j : F\psi^{-1} \to BU$. (Note the analogy between this construction and the previous one.)

Our next points concern Galois groups. Let $\mu_\infty(E)$ respectively μ_∞ be the group of 2-primary roots of unity in E respectively $\mathbb{C}$, and $\Gamma'_E = \mathrm{Gal}\,(E(\mu_\infty)/E)$. The natural action of Γ'_E on μ_∞ gives a monomorphism

(1) $$\phi : \Gamma'_E \to Aut(\mu_\infty) \cong \mathbb{Z}_2\hat{} \oplus \mathbb{Z}/2$$

– compare §1 of [6]. By considering $\pi_1(\mathrm{Spec}\, R_E)$ as the Galois group of the maximal unramified extension of R_E, we see that the action of $\pi_1(\mathrm{Spec}\, R_E)$ on μ_∞ factors through Γ'_E to give a composite:

(2) $$\pi_1(\mathrm{Spec}\, R_E) \to \Gamma'_E \xrightarrow{\phi} Aut(\mu_\infty)$$

which we shall call $\hat{\phi}$. Clearly the images of ϕ and $\hat{\phi}$ in $Aut(\mu_\infty)$ are the same subgroup, say Λ. If $E_0 = E(\sqrt{-1})$, it is a consequence of the 'exceptional' condition on E that $\Gamma'_E = \Gamma_E \times \mathbb{Z}/2$, where $\Gamma_E = \mathrm{Gal}\,(E(\mu_\infty)/E_0)$. Still following [6], set $a_E = \nu_2(|\mu_\infty(E_0)|)$ (the 2-adic valuation). Then Λ is (topologically) generated by elements q, σ where σ (order 2) is conjugation and $q \in \mathbb{Z}_2\hat{}$ is represented by any integer such that q is $\equiv \pm 1 \bmod 2^{a_E}$ but not $\bmod 2^{a_E+1}$. By Čebotarev's theorem we can always choose a prime $\mathfrak{P}$ in R_E such that the order q of the finite field $R_E/(\mathfrak{P})$ is an integer with these properties.

For future reference, we define numbers $w_m = w_m(E)$ by: $w_m = 2$ (m odd), $w_m = 2^{a_E + \nu_2(m)}$ (m even). (Compare, e.g., the definition in [9], which is equivalent in the exceptional case. Mitchell [6] writes w_i for the exponents, rather than the powers of 2.)

Our main result is as follows:

Theorem 1.1. *With the preceding notation, let E be 2-regular and exceptional. Then at the prime 2:*

(i) *If E is totally imaginary ($r_1 = 0$), $K(R_E)$ is homotopy equivalent to the product*
$$J_c(q) \times \prod^{r_2 - 1} U$$
($r_2 - 1$ factors in the product).

(ii) *If $r_1 > 0$, $K(R_E)$ is homotopy equivalent to the product*
$$JK(q) \times \prod^{r_2} U \times \prod^{r_1 - 1} (U/O).$$

Our strategy in proving this result is based on ideas present in [3], and is as follows. We find a space X, and map $f : X \to \operatorname{Spec} R_E$ which induces an isomorphism on mod 2 homology. The space X is a wedge of $r_2 + 1$ circles and r_1 copies of the infinite projective space RP^∞. The key question is then how the wedge components of X map via the composite:

$$(3) \qquad \pi_1(X) \xrightarrow{f_*} \pi_1(\operatorname{Spec} R_E) \xrightarrow{\hat{\phi}} Aut(\mu_\infty).$$

Specifically, the condition required is that $Im(\hat{\phi} \circ f_*) = Im(\hat{\phi}) = \Lambda$. According to [3], in the given circumstances, $K(R_E)$ is homotopy equivalent to a space called $K(X)$; and to find $K(X)$ we need to know the wedge components, and the way that their fundamental groups map into $Aut(\mu_\infty)$. We therefore find these, simplify as much as possible, and our theorem will follow from Dwyer and Friedlander's result. As can be seen from the statement of the theorem, cases (i) and (ii) need separate treatment; they are dealt with in Sections 2, 3 respectively.

It should be noted that all of these computations could have been done at the time of the original paper [3], modulo replacing the K-theory spaces by their étale topological versions; indeed, our reliance on the methods of Dwyer and Friedlander is substantial. However, the étale descent results we mentioned in the beginning make it possible to state the results for the K-theory spaces themselves.

A general solution to the problem of finding K-theory spectra of number rings at the prime 2 has been undertaken by Mitchell [5, 6]. The results are fuller, and give in fact the 2-adic homotopy type when combined with [10] and [14]. But their interpretation requires knowledge of the 'Iwasawa module', which appears difficult in general. Mitchell has kindly pointed out to us that our main result for real number fields can be deduced from results in [6] and [14], see Example 2 in [6]. The deduction itself will not be reproduced here since it requires technology which is not part of this paper.

2. The totally imaginary case

In this section, E will denote a number field which is 2-regular, exceptional, and totally imaginary. We can deduce immediately (cf. [11]):

Lemma 2.1. *The mod 2 homology of R_E is given by:*

$$H_1(R_E;\mathbb{Z}/2) = (r_2+1).\mathbb{Z}/2$$
$$H_i(R_E;\mathbb{Z}/2) = 0 \quad (i > 1).$$

However, in this case we can do better, since the 2-completed homology is also simple; for this we write $H_i(\ )$, omitting the coefficients.

Lemma 2.2. *The 2-completed homology of R_E is given by:*

$$H_1(R_E) = (r_2+1).\mathbb{Z}_2\hat{}$$
$$H_i(R_E) = 0 \quad (i > 1).$$

Proof. Let α be the unique prime above 2 in $\mathcal{O}_E$. Then by Theorem 2.2 of [3], $H_1(R_E)$ is a quotient of $(\mathcal{O}_E)^{*\,\hat{}}_{\alpha}$ (since the narrow Picard group vanishes); that is, of a direct sum of copies of $\mathbb{Z}_2\hat{}$ and 2-torsion groups. From the description of the mod 2 homology in Lemma 2.1, there is no 2-torsion, so H_1 is a sum of copies of $\mathbb{Z}_2\hat{}$, and the rank, again from Lemma 2.1, is $r_2 + 1$. □

Next, recall from §1 the homomorphism $\hat{\phi} : \pi_1(\text{Spec}\, R_E) \to Aut(\mu_\infty)$. In [15] (following a model from [3]) it was shown that in the non-exceptional case there is a map, $f : X \to \text{Spec}\, R_E$, inducing an isomorphism on mod 2 homology, where:

(i) X is a wedge of $r_2 + 1$ circles;
(ii) The circles can be chosen in such a way that the first one maps to a topological generator of $\mathbb{Z}_2\hat{}$, and the rest map trivially.

Our first result in this section is parallel to this, if slightly more complicated. It states:

Proposition 2.3. *In the exceptional totally imaginary case, there is a map $f : X \to$ Spec R_E such that:*

(i) *X is a wedge of $r_2 + 1$ circles;*
(ii) *The first circle, considered as an element of $\pi_1(X)$, maps under $\hat{\phi} \circ f_*$ to $q \in \Lambda$; the second to the non-trivial element σ of order 2 in Λ; and the rest (if any) map trivially;*
(iii) *f induces an isomorphism on mod 2 homology.*

Proof. This is essentially elementary topology, using what we know of the homology of Spec R_E. In fact, we can clearly choose $r_2 + 1$ maps from S^1 to Spec R_E whose images under the Hurewicz map generate $H_1(R_E)$ ($\mathbb{Z}_2\hat{}$ or $\mathbb{Z}/2$ coefficients). The key adjustments to be made concern their images under ϕ. We know the structure of $Im(\hat{\phi}) = \Lambda$, and since the latter is Abelian, $\hat{\phi}$ factors through $H_1(R_E)$. Hence we can find maps $f_1, f_2 : S^1 \to \text{Spec}\, R_E$ which represent generators of H_1, such that f_1 maps under ϕ to q, and f_2 maps to σ.

Now let $f_3, \dots, f_{r_2+1} : S^1 \to \text{Spec}\, R_E$ represent the remaining generators of H_1. We can multiply (in π_1) by suitable powers of f_1, f_2 to obtain maps $g_3, \dots, g_{r_2+1}$ which still define a basis of H_1 together with f_1, f_2, and map trivially under $\hat{\phi}$. We can now use $f_1, f_2, g_3, \dots, g_{r_2+1}$ to construct a map f from the wedge X of $r_2 + 1$ circles to Spec R_E which has the properties claimed in Proposition 2.3. □

Now Proposition 3.2 of [3] tells us that f induces a homotopy equivalence from $K(R_E)$ to a space $K(X)$, whose definition strictly depends not on X as space but on the composite map

$$X \xrightarrow{f} \text{Spec}\, R_E \to \text{Spec}\, \mathbb{Z}[\frac{1}{2}].$$

Again using the methods of [3], we write $X = V_1 \vee V_2 \vee W$, where V_1, V_2 are the first two circles and W is the wedge of the remaining ones. There is a fibre square

$$\begin{array}{ccc} K(X) & \longrightarrow & K(V_1 \vee V_2) \\ \downarrow & & \downarrow \\ K(W) & \longrightarrow & K(*) = BU \end{array}$$

and $K(W)$, by the arguments of [3] Proposition 4.5, is the unpointed function space $BU^W = BU \times \prod^{r_2-1} U$. It follows easily that:

Proposition 2.4. *The space $K(X)$ is homotopy equivalent to the product:*

$$K(V_1 \vee V_2) \times \prod^{r_2-1} U.$$

And our main challenge is to identify the space $K(V_1 \vee V_2)$. For this we have a second fibre square:

$$\begin{array}{ccc} K(V_1 \vee V_2) & \longrightarrow & K(V_1)\ . \\ \downarrow & & \downarrow \\ K(V_2) & \longrightarrow & BU \end{array}$$

The space $K(V_1)$ is essentially well known, and often used, being the 'finite fields' K-theory space $F\psi^q$ of Quillen [8]. It follows from the choice of the prime $\mathfrak{P}$ and the integer q in §1 that the composite

$$S^1 \to \operatorname{Spec} R_E/(\mathfrak{P}) \to \operatorname{Spec} R_E$$

where the generator of $\pi_1(S^1)$ is mapped into the Frobenius of the finite field, can be taken for the inclusion of V_1 – its image under $\hat{\phi}$ is precisely q. We can therefore identify $K(V_1)$ with the 2-adic K-theory space of the finite field; and this is $F\psi^q$, the fibre of $\psi^q - 1 : BU \to BU$.

The following result is now not surprising, given its similarity to the foregoing:

Lemma 2.5. *The space $K(V_2)$ is homotopy equivalent to the fibre $F\psi^{-1}$ of $\psi^{-1} - 1 : BU \to BU$.*

Proof. We give a rather ad hoc proof of what is probably a special case of a larger result. Let R denote the 'ground ring' $\mathbb{Z}[\frac{1}{2}]$. The map $f_2 : V_2 \to \operatorname{Spec} R$, which defines $K(V_2)$ (see above), has an obvious double cover $\tilde{f}_2 : \tilde{V}_2 \to \operatorname{Spec} R$. However, $\tilde{f}_2$ is trivial on π_1, so that the corresponding space $K(\tilde{V}_2)$ is homotopy equivalent to $BU^{\tilde{V}_2}$ as before. ($\tilde{V}_2$ is still a circle, of course.) Now the theory of [4] tells us that $K(V_2)$ can be identified with the 'homotopy fixed points' $Hom^{\Sigma}(E\Sigma, K(\tilde{V}_2))$, where Σ is the covering group ($=\mathbb{Z}/2$). In our case these are equivalent to the fixed points, i.e., to the equivariant maps $\lambda : \tilde{V}_2 \to BU$; where Σ acts on $\tilde{V}_2$ by the antipodal map, and on BU by complex conjugation ψ^{-1}.

Clearly such a map λ is determined by its values on the upper half-circle, and a standard argument shows that we can identify the space of such maps with the space of maps $\lambda' : [0,1] \to BU$ satisfying $\lambda'(1) = \psi^{-1}(\lambda'(0))$. But this is (one definition of) the homotopy fibre $F\psi^{-1}$. □

Now we can compute $K(V_1 \vee V_2)$. In fact, again following [3], the wedge product gives rise to a fibre square, which we extend to the right:

$$\begin{array}{ccccc} K(V_1 \vee V_2) & \longrightarrow & K(V_2) & \longrightarrow & BU\,. \\ \Big\downarrow & & \Big\downarrow{\scriptstyle j} & & \Big\| \\ K(V_1) & \longrightarrow & BU & \xrightarrow{\psi^q - 1} & BU \end{array}$$

From this it is clear that $K(V_1 \vee V_2)$ is the fibre of the composite $(\psi^q - 1) \circ j : F\psi^{-1} \to BU$, which is the space denoted $J_c(q)$ in §1. This identification, together with Proposition 2.4, give us case (i) of Theorem 1.1.

3. The case where $r_1 > 0$

The case where E admits real embeddings is (contrary to what one might expect) simplified by the fact that the '$\mathbb{Z}/2$ part' of the model space X can be absorbed into the real embeddings and dealt with there. In this section we suppose E 2-regular, with $r_1 > 0$; again we begin by producing a space X which will serve as a model for the homology of Spec R_E. The homology (again adapted from [11]) is described by:

Lemma 3.1. *The mod 2 homology of R_E is as follows:*

$$H_j(R_E; \mathbb{Z}/2) = \begin{cases} (r_1 + r_2 + 1).\mathbb{Z}/2 & j = 1 \\ r_1.\mathbb{Z}/2 & j > 1. \end{cases}$$

The r_1 real embeddings $R_E \xrightarrow{f_i} \mathbb{R}$ $(i = 1, \ldots, r_1)$ induce isomorphisms from the group $H_j(R_E; \mathbb{Z}/2)$ to $\oplus^{r_1} H_j(\mathbb{R}; \mathbb{Z}/2)$ in dimensions $j > 1$.

The last sentence in the above statement is Tate's theorem: the higher homology of R_E comes solely from the real embeddings. The fact that E is 2-regular makes it possible to extend the statement (true in general only for $j > 2$) to the case $j = 2$.

As is the previous section, we can extend the statement to the $\hat{\mathbb{Z}_2}$ homology:

Lemma 3.2.

(i) *The 2-completed homology of R_E is as follows:*

$$H_1(R_E) = (r_2 + 1).\hat{\mathbb{Z}_2} \oplus r_1.\mathbb{Z}/2$$
$$H_{2j}(R_E) = 0 \quad (j > 0)$$
$$H_{2j+1}(R_E) = r_1.\mathbb{Z}/2 \quad (j > 0).$$

(ii) *The r_1 real embeddings $R_E \xrightarrow{f_i} \mathbb{R}$ $(i = 1, \ldots, r_1)$ induce isomorphisms from the group $\oplus^{r_1} H_j(\mathbb{R})$ to $H_j(R_E)$ in dimensions $j > 1$.*

(iii) *For* $i = 1, \ldots r_1$, *the image of the non-trivial element of* $H_1(\mathbb{R})$ *under* $(f_i)_*$ *maps to* σ *under* $\hat{\phi}$.

Proof. Parts (i), (ii) require essentially the same as the proof of Lemma 2.2. Part (iii) is a consequence of the statement that $\pi_1(\operatorname{Spec}\mathbb{R}) = \mathbb{Z}/2$ is generated by complex conjugation, which corresponds to σ. □

Parallel to Proposition 2.3 we have the following result setting up a space X for this case.

Proposition 3.3. *There is a map* $f : X \to \operatorname{Spec} R_E$ *such that:*

(i) X *is a wedge of* $r_2 + 1$ *circles and* r_1 *copies of* $\operatorname{Spec}\mathbb{R}$;
(ii) *The first circle, considered as an element of* $\pi_1(X)$, *maps to* q *under the composite* $\hat{\phi} \circ f_*$, *and the rest (if any) map trivially;*
(iii) *For each copy of* $\operatorname{Spec}\mathbb{R}$, *the non-trivial element of* $\pi_1(\operatorname{Spec}\mathbb{R})$ *maps to* $\sigma \in (\mathbb{Z}_{\hat{2}})^*$;
(iv) f *induces an isomorphism on mod* 2 *homology.*

Note. The copies of Spec $\mathbb{R}$ can be replaced by the infinite projective space $RP^\infty \sim B\mathbb{Z}/2$ (cf. [3]), but this is not particularly useful for our purposes.

Proof. First note that the copies of Spec $\mathbb{R}$ are already taken care of by Lemma 3.2. Now as before consider the homomorphism from $H_1(R_E)$ onto $\Lambda \subset Aut(\mu_\infty)$. Since the 2-torsion subgroup maps onto the subgroup generated by $\sigma \in Aut(\mu_\infty)$, we can, by subtracting 2-torsion elements if necessary, choose a free direct summand $(r_2+1).\mathbb{Z}_{\hat{2}}$ which maps entirely into the $\mathbb{Z}_{\hat{2}}$ part of Λ. By choosing generators for this summand as in the proof of Proposition 2.3, we can ensure that the first maps to q and the others to zero.

Now represent these (r_2+1) generators by maps $S^1 \to \operatorname{Spec} R_E$. The wedge of these, and of our r_1 copies of Spec $\mathbb{R}$, with the given mappings into Spec R_E, is the required space X. □

We now have once again to find the space $K(X)$, representing wedges by fibred products over BU. However, we can speed up this process by dealing with all the copies of Spec $\mathbb{R}$ at once.

Proposition 3.4. *Let* $X_1, \ldots, X_k$ *be copies of* $\operatorname{Spec}\mathbb{R}$ *mapping into* $\operatorname{Spec} R_E$ *by maps* $f_1, \ldots, f_k$ *which arise from real embeddings. Then* $K(X_1 \vee X_2 \cdots \vee X_k)$ *is homotopy equivalent to the product* $BO \times \prod^{k-1}(U/O)$; *where the structure map from* $K(X_1 \vee X_2 \cdots \vee X_k)$ *to* BU *is the complexification on* BO, *and the trivial map on each factor* U/O.

Proof. First, by Proposition 4.1 of [3], we can identify $K(\mathbb{R})$ with BO, mapped into BU in the usual way. We therefore need to find the fibred product of k such copies of BO. To do this, replace the maps $BO \to BU$ by principal fibrations $p_1, \ldots, p_k$, with fibre U/O. A point of the fibred product is a sequence $(x_1, \ldots x_k)$ with $p_1(x_1) = \cdots = p_k(x_k)$. It follows that there are unique elements $g_2, \ldots g_k$ of U/O such that $x_i = x_1.g_i$ $(i = 2, \ldots, k)$; and that the map $(x_1, \ldots, x_k) \mapsto$

$(x_1, g_2, \ldots, g_k)$ is a homeomorphism from the fibred product to $BO \times \prod^{k-1}(U/O)$. The statement about the structure map is now obvious. $\square$

It is now relatively easy to complete the proof of Theorem 1.1. Let us write V for the wedge of the circles in the space X, and W for the wedge of the $\operatorname{Spec}\mathbb{R}$'s. We know that the circles, as in the previous section (but here simplified by the absence of a $\mathbb{Z}/2$-component) combine to give a space $K(V) = F\psi^q \times \prod^{r_2} U$; while $K(W)$ is given by the preceding result. To find the fibred product of $K(V)$ and $K(W)$ we need only look at those parts which are non-trivial over BU, i.e., at the $F\psi^q$ in $K(V)$ and the BO in $K(W)$. Now the fibred product of these is again well known, going back to Bökstedt and Dwyer-Friedlander; from the diagram

$$\begin{array}{ccccc} K(\operatorname{Spec}\mathbb{F}_q \vee \operatorname{Spec}\mathbb{R}) & \longrightarrow & K(\mathbb{R}) = BO & \longrightarrow & BU \\ \downarrow & & \downarrow c & & \| \\ K(\mathbb{F}_q) = F\psi^q & \longrightarrow & BU & \xrightarrow{\psi^q - 1} & BU \end{array}$$

we see that it is precisely the fibre of the composite

$$BO \xrightarrow{c} BU \xrightarrow{\psi^q - 1} BU$$

or what we have called $JK(q)$. Theorem 1.1 in the case of real embeddings follows immediately. $\square$

4. The homotopy groups

As a check on the correctness of the preceding results, it makes sense to show that they imply the results on $K_i(R_E) = \pi_i(K(R_E))$ proved in [11]. Some attention to details is needed since the homotopy groups of a fibre may not be uniquely determined by the exact sequence of the fibration (there may be extensions).

Again, we have to separate the two cases. In the totally imaginary case, since $\pi_i(\prod^{r_2 - 1} U)$ is well known to be $(r_2 - 1).\mathbb{Z}_2\hat{}$ for i odd and 0 for i even, the question reduces to finding the homotopy groups of the factor we have called $J_c(q)$, the homotopy fibre of the composite

$$F\psi^{-1} \xrightarrow{j} BU \xrightarrow{\psi^q - 1} BU$$

where j is the fibre inclusion. The homotopy groups of $F\psi^{-1}$ (the classifying space of self-conjugate K-theory) were computed by D. Anderson in his thesis from 1963, but see also Atiyah's paper 'K-theory and Reality' [1].

Lemma 4.1. *The homotopy groups of $F\psi^{-1}$ are:*

$$\pi_{4k+1} = \mathbb{Z}/2 \quad \pi_{4k+2} = 0$$
$$\pi_{4k+3} = \mathbb{Z}_2\hat{} \quad \pi_{4k} = \mathbb{Z}_2\hat{}$$

and the fibre inclusion from $F\psi^{-1}$ to BU induces an isomorphism on π_{4k}, and zero in all other cases.

Clearly we can deduce that $((\psi^q - 1) \circ j)_*$ is zero from $\pi_i(F\psi^q)$ to $\pi_i(BU)$ unless $i = 4k$, in which case it is multiplication by $q^{2k} - 1$ – this from the known action of the ψ's on the homotopy groups of BU.

It should be noted that $F\psi^{-1}$, or 'KSC' as it is traditionally called, is one of a family of K-spaces, and admits Adams operations ψ^q for $q > 1$ and odd, say. The fibre of $\psi^q - 1 : F\psi^{-1} \to F\psi^{-1}$ will be called $JSC(q)$, by an obvious analogy. Similarly, we use $JO(q)$ to denote the fibre of $\psi^q - 1 : BO \to BO$.

From the above, we can deduce:

Proposition 4.2. *The homotopy groups of $J_c(q)$ are zero in even dimensions; for the odd ones we have:*

$$\pi_{4k-1}(J_c(q)) = \mathbb{Z}_2\hat{} \oplus (\mathbb{Z}/(q^{2k} - 1))_2$$

and there is a split exact sequence

$$0 \to \mathbb{Z}_2\hat{} \to \pi_{4k+1}(J_c(q)) \to \mathbb{Z}/2 \to 0.$$

Proof. The kernel of $((\psi^q - 1) \circ j)_*$ is $\mathbb{Z}/2$ for π_{4k+1}, $\mathbb{Z}_2\hat{}$ for π_{4k-1}, and zero otherwise, from the lemma; the cokernel is $\mathbb{Z}_2\hat{}$ for π_{4k+2}, $(\mathbb{Z}/(q^{2k} - 1))_2$ for π_{4k}, and zero otherwise. Hence it remains to prove that the exact sequence is split. Consider the diagram of fibre sequences:

$$\begin{array}{ccccc} JSC(q) & \longrightarrow & F\psi^{-1} & \xrightarrow{\psi^q - 1} & F\psi^{-1} . \\ \downarrow & & \| & & \downarrow \\ J_c(q) & \longrightarrow & F\psi^{-1} & \longrightarrow & BU \end{array}$$

The map $\pi_{4k+1}(JSC(q)) \to \pi_{4k+1}(F\psi^{-1})$ is an isomorphism by the choice of q, so we get the claimed splitting. □

How does this fit with the results of [11]? Once we have added $(r_2 - 1)$ copies of $\pi_i(U)$ as required, we have an obvious agreement in the even dimensions. In dimension $4k + 1$, [11] give:

$$\pi_{4k+1}(K(R_E)) = r_2.\mathbb{Z}_2\hat{} \oplus \mathbb{Z}/w_{2k+1}$$

where $w_{2k+1} = 2$ is as defined in §1; hence our result agrees. In dimension $4k + 3$, their result is

$$\pi_{4k-1}(K(R_E)) = r_2.\mathbb{Z}_2\hat{} \oplus \mathbb{Z}/w_{2k}.$$

This agrees with our result provided we can identify w_{2k} with the 2-part of $q^{2k} - 1$. However, from the requirements on q and on w_i in §1, this is immediately evident.

We now consider the case where $r_1 > 0$. Here we need the homotopy groups of U/O; these are periodic of period 8, as follows:

$$\pi_{8k}(U/O) = 0, \quad \pi_{8k+1}(U/O) = \mathbb{Z}_2\hat{}, \quad \pi_{8k+2}(U/O) = \mathbb{Z}/2, \quad \pi_{8k+3}(U/O) = \mathbb{Z}/2$$

$$\pi_{8k+4}(U/O) = 0, \quad \pi_{8k+5}(U/O) = \mathbb{Z}_2\hat{}, \quad \pi_{8k+6}(U/O) = 0, \quad \pi_{8k+7}(U/O) = 0.$$

(These groups are well known in the study of Bott periodicity; they also follow from identifying U/O with $B(\hat{\mathbb{Z}}_2 \times BO)$.)

The homotopy groups $\pi_i(K(R_E)) = K_i(R_E)$ are the sum of $(r_1 - 1)$ copies of these, r_2 copies of $\pi_i(U)$, and $\pi_i(JK(q))$, which we must now determine. From the fibration

$$JK(q) \to BO \overset{(\psi^q - 1)\circ c}{\longrightarrow} BU$$

we derive an exact homotopy sequence. All homomorphisms from $\pi_i(BO)$ to $\pi_i(BU)$ are necessarily zero (torsion to zero or free) except when $i = 4k$. It is known that c_* is an isomorphism for $i = 8k$ and multiplication by 2 for $i = 8k+4$, and hence $((\psi^q - 1)\circ c)_*$ is multiplication by $q^{4k} - 1$ for $i = 8k$ and by $2(q^{4k+2} - 1)$ for $i = 8k + 4$. From this (using the relation between q and the w_i's as before) we deduce the following.

Proposition 4.3. *The homotopy groups of $JK(q)$ are as follows:*

$$\pi_{8k} = 0, \quad \pi_{8k+2} = \mathbb{Z}/2, \quad \pi_{8k+3} = \mathbb{Z}/2w_{4k+2}, \quad \pi_{8k+4} = 0$$

$$\pi_{8k+5} = \hat{\mathbb{Z}}_2, \quad \pi_{8k+6} = 0, \quad \pi_{8k+7} = \mathbb{Z}/w_{4k+4}$$

and there is a split short exact sequence

$$0 \to \hat{\mathbb{Z}}_2 \to \pi_{8k+1} \to \mathbb{Z}/2 \to 0.$$

Once again, these give exactly the results computed in [11].

Proof. Using the above remarks, we see:

(a) that the cokernel of $((\psi^q - 1)\circ c)_*$ is zero on odd dimensions, $\hat{\mathbb{Z}}_2$ in dimension $4k + 2$, $\mathbb{Z}/2w_{4k+2}$ in dimension $8k + 4$, and $\mathbb{Z}/w_{4k+4}$ in dimension $8k + 8$;

(b) that the kernel is $\mathbb{Z}/2$ in dimensions $8k + 1$, $8k + 2$, and zero otherwise.

Hence in the short exact sequence

$$0 \to Coker((\psi^q - 1)\circ c)_*(i+1) \to \pi_i(JK(q)) \to Ker((\psi^q - 1)\circ c)_*(i) \to 0$$

either the kernel or cokernel vanishes except for $i = 8k + 1$. This therefore gives the only extension problem, and the other calculations are immediate. To answer the extension question we use an argument which the second author learned from J. Rognes. Consider the diagram of fibre sequences:

$$\begin{array}{ccccc} JO(q) & \longrightarrow & BO & \xrightarrow{\psi^q - 1} & BO\,. \\ \downarrow & & \| & & \downarrow c \\ JK(q) & \longrightarrow & BO & \longrightarrow & BU \end{array}$$

The map $\pi_{8k+1}(JO(q)) \to \pi_{8k+1}(BO)$ is a split surjection from $\mathbb{Z}/2 \oplus \mathbb{Z}/2$ to $\mathbb{Z}/2$, and we can conclude. □

References

[1] M.F. Atiyah: *K-theory*, Benjamin, New York, 1967.

[2] S. Bloch and S. Lichtenbaum: *A spectral sequence for motivic cohomology*, preprint 1995.

[3] W.G. Dwyer and E.M. Friedlander: *Conjectural calculations of general linear group homology*, Contemp. Math. **55** part I, Amer. Math. Soc., Providence (1986), 135–147.

[4] W.G. Dwyer and E.M. Friedlander: *Algebraic and etale K-theory*, Trans. Amer. Math. Soc. **292** (1985), 247–280.

[5] S.A. Mitchell: *Topological K-theory of algebraic K-theory spectra*, K-Theory **21** (2000), 229–247.

[6] S.A. Mitchell: *K-theory hypercohomology spectra of number rings at the prime 2*, Contemp. Math. **265**, Amer. Math. Soc., Providence (2000), 129–157.

[7] D.G. Quillen: *Higher algebraic K-theory I*, Lecture Notes in Math. **342**, Springer, New York (1973), 85–147.

[8] D.G. Quillen: *On the cohomology and K-theory of the general linear group over finite fields*, Annals of Math. **96** (1972), 552–586.

[9] J. Rognes and C.A. Weibel: *Two-primary algebraic K-theory of rings of integers in number fields*, J. of Amer. Math. Soc. **13** (2000), 1–54.

[10] J. Rognes and C.A. Weibel: *Étale descent for two-primary algebraic K-theory of totally imaginary number rings*, K-Theory **15** (1998), 101–104.

[11] J. Rognes and P.A. Østvær: *K-theory of 2-regular number fields*, Math. Z. **233** (2000), 251–263.

[12] A. Suslin and V. Voevodsky: *Bloch-Kato conjecture and motivic cohomology with finite coefficients*, The arithmetic and geometry of algebraic cycles, Proc. Conf., Banff 1998, Nato Science Series C **548** Kluwer (2000), 117–189.

[13] V. Voevodsky: *The Milnor conjecture*, preprint 1996.

[14] P.A. Østvær: *Étale descent for real number fields*, Topology **42** (2003), 197–225.

[15] P.A. Østvær: *Algebraic K-theory spectra of non-exceptional two-regular number fields*, Contemp. Math. **265** Amer. Math. Soc., Providence (2000), 187–192.

Luke Hodgkin
Department of Mathematics
King's College, London, United Kingdom
e-mail: luke.hodgkin@kcl.ac.uk

Paul Arne Østvær
Department of Mathematics
University of Oslo, Oslo, Norway
e-mail: paularne@math.uio.no

Progress in Mathematics, Vol. 215
Algebraic Topology: Categorical Decomposition Techniques, 179–198

Strict Model Structures for Pro-categories

Daniel C. Isaksen

Abstract. We show that if $\mathcal{C}$ is a proper model category, then the pro-category pro-$\mathcal{C}$ has a proper model structure in which the weak equivalences are the levelwise weak equivalences. The structure is simplicial when $\mathcal{C}$ is simplicial. This is related to a major result of [10] and is the starting point for many homotopy theories of pro-objects such as those described in [5], [17], and [19].

1. Introduction

If $\mathcal{C}$ is a category, then the category pro-$\mathcal{C}$ has as objects all cofiltered diagrams in $\mathcal{C}$ and has morphisms defined by

$$\mathrm{Hom}_{\text{pro-}\mathcal{C}}(X, Y) = \lim_t \operatorname{colim}_s \mathrm{Hom}_{\mathcal{C}}(X_s, Y_t).$$

Pro-categories have found many uses over the years in fields such as algebraic geometry [2], shape theory [20], geometric topology [6], and possibly even applied mathematics [7, Appendix].

When working with pro-categories, one would frequently like to have a homotopy theory of pro-objects. The first attempts at this appear in [2] and [26] in which pro-objects in homotopy categories are considered. The difficulty with this approach is that the diagrams commute only up to homotopy, and this makes it virtually impossible to make sense of most of the standard notions of homotopy theory in this context.

Much better is to first consider actually commuting cofiltered diagrams (of spaces or simplicial sets or spectra or whatever) and then to define a notion of weak equivalence between such pro-objects. This approach was first taken by [12] in a restricted context. See also [22] and [23] for early but incomplete attempts.

This framework was applied much more generally in [10]. The idea is to start with a model structure (*i.e.*, a homotopy theory) on a category $\mathcal{C}$ and then to construct a *strict model structure* on pro-$\mathcal{C}$ in which the weak equivalences are more or less just the levelwise weak equivalences. The resulting homotopy theory is precisely suited to study homotopy limits [4, Ch. XI] of cofiltered diagrams. The

Received by the editors November 6, 2001.
2000 *Mathematics Subject Classification.* 18G55, 55U35.
Key words and phrases. Closed model category, pro-homotopy theory, pro-space.
The author was supported by an NSF Postdoctoral Research Fellowship.

strict model structure is a starting point for several other model structures such as those described in [5], [17], and [19].

The strict model structure on pro-$\mathcal{C}$ does not seem to exist for a completely arbitrary model category $\mathcal{C}$. A niceness hypothesis was required in [10, p. 45] (which was weakened in [13]). Unfortunately, many important examples of model categories, such as the usual model for topological spaces or any of the usual models for spectra, do not satisfy this hypothesis. The main purpose of this paper is to prove that the strict structure on pro-$\mathcal{C}$ exists whenever $\mathcal{C}$ is a proper model category. Almost all of the most important examples of model categories are proper.

Theorem 1.1. *If $\mathcal{C}$ is a proper model category, then* pro-$\mathcal{C}$ *has a proper model structure in which the weak equivalences (*resp.*, cofibrations) are the essentially levelwise weak equivalences (*resp.*, cofibrations) and the fibrations are defined by the right lifting property. If $\mathcal{C}$ is a simplicial model category, then so is* pro-$\mathcal{C}$.

The terminology used in the statement of the theorem is explained in Sections 2, 3, and 4. Beware that the strict structure on pro-$\mathcal{C}$ does not have functorial factorizations, even when the original model category $\mathcal{C}$ does.

Another problem with [10] is that a non-standard set of axioms for model structures is used. From a modern perspective, it is harder to comprehend the technical details of [10] than to simply work out new proofs from scratch. The secondary goal of this paper is to give these new more modern proofs. Oddly, the two-out-of-three axiom is the most difficult part of the proof; in most model structures, it is automatic from the definition of weak equivalence.

The last goal of the paper is to consider whether strict model structures on pro-categories are fibrantly generated. It is already known that these model structures are not cofibrantly generated in general, even when $\mathcal{C}$ is [17, §19]. We produce reasonable collections of generating fibrations and generating acyclic fibrations that have cosmall codomains, but these collections are not sets. In fact, the strict model structure for pro-simplicial sets is not fibrantly generated. We show that if this strict structure were fibrantly generated, then in the category of simplicial sets there would exist a set of fibrations that detect acyclic cofibrations.

The paper is organized as follows. First we introduce the language of pro-categories and give some background results. Then we define the strict weak equivalences and prove that they satisfy the two-out-of-three axiom when $\mathcal{C}$ is proper. Next we prove that the strict model structure exists when $\mathcal{C}$ is proper. Finally, we consider whether the strict model structure is fibrantly generated.

1.1. Model categories

We assume that the reader is familiar with model categories. The original reference is [25], but we follow the notation and terminology of [14] as closely as possible. Other references include [9] and [15].

We do not assume that model structures have functorial factorizations. The reason for this choice is that the strict model structure on pro-$\mathcal{C}$ does not have functorial factorizations.

It does not really matter whether we assume that $\mathcal{C}$ has small limits and colimits or just finite limits and colimits. If $\mathcal{C}$ has small limits and colimits, then so does pro-$\mathcal{C}$ [17, Prop. 11.1]. On the other hand, if $\mathcal{C}$ has finite limits and colimits, then so does pro-$\mathcal{C}$ [2, App. 4.2]. For the sake of generality, we assume only that model structures have finite limits and colimits.

2. Preliminaries on pro-categories

We begin with a review of the necessary background on pro-categories. This material can be found in [1], [2], [8], [10], and [18].

2.1. Pro-categories

Definition 2.1. For a category $\mathcal{C}$, the category **pro-$\mathcal{C}$** has objects all cofiltering diagrams in $\mathcal{C}$, and

$$\mathrm{Hom}_{\text{pro-}\mathcal{C}}(X,Y) = \lim_s \operatorname*{colim}_t \mathrm{Hom}_{\mathcal{C}}(X_t, Y_s).$$

Composition is defined in the natural way.

A category I is **cofiltering** if the following conditions hold: it is non-empty and small; for every pair of objects s and t in I, there exists an object u together with maps $u \to s$ and $u \to t$; and for every pair of morphisms f and g with the same source and target, there exists a morphism h such that fh equals gh. Recall that a category is **small** if it has only a set of objects and a set of morphisms. A diagram is said to be **cofiltering** if its indexing category is so. Beware that some material on pro-categories, such as [2] and [21], consider cofiltering categories that are not small. All of our pro-objects will be indexed by small categories.

Objects of pro-$\mathcal{C}$ are functors from cofiltering categories to $\mathcal{C}$. We use both set theoretic and categorical language to discuss indexing categories; hence "$t \geq s$" and "$t \to s$" mean the same thing when the indexing category is actually a cofiltering partially ordered set.

The word *pro-object* refers to objects of pro-categories. A **constant** pro-object is one indexed by the category with one object and one (identity) map. Let $c : \mathcal{C} \to \text{pro-}\mathcal{C}$ be the functor taking an object X to the constant pro-object with value X. Note that this functor makes $\mathcal{C}$ a full subcategory of pro-$\mathcal{C}$. The limit functor $\lim : \text{pro-}\mathcal{C} \to \mathcal{C}$ is the right adjoint of c. To avoid confusion, we write $\lim^{\text{pro}}$ for limits computed within the category pro-$\mathcal{C}$.

Let $Y : I \to \mathcal{C}$ and $X : J \to \mathcal{C}$ be arbitrary pro-objects. We say that X is **cofinal** in Y if there is a cofinal functor $F : J \to I$ such that X is equal to the composite YF. This means that for every s in I, the overcategory $F \downarrow s$ is cofiltered. In the case when F is an inclusion of cofiltering partially ordered sets, F is cofinal if and only if for every s in I there exists t in J such that $t \geq s$. The importance of this definition is that X is isomorphic to Y in pro-$\mathcal{C}$.

2.2. Level representations

A **level representation** of a map $f : X \to Y$ is: a cofiltered index category I; pro-objects $\tilde{X}$ and $\tilde{Y}$ indexed by I and pro-isomorphisms $X \to \tilde{X}$ and $Y \to \tilde{Y}$; and a collection of maps $f_s : \tilde{X}_s \to \tilde{Y}_s$ for all s in I such that for all $t \to s$ in I, there is a commutative diagram

$$\begin{array}{ccc} \tilde{X}_t & \longrightarrow & \tilde{Y}_t \\ \downarrow & & \downarrow \\ \tilde{X}_s & \longrightarrow & \tilde{Y}_s \end{array}$$

and such that the maps f_s represent a map $\tilde{f} : \tilde{X} \to \tilde{Y}$ belonging to a commutative square

$$\begin{array}{ccc} X & \xrightarrow{f} & Y \\ \cong\downarrow & & \downarrow\cong \\ \tilde{X} & \xrightarrow[\tilde{f}]{} & \tilde{Y} \end{array}$$

in pro-$\mathcal{C}$. That is, a level representation is just a natural transformation such that the maps f_s represent the element f of

$$\lim_s \operatorname{colim}_t \operatorname{Hom}_{\mathcal{C}}(X_t, Y_s) \cong \lim_s \operatorname{colim}_t \operatorname{Hom}_{\mathcal{C}}(\tilde{X}_t, \tilde{Y}_s).$$

Every map has a level representation [2, App. 3.2] [21].

More generally, suppose given any diagram $A \to$ pro-$\mathcal{C}$: $a \mapsto X^a$. A **level representation** of X is: a cofiltered index category I; a functor $\tilde{X} : A \times I \to \mathcal{C}$: $(a, s) \mapsto \tilde{X}^a_s$; and pro-isomorphisms $X^a \to \tilde{X}^a$ such that for every map $\phi : a \to b$ in A, $\tilde{X}^\phi$ is a level representation for X^ϕ. In other words, $\tilde{X}$ is a uniform level representation for all the maps in the diagram X.

Not every diagram of pro-objects has a level representation. However, finite diagrams without loops do have level representations. This makes computations of limits and colimits of such diagrams in pro-$\mathcal{C}$ relatively straightforward. To compute this limit or colimit, just take the levelwise limit or colimit of the level representation [2, App. 4.2].

A pro-object X satisfies a certain property **levelwise** if each X_s satisfies that property, and X satisfies this property **essentially levelwise** if it is isomorphic to another pro-object satisfying this property levelwise. Similarly, a level representation $X \to Y$ satisfies a certain property **levelwise** if each $X_s \to Y_s$ has this property. A map of pro-objects satisfies this property **essentially levelwise** if it has a level representation satisfying this property levelwise. The following surprisingly general and very useful proposition about retracts of essentially levelwise maps is proved in [18, Thm. 5.5].

Proposition 2.2. *Let C be any class of maps in a category $\mathcal{C}$. Then retracts preserve the class of maps in* pro-$\mathcal{C}$ *that belong to C essentially levelwise.*

We make an observation concerning level representations of commutative squares that will be used repeatedly.

Lemma 2.3. *Suppose given a square*

$$\begin{array}{ccc} W & \longrightarrow & X \\ {\scriptstyle f}\downarrow & & \downarrow{\scriptstyle g} \\ Y & \longrightarrow & Z \end{array}$$

of pro-maps such that f belongs to some class C essentially levelwise and g belongs to some class D essentially levelwise. Then we may find a level representation for the square such that the maps $\tilde{f} : \tilde{W} \to \tilde{Y}$ and $\tilde{g} : \tilde{X} \to \tilde{Z}$ belong levelwise to C and D respectively.

Proof. We use the fact that the pro-category pro-Ar($\mathcal{C}$) of the category of morphisms Ar($\mathcal{C}$) of $\mathcal{C}$ is equivalent to the category of morphisms Ar(pro-$\mathcal{C}$) [21]. Then the square is just a pro-map $f \to g$, where f and g are pro-objects belonging levelwise to C and D respectively. Using the method of [2, App. 3.2], this pro-map has a level representation in which the source and target still belong levelwise to C and D. □

2.3. Cofiniteness

A partially ordered set $(I, \leq)$ is **directed** if for every s and t in I, there exists u such that $u \geq s$ and $u \geq t$. A directed set $(I, \leq)$ is **cofinite** if for every t, the set of elements s of I such that $s \leq t$ is finite. A pro-object or level representation is **cofinite directed** if it is indexed by a cofinite directed set.

For every cofiltered category I, there exists a cofinite directed set J and a cofinal functor $J \to I$ [10, Th. 2.1.6] (or [1, Exposé 1, 8.1.6]). Therefore, every pro-object is isomorphic to a cofinite directed pro-object. Similarly, every map has a cofinite directed level representation. Thus, it is possible to restrict the definition of a pro-object to only consider cofinite directed sets as index categories, but we find this unnatural for general definitions and constructions. On the other hand, we find it much easier to work with cofinite directed pro-objects in practice. Thus, most of our results start by assuming without loss of generality that a pro-object is indexed by a cofinite directed set. Cofiniteness is critical because many arguments and constructions proceed inductively.

Definition 2.4. Let $f : X \to Y$ be a cofinite directed level representation of a map in a pro-category. For every index t, the **relative matching map** $\boldsymbol{M_t f}$ is the map

$$X_t \to \lim_{s<t} X_s \times_{\lim_{s<t} Y_s} Y_t.$$

The terminology is motivated by the fact that these maps appear in Reedy model structures [14, Defn. 15.3.2]. The similarity is not coincidental. The strict model structure is closely linked to the Reedy model structures for each fixed cofinite directed index category [10, §3.2].

3. Strict weak equivalences

We now study strict weak equivalences for pro-categories as described in [10] [22] [23]. The niceness hypothesis of [10, p. 45] is not satisfied by many categories of interest. These include the standard model structure for topological spaces and many of the standard models for spectra, such as Bousfield-Friedlander spectra [3], symmetric spectra [16], or S-modules [11]. We shall study strict weak equivalences in pro-$\mathcal{C}$ whenever $\mathcal{C}$ is a proper model category.

In [10, §3.3], the weak equivalences of pro-$\mathcal{C}$ were defined to be the compositions of essentially levelwise acyclic cofibrations and essentially levelwise acyclic fibrations. We make the following simpler definition of weak equivalences in pro-$\mathcal{C}$. We shall see in Proposition 4.15 that the two definitions are actually the same. Therefore, we have a much simpler description of the same class of maps.

Definition 3.1. The **strict weak equivalences** of pro-$\mathcal{C}$ are the essentially levelwise weak equivalences.

It is not obvious from the definition that the strict weak equivalences satisfy the two-out-of-three axiom. The next few lemmas prove this axiom. These proofs are the technical heart of the paper. They are the reason that we must assume that $\mathcal{C}$ is proper. The basic complication is that given a diagram of strict weak equivalences, it is not necessarily possible to find a level representation for the diagram such that the level representations of all maps are levelwise weak equivalences. Independently, each map has a level representation that is a levelwise weak equivalence, but the reindexing for these level representations may be different.

Lemma 3.2. *Let $\mathcal{C}$ be a model category, and let $f : X \to Y$ be a level representation of an isomorphism in* pro-$\mathcal{C}$. *After reindexing along a cofinal functor, f can be factored in either of the following two ways:*

1. *as a levelwise cofibration followed by a levelwise acyclic fibration, both of which are pro-isomorphisms.*
2. *as a levelwise acyclic cofibration followed by a levelwise fibration, both of which are pro-isomorphisms.*

Remark 3.3. The model structure on $\mathcal{C}$ is not really necessary. We just need a category $\mathcal{C}$ and two classes of morphisms C and F such that each morphism of $\mathcal{C}$ can be factored (not necessarily functorially) into an element of C followed by an element of F.

Proof. We prove only the first claim; the proof of the second is identical with the obvious substitutions.

We may assume that f is indexed by a cofinite directed set I. Since f is an isomorphism, for every s in I, there exists $t > s$ and a map $h_{ts} : Y_t \to X_s$

belonging to a commutative diagram

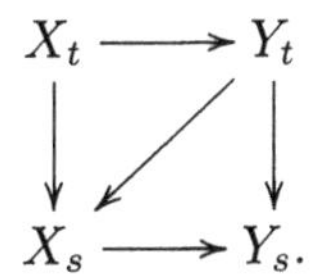

In effect, the maps h_{ts} represent the inverse of f. Let I' be the set of all pairs (s,t) in I for which $t > s$ and such an h_{ts} exists. We choose a map $h_{ts} : Y_t \to X_s$ for each element (t,s) of I' making the above diagram commute but require no compatibility between these choices.

Factor each map $f_s : X_s \to Y_s$ into a cofibration $X_s \to Z_s$ followed by an acyclic fibration $Z_s \to Y_s$. We shall define structure maps making Z into a pro-object.

Define a category J whose objects are the elements of I and whose morphisms $t \to s$ are finite chains $t = u_0 > u_1 > \cdots > u_n = s$ such that each pair (u_i, u_{i-1}) belongs to I'. Composition is defined by concatenation of chains. Note that J is not a directed set; it is not even cofiltered.

For every morphism $t = u_0 > u_1 > \cdots > u_n = s$ in J, define a map $Z_t \to Z_s$ by the composition

$$Z_t \longrightarrow Y_t = Y_{u_0} \xrightarrow{h_{u_0 u_1}} X_{u_1} \longrightarrow X_{u_2} \longrightarrow \cdots \longrightarrow X_{u_n} = X_s \longrightarrow Z_s.$$

A diagram chase shows that this makes Z into a diagram indexed by J.

There may be more than one map from a given Z_t to a given Z_s, but another diagram chase shows that they become equal after composition with some map $Z_u \to Z_t$. Consider the category K defined to be a quotient of J as follows. The objects of K are the same as the objects of J, and two morphisms from t to s in J are identified in K if the corresponding maps from Z_t to Z_s are equal in $\mathcal{C}$. Now K is a cofiltered category, and we may consider it as the indexing category of Z.

The projection functor $K \to I$ is cofinal, so we may reindex X and Y along this functor. More diagram chases show that the maps $X_s \to Z_s$ and $Z_s \to Y_s$ assemble into level representations $X \to Z$ and $Z \to Y$. It remains only to show that these maps are pro-isomorphisms. This follows from the commutative diagrams

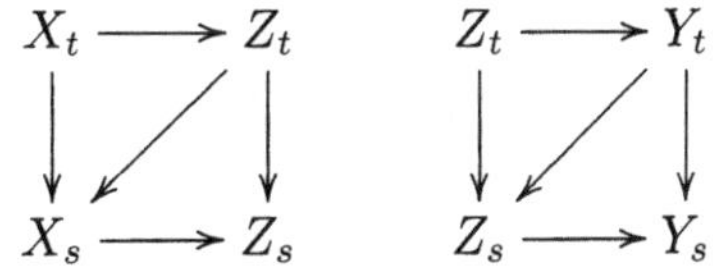

for every pair (s,t) belonging to I'. The diagonal maps in the above diagrams are the compositions $Z_t \to Y_t \to X_s$ and $Y_t \to X_s \to Z_s$ respectively. □

Remark 3.4. There is a more obvious argument where a level representation $X \to Y$ is functorially factored into a levelwise cofibration $X \to Z$ followed by a levelwise fibration $Z \to Y$. This does not give the desired factorization; the maps $X \to Z$ and $Z \to Y$ are not necessarily isomorphisms.

The following lemma appears in [17, Prop. 10.4], but the previous lemma makes the technical details of that proof clearer.

Lemma 3.5. *If $\mathcal{C}$ is a proper model category, then the strict weak equivalences of* pro-$\mathcal{C}$ *are closed under composition.*

Proof. It suffices to assume that there is a cofinite directed level representation for the diagram

$$X \xrightarrow[\sim]{f} Y \xleftarrow[\cong]{h} Z \xrightarrow[\sim]{g} W$$

in which f and g are levelwise weak equivalences while h is a pro-isomorphism (but not a levelwise isomorphism). This requires an argument similar to the proof of Lemma 2.3. We must construct a levelwise weak equivalence isomorphic to the composition $gh^{-1}f$.

By Lemma 3.2, after reindexing we can factor $h : Z \to Y$ into a levelwise cofibration $Z \to A$ followed by a levelwise fibration $A \to Y$ such that

$$X \xrightarrow{\sim} Y \twoheadleftarrow^{\cong} A \xleftarrow{\cong} Z \xrightarrow{\sim} W$$

is a level representation in which the first and fourth maps are levelwise weak equivalences, and the second and third are pro-isomorphisms.

Let B be the pullback $X \times_Y A$, and let C be the pushout $A \amalg_Z W$. Since pushouts and pullbacks can be constructed levelwise, the maps $B \to A$ and $A \to C$ are levelwise weak equivalences. Here we use that the model structure is left and right proper. Moreover, the maps $B \to X$ and $W \to C$ are pro-isomorphisms since base and cobase changes preserve isomorphisms. Hence the composition $B \to C$ is the desired levelwise weak equivalence. □

Lemma 3.6. *Suppose that $\mathcal{C}$ is a proper model category, and let f and g be two composable maps in pro-$\mathcal{C}$. If g and gf are strict weak equivalences, then f is a strict weak equivalence. If f and gf are strict weak equivalences, then g is a strict weak equivalence.*

Proof. We first prove the first claim. Apply Lemma 2.3 to the square

$$\begin{array}{ccc} X & \xrightarrow{f} & Y \\ {\scriptstyle gf}\downarrow & & \downarrow{\scriptstyle g} \\ Z & \xrightarrow[=]{} & Z \end{array}$$

to obtain a cofinite directed level representation

$$\begin{array}{ccc} X & \xrightarrow{f} & Y \\ {\scriptstyle \sim}\downarrow & & \downarrow{\scriptstyle \sim\; g} \\ W & \xrightarrow[\cong]{} & Z \end{array}$$

where the vertical maps are levelwise weak equivalences and the bottom horizontal map is a pro-isomorphism (but not a levelwise isomorphism). We want to show that the top horizontal map is an essentially levelwise weak equivalence.

By Lemma 3.2, after reindexing there exists a level representation

$$\begin{array}{ccccc} X & \longrightarrow & B & \longrightarrow & Y \\ {\scriptstyle\sim}\downarrow & & \downarrow & & \downarrow{\scriptstyle\sim} \\ W & \underset{\cong}{\overset{\sim}{\rightarrowtail}} & A & \underset{\cong}{\twoheadrightarrow} & Z \end{array}$$

such that B is the levelwise pullback $A \times_Z Y$; the first and third vertical maps are levelwise weak equivalences; the map $W \to A$ is a levelwise acyclic cofibration and a pro-isomorphism; and the map $A \to Z$ is a levelwise fibration and a pro-isomorphism. Because of right properness, the map $B \to A$ is also a levelwise weak equivalence. By the two-out-of-three axiom in $\mathcal{C}$, the induced map $X \to B$ is a levelwise weak equivalence.

On the other hand, the map $B \to Y$ is an isomorphism because base changes preserve isomorphisms. Hence $X \to B$ is isomorphic to f.

The proof of the second claim is similar. Again using Lemma 2.3, we start with a cofinite directed level representation

$$\begin{array}{ccc} X & \overset{\cong}{\longrightarrow} & W \\ f\downarrow{\scriptstyle\sim} & & \downarrow{\scriptstyle\sim} \\ Y & \underset{g}{\longrightarrow} & Z \end{array}$$

where the vertical maps are levelwise weak equivalences and the top horizontal map is a pro-isomorphism. We want to show that the bottom horizontal map is an essentially levelwise weak equivalence. We produce a level representation

$$\begin{array}{ccccc} X & \overset{\cong}{\rightarrowtail} & A & \underset{\sim}{\overset{\cong}{\twoheadrightarrow}} & W \\ {\scriptstyle\sim}\downarrow & & \downarrow & & \downarrow{\scriptstyle\sim} \\ Y & \longrightarrow & B & \longrightarrow & Z \end{array}$$

such that B is the levelwise pushout $A \amalg_X Y$; the first and third vertical maps are levelwise weak equivalences; the map $A \to W$ is a levelwise acyclic fibration and a pro-isomorphism; and the map $X \to A$ is a levelwise cofibration and a pro-isomorphism. The map $B \to Z$ is the desired level representation. □

4. Strict model structures

Beginning with a proper model structure on a category $\mathcal{C}$, we now establish a model structure on the category pro-$\mathcal{C}$.

Definition 4.1. The **strict cofibrations** of pro-$\mathcal{C}$ are the essentially levelwise cofibrations.

Definition 4.2. A map in pro-$\mathcal{C}$ is a **special fibration** if it has a cofinite directed level representation p for which every relative matching map $M_s p$ is a fibration. A map in pro-$\mathcal{C}$ is a **strict fibration** if it is a retract of a special fibration.

In order to help us understand these definitions, we need some auxiliary notions.

Definition 4.3. A map in pro-$\mathcal{C}$ is a **special acyclic fibration** if it has a cofinite directed level representation p for which every relative matching map $M_s p$ is an acyclic fibration.

Note that every special acyclic fibration is a special fibration, so it is also a strict fibration.

We shall use these definitions to establish the strict model structure on pro-$\mathcal{C}$. The argument is organized as follows. We shall show that the strict acyclic cofibrations, *i.e.*, the maps that are both strict cofibrations and strict weak equivalences, are the same as the essentially levelwise acyclic cofibrations. This does not follow immediately from the definitions. If a map is isomorphic to a levelwise weak equivalence and to a levelwise cofibration, it may be necessary to use two different index categories to obtain the two level representations.

Similarly, the strict acyclic fibrations, *i.e.*, the maps that are both strict fibrations and strict weak equivalences, are the retracts of special acyclic fibrations. Using these descriptions of the strict acyclic cofibrations and the strict acyclic fibrations, the proofs of the model structure axioms are relatively straightforward.

Lemma 4.4. *Special acyclic fibrations are essentially levelwise acyclic fibrations. In particular, they are strict acyclic fibrations.*

Proof. Special acyclic fibrations are special fibrations, so they are strict fibrations by definition. This means that the second statement follows from the first.

Suppose given a cofinite directed level representation $p : X \to Y$ for which each $M_s p$ is an acyclic fibration. The map $p_s : X_s \to Y_s$ factors as

$$X_s \xrightarrow{M_s p} Y_s \times_{\lim_{t<s} Y_t} \lim_{t<s} X_t \xrightarrow{q_s} Y_s.$$

Since compositions and base changes preserve acyclic fibrations, it suffices to show that $\lim_{t<s} p_t : \lim_{t<s} X_t \to \lim_{t<s} Y_t$ is an acyclic fibration for every s.

Let S be the set of indices t such that $t < s$. Recall that a subset T of S is *initial* if u belongs to T whenever $u < t$ and t belongs to T. Since S is finite, we may choose a sequence

$$\emptyset = S_0 \subset S_1 \subset \cdots \subset S_n = S$$

such that each S_i is initial and such that S_i is obtained from S_{i-1} by adding a single element t_i.

For all initial subsets T of S, define $X_T = \lim_{t \in T} X_t$ and $Y_T = \lim_{t \in T} Y_t$. Also, let $Z_T = X_T \times_{Y_T} Y_S$. Note that the desired map $\lim_{t<s} p_t$ is the composition

$$\lim_{t<s} X_t = Z_{S_n} \to Z_{S_{n-1}} \to \cdots \to Z_{S_0} = \lim_{t<s} Y_t,$$

so we need only show that each map $Z_{S_i} \to Z_{S_{i-1}}$ is an acyclic fibration.

Observe that X_{S_i} is equal to

$$X_{S_{i-1}} \times_{\lim_{u<t_i} X_u} X_{t_i}$$

and that Y_{S_i} is equal to

$$Y_{S_{i-1}} \times_{\lim_{u<t_i} Y_u} Y_{t_i}$$

because $S_i = S_{i-1} \amalg \{t_i\}$. A diagram chase shows that $Z_{S_i} \to Z_{S_{i-1}}$ is a base change of the map $M_{t_i} p$, so it is an acyclic fibration. □

Lemma 4.5. *Strict fibrations are essentially levelwise fibrations.*

Proof. The proof of Lemma 4.4 shows that special fibrations are essentially levelwise fibrations. Retracts of special fibrations are also essentially levelwise fibrations by Proposition 2.2. □

Lemma 4.6. *Every map $f : X \to Y$ in* pro-$\mathcal{C}$ *factors as a strict cofibration $i : X \to Z$ followed by a special acyclic fibration $p : Z \to Y$.*

Proof. We may suppose that f is a level representation indexed by a cofinite directed set. Suppose for induction that the maps $i_t : X_t \to Z_t$ and $p_t : Z_t \to Y_t$ have already been defined for $t < s$. Consider the map

$$X_s \to Y_s \times_{\lim_{t<s} Y_t} \lim_{t<s} Z_t.$$

Factor it into a cofibration $i_s : X_s \to Z_s$ followed by an acyclic fibration

$$p_s : Z_s \to Y_s \times_{\lim_{t<s} Y_t} \lim_{t<s} Z_t.$$

This extends the factorization to level s. □

Lemma 4.7. *Every map $f : X \to Y$ in* pro-$\mathcal{C}$ *factors as an essentially levelwise acyclic cofibration $i : X \to Z$ followed by a special fibration $p : Z \to Y$.*

Proof. The proof is identical to the proof of Lemma 4.6, except that we factor the map

$$X_s \to Y_s \times_{\lim_{t<s} Y_t} \lim_{t<s} Z_t$$

into an acyclic cofibration followed by a fibration. □

Remark 4.8. The previous two lemmas do not give functorial factorizations. The problem is that the reindexing into cofinite directed level representations is not functorial. This is the reason that strict model structures do not have functorial factorizations.

Next we show that the classes of strict cofibrations and retracts of special acyclic fibrations determine each other by lifting properties. Similarly, the classes of essentially levelwise acyclic cofibrations and strict fibrations determine each other.

Lemma 4.9. *A map is a strict cofibration if and only if it has the left lifting property with respect to all retracts of special acyclic fibrations. Also, a map is a retract of a special acyclic fibration if and only if it has the right lifting property with respect to all strict cofibrations.*

Proof. Suppose given a square

$$\begin{array}{ccc} A & \longrightarrow & X \\ {\scriptstyle i}\downarrow & & \downarrow{\scriptstyle p} \\ B & \longrightarrow & Y \end{array}$$

in which i is a strict cofibration and p is a special acyclic fibration. By reindexing, we may assume that the diagram is a cofinite directed level representation such that each $i_s : A_s \to B_s$ is a cofibration and such that each map $M_s p$ is an acyclic fibration.

Assume that for all $t < s$, we have already constructed maps $B_t \to X_t$ belonging to a commuting diagram

$$\begin{array}{ccc} A_t & \longrightarrow & X_t \\ \downarrow & \nearrow & \downarrow \\ B_t & \longrightarrow & Y_t \times_{\lim_{u<t} Y_u} \lim_{u<t} X_u . \end{array}$$

Thus we have a diagram

$$\begin{array}{ccc} A_s & \longrightarrow & X_s \\ {\scriptstyle i_s}\downarrow & & \downarrow{\scriptstyle \sim\; M_s p} \\ B_s & \longrightarrow & Y_s \times_{\lim_{t<s} Y_s} \lim_{t<s} X_s . \end{array}$$

A lift exists in this diagram since i_s is a cofibration and $M_s p$ is an acyclic fibration. By induction on s, we construct a lift $B \to X$.

We have shown that strict cofibrations lift with respect to special acyclic fibrations. Formally, it follows that strict cofibrations lift with respect to retracts of special acyclic fibrations.

Now suppose that a map $i : A \to B$ has the left lifting property with respect to all special acyclic fibrations. Use Lemma 4.6 to factor i as a strict cofibration $i' : A \to B'$ followed by a special acyclic fibration $p : B' \to B$. Then we have a square

$$\begin{array}{ccc} A & \xrightarrow{i'} & B' \\ {\scriptstyle i}\downarrow & & \downarrow{\scriptstyle p} \\ B & \xrightarrow[=]{} & B, \end{array}$$

and a lift exists in this square by assumption. Hence i is a retract of i'. But retracts preserve strict cofibrations by Proposition 2.2, so i is again a strict cofibration.

Finally, suppose that $p : X \to Y$ has the right lifting property with respect to all strict cofibrations. Use Lemma 4.6 to factor p as a strict cofibration $i : X \to X'$ followed by a special acyclic fibration $p' : X' \to Y$. Similarly to the previous paragraph, p is a retract of p'. □

Lemma 4.10. *A map is an essentially levelwise acyclic cofibration if and only if it has the right lifting property with respect to all strict fibrations. Also, a map is a strict fibration if and only if it has the right lifting property with respect to all essentially levelwise acyclic cofibrations.*

Proof. The proof is the same as the proof of Lemma 4.9, except that the roles of cofibrations and acyclic fibrations are replaced by acyclic cofibrations and fibrations respectively. Lemma 4.7 is relevant instead of Lemma 4.6. □

Proposition 4.11. *A map is a strict acyclic cofibration if and only if it is an essentially levelwise acyclic cofibration.*

Proof. One implication follows from the definitions. For the other implication, let $i : A \to B$ be a strict weak equivalence and strict cofibration. We may assume that i is a level representation that is a levelwise weak equivalence. Then we may use the argument of the proof of Lemma 4.6 to factor i as an essentially levelwise acyclic cofibration $i' : A \to B'$ followed by a special acyclic fibration $p : B' \to B$. Hence we have a square

$$\begin{array}{ccc} A & \xrightarrow{i'} & B' \\ {\scriptstyle i}\downarrow & & \downarrow{\scriptstyle p} \\ B & \xrightarrow[=]{} & B, \end{array}$$

and a lift exists in this square by Lemma 4.9 because i is a strict cofibration and p is a special acyclic fibration. Therefore, i is a retract of i'. But retracts preserve essentially levelwise acyclic cofibrations by Proposition 2.2, so i is also an essentially levelwise acyclic cofibration. □

Proposition 4.12. *A map is a strict acyclic fibration if and only if it is a retract of a special acyclic fibration.*

Proof. A special acyclic fibration is a strict acyclic fibration by Lemma 4.4. For the other direction, let $p : X \to Y$ be a strict weak equivalence and strict fibration. We may assume that p is a level representation that is a levelwise weak equivalence.

Then we may use the argument of the proof of Lemma 4.6 to factor p as an essentially levelwise acyclic cofibration $i : X \to X'$ followed by a special acyclic fibration $p' : X' \to Y$. Similarly to the proof of Proposition 4.11, p is a retract of p'. □

Theorem 4.13. *Let $\mathcal{C}$ be a proper model category. Then the classes of strict weak equivalences, strict cofibrations, and strict fibrations define a proper model structure on* pro-$\mathcal{C}$.

Proof. The existence of finite limits and colimits is demonstrated in [2, App. 4.2]. The two-out-of-three axiom is proved in Lemmas 3.5 and 3.6. The retract axiom for strict cofibrations and strict weak equivalences follows from Proposition 2.2, where it is shown that any class of essentially levelwise maps is always closed under retract. The retract axiom for strict fibrations is true by definition.

Using Propositions 4.11 and 4.12, the factorization axiom is given in Lemmas 4.6 and 4.7. Similarly, the lifting axiom follows from Lemmas 4.9 and 4.10.

This finishes the proofs of the basic model structure axioms. It remains to consider properness. We shall show that the strict model structure is right proper; the proof of left properness is dual.

Let $p : X \to Y$ be a strict fibration, and let $f : Z \to Y$ be a strict weak equivalence. By Lemma 4.5, we know that p is an essentially levelwise fibration.

We do not have a level representation

$$Z \xrightarrow{f} Y \xleftarrow{p} X$$

because we cannot necessarily represent f by a levelwise weak equivalence and p by a levelwise fibration with the same indexing category. Nevertheless, we do have a level representation

$$Z \xrightarrow[\sim]{f} W \xrightarrow[\cong]{g} Y \twoheadleftarrow^{p} X$$

in which p is a levelwise fibration, f is a levelwise weak equivalence, and g is a pro-isomorphism (but not necessarily a levelwise isomorphism). This gives us a level representation

$$\begin{array}{ccccc} Z \times_Y X & \xrightarrow{f'} & W \times_Y X & \xrightarrow[\cong]{g'} & X \\ \downarrow & & \downarrow{\scriptstyle p'} & & \downarrow{\scriptstyle p} \\ Z & \xrightarrow[f]{\sim} & W & \xrightarrow[g]{\cong} & Y, \end{array}$$

where the pullbacks are computed levelwise. Because g and therefore g' are isomorphisms, f' is a level representation for $Z \times_Y X \to X$. Hence it suffices to show that f' is a levelwise weak equivalence.

Since pullbacks preserve fibrations in $\mathcal{C}$, we know that p' is a levelwise fibration. Now the map f' is a levelwise pullback of a weak equivalence along a fibration, so it is a levelwise weak equivalence because $\mathcal{C}$ is right proper. □

Remark 4.14. The proof of properness in [17, Prop. 17.1] is incorrect, but the techniques described in the above proof can be used to fix it.

We are now ready to prove that the weak equivalences of [10, §3.3] are the same as the strict weak equivalences.

Proposition 4.15. *When $\mathcal{C}$ is a proper model category, the strict weak equivalences of Definition* 3.1 *agree with the weak equivalences of* [10, §3.3].

Proof. The weak equivalences of [10, §3.3] are by definition compositions of essentially levelwise weak equivalences. By Lemma 3.5, these compositions are again essentially levelwise weak equivalences. On the other hand, the proof of Lemma 4.6 (and Lemma 4.4) shows that every levelwise weak equivalence can be factored into a levelwise acyclic cofibration followed by a levelwise acyclic fibration. Therefore, every levelwise weak equivalence is a weak equivalence in the sense of [10, §3.3]. □

The preceding proposition is closely related to the main result of [24]. However, we make a useful observation missed there. Namely, it is not necessary to saturate the essentially levelwise weak equivalences; they are already saturated when $\mathcal{C}$ is proper.

4.1. Simplicial model structures

If $\mathcal{C}$ is a simplicial category, then pro-$\mathcal{C}$ is again a simplicial category. For any two pro-objects X and Y, the **simplicial mapping space $\mathbf{Map}(X, Y)$** is defined to be $\lim_s \operatorname{colim}_t \operatorname{Map}_{\mathcal{C}}(X_t, Y_s)$.

Beware that the definitions of tensor and cotensor are straightforward for finite simplicial sets but are slightly subtle in general. If X is a pro-object and K is a finite simplicial set, then $X \otimes K$ is defined by tensoring levelwise. Similarly, X^K is defined by cotensoring levelwise.

For an arbitrary simplicial set K, write it as a colimit $\operatorname{colim}_s K_s$, where each K_s is a finite simplicial set. Then define $X \otimes K$ to be $\operatorname{colim}_s^{\text{pro}}(X \otimes K_s)$, where the colimit is computed in the category pro-$\mathcal{C}$. This is not the same as tensoring levelwise with K. Similarly, X^K is defined to be $\lim_s^{\text{pro}}(X^{K_s})$, where the limit is computed in the category pro-$\mathcal{C}$. Again, this is not the same as cotensoring levelwise with K.

Theorem 4.16. *If $\mathcal{C}$ is a proper simplicial model category, then the strict structure on* pro-$\mathcal{C}$ *is also simplicial.*

Proof. Most of the axioms are obvious or follow formally from the simplicial structure on $\mathcal{C}$. See [17, §16] for more details. We shall show that if $i : K \to L$ is a cofibration of finite simplicial sets and $j : A \to B$ is a strict cofibration in pro-$\mathcal{C}$, then

$$A \otimes L \amalg_{A \otimes K} B \otimes K \to B \otimes L$$

is a strict cofibration, and it is acyclic if either i or j is. By adjointness [14, Lem. 10.3.6], this is equivalent to the usual SM7 axiom for simplicial model categories.

We may assume that j is a levelwise cofibration. If j is also acyclic, then we may assume by Proposition 4.11 that j is a levelwise acyclic cofibration. Because $\mathcal{C}$ is a simplicial model category, the map

$$A_s \otimes L \amalg_{A_s \otimes K} B_s \otimes K \to B_s \otimes L$$

is a cofibration, and it is acyclic if either i or j_s is acyclic. Because K and L are finite, this shows that

$$A \otimes L \amalg_{A \otimes K} B \otimes K \to B \otimes L$$

is a levelwise cofibration and that it is a levelwise acyclic cofibration if either i or j is acyclic. □

5. Non-fibrantly generated model categories

In [17, Cor. 19.3], it was shown that strict model structures are not always cofibrantly generated [14, Defn. 11.1.2], even if $\mathcal{C}$ is cofibrantly generated. For example, the strict model structure for pro-simplicial sets is not cofibrantly generated. In this section, we study whether strict model structures are fibrantly generated.

Let λ be any ordinal. Then λ is the partially ordered set of all ordinals strictly less than λ. A **λ-tower** Z in a category is a contravariant functor from λ such that for all limit ordinals β, the object Z_β is isomorphic to $\lim_{\alpha<\beta} Z_\alpha$. In other words, it is a diagram

$$\cdots \to Z_\beta \to \cdots \to Z_1 \to Z_0$$

of length λ. Note that Z_β is defined only for $\beta < \lambda$, not for $\beta = \lambda$. This definition is dual to the notion of λ-sequence that arises in discussions of the small object argument [14, Defn. 10.2.1].

Definition 5.1. Let C be a class of maps in a category. A map $p : X \to Y$ is a ***C*-principal cocomplex** if there exists a λ-tower Z such that each map $Z_{\beta+1} \to Z_\beta$ is a base change of a map belonging to C and such that the projection $\lim_{\beta<\lambda} Z_\beta \to Z_0$ is isomorphic to the map p.

This definition is dual to the usual notion of C-cell complex [14, Defn. 10.5.8].

Proposition 5.2. *Let C be any class of maps in a category $\mathcal{C}$, and let cC be the image in* pro-$\mathcal{C}$ *of the class C under the constant functor. If $f : X \to Y$ is a cofinite directed level representation such that each map $M_s f$ belongs to C, then f is a cC-principal cocomplex.*

Proof. Let I be the cofinite directed indexing category for f. Choose a well-ordering ϕ of I that respects the ordering on I; thus ϕ is a bijection from I to some ordinal λ such that $\phi(s) \geq \phi(t)$ whenever $s \geq t$.

We define a λ-tower Z as follows. Set Z_0 equal to Y. For every successor ordinal $\beta + 1$, let s be the unique element of I such that $\phi(s) = \beta$. Define $Z_{\beta+1}$ by the pullback square

$$\begin{array}{ccc} Z_{\beta+1} & \longrightarrow & Z_\beta \\ \downarrow & & \downarrow \\ cX_s & \longrightarrow & cY_s \times_{\lim^{\text{pro}}_{t<s} cY_t} \lim^{\text{pro}}_{t<s} cX_t. \end{array}$$

Note that the bottom horizontal map is equal to $cM_s f$ because c commutes with finite limits. Also note that by transfinite induction, each Z_β comes equipped with a map

$$Z_\beta \to cY_s \times_{\lim^{\text{pro}}_{t<s} cY_t} \lim^{\text{pro}}_{t<s} cX_t.$$

For limit ordinals β, define Z_β to be $\lim^{\text{pro}}_{\alpha<\beta} Z_\alpha$. Thus we have constructed a cC-principal cocomplex $Z \to Y$. A consideration of universal properties shows that $\lim^{\text{pro}}_{\beta<\lambda} Z_\beta$ is isomorphic to X; here we use that X is isomorphic to $\lim^{\text{pro}}_s cX_s$. Also, the projection $\lim^{\text{pro}}_{\beta<\lambda} Z_\beta \to Z_0$ is equal to f. □

Proposition 5.3. *Let $\mathcal{C}$ be a proper model category, and let I and J be the classes of fibrations and acyclic fibrations in $\mathcal{C}$ respectively. The class of strict fibrations is equal to the class of retracts of cI-principal cocomplexes, and the class of strict acyclic fibrations is equal to the class of retracts of cJ-principal cocomplexes.*

Proof. We prove only the first statement; the proof of the second uses Proposition 4.12 but is otherwise identical.

One implication follows immediately from Proposition 5.2. For the other direction, first observe that maps in cI are special fibrations. Base changes and compositions along λ-towers preserve right lifting properties, so all cI-principal cocomplexes are strict fibrations. Finally, retracts preserve lifting properties, so retracts of cI-principal cocomplexes are strict fibrations. □

Remark 5.4. One direction of the previous proof uses the lifting property characterization of strict fibrations, which we know because of Theorem 4.13, but the other direction does not. It seems plausible that for any category $\mathcal{C}$ and any class C of maps, a map in pro-$\mathcal{C}$ should be a retract of a cC-principal cocomplex if and only if it is a retract of a map that has a cofinite directed level representation for which all the relative matching maps belong to C. However, we have not been able to prove this claim in such generality.

Proposition 5.5. *Let $\mathcal{C}$ be a proper model category, and let I and J be the classes of fibrations and acyclic fibrations in $\mathcal{C}$ respectively. The class of strict cofibrations is determined by the left lifting property with respect to cJ, and the class of strict acyclic cofibrations is determined by the left lifting property with respect to cI.*

Proof. We only prove the first claim; the proof of the second is identical.

Every element of cJ is a special acyclic fibration, so strict cofibrations lift with respect to them by Lemma 4.9. Conversely, if a map i has the left lifting property with respect to every element of cJ, then i has the left lifting property with respect to retracts of cJ-principal cocomplexes. By Proposition 5.3, i lifts with respect to all strict acyclic fibrations, so it is a strict cofibration. □

Every object of every pro-category is cosmall [5]. Therefore, Proposition 5.5 almost shows that the strict model structure on pro-$\mathcal{C}$ is fibrantly generated. The problem is that the collections cI and cJ of generating strict fibrations and generating strict acyclic fibrations are *not* sets.

Proposition 5.6. *The strict structure on pro-simplicial sets is not fibrantly generated.*

Proof. Suppose that the strict structure on pro-simplicial sets were fibrantly generated. Let I be a set of generating strict fibrations. By the dual to the usual small object argument, every strict fibration is a retract of an I-principal cocomplex.

Now let $\lim I$ be the image of I under the limit functor. Since the constant functor c and the limit functor form a Quillen pair between $\mathcal{C}$ and pro-$\mathcal{C}$, every element of $\lim I$ is a fibration of simplicial sets.

Every fibration of simplicial sets is the image of a strict fibration under the limit functor, so every fibration is a retract of a $(\lim I)$-principal cocomplex. This means that the set $\lim I$ of fibrations detects acyclic cofibrations of simplicial sets. No such set exists, so we have obtained a contradiction. □

Remark 5.7. The fact that no set of fibrations detects acyclic cofibrations of simplicial sets seems obvious, but the proof is not elementary. Bill Dwyer has shown us a proof, but it is too lengthy to reproduce here.

We describe one way to obtain a fibrantly generated model structure for pro-simplicial sets. Choose an infinite cardinal κ. We say that a simplicial set is κ-bounded if it has fewer than κ simplices.

The category of κ-bounded simplicial sets equipped with the usual notions of weak equivalence, cofibration, and fibration is a model category. Only the factorization axiom has a non-obvious proof. One must check that the usual small object argument factorizations produce κ-bounded simplicial sets.

Now the category of κ-bounded simplicial sets is small and has all finite limits. This means that the category pro-(κ-bounded simplicial sets) has all small limits and colimits [2, App. 4.3 and App. 4.4]. We obtain a strict model structure for pro-(κ-bounded simplicial sets), and it is fibrantly generated. As shown in Proposition 5.5, the constant fibrations of κ-bounded simplicial sets form a set of generating fibrations, and the constant acyclic fibrations of κ-bounded simplicial sets form a set of generating acyclic fibrations.

Therefore, the strict model structure on pro-(κ-bounded simplicial sets) is fibrantly generated. In some applications of pro-simplicial sets, it suffices to choose a κ larger than any of the simplicial sets occurring in the application. Thus, it is possible sometimes to use a fibrantly generated model structure to study the homotopy theory of pro-simplicial sets.

References

[1] M. Artin, A. Grothendieck, and J.L. Verdier, *Théorie des topos et cohomologie étale des schémas*, Lecture Notes in Mathematics, vol. 269, Springer Verlag, 1972.

[2] M. Artin and B. Mazur, *Etale homotopy*, Lecture Notes in Mathematics, vol. 100, Springer Verlag, 1969.

[3] A.K. Bousfield and E.M. Friedlander, *Homotopy theory of Γ-spaces, spectra, and bisimplicial sets*, Geometric Applications of Homotopy Theory, vol. II (Proc. Conf., Evanston, IL, 1977), Lecture Notes in Mathematics, vol. 658, Springer Verlag, 1978, pp. 80–130.

[4] A.K. Bousfield and D. Kan, *Homotopy limits, completions, and localizations*, Lecture Notes in Mathematics, vol. 304, Springer Verlag, 1972.

[5] J.D. Christensen and D.C. Isaksen, *Duality and pro-spectra*, preprint.

[6] R.L. Cohen, J.D.S. Jones, and G.B. Segal, *Floer's infinite-dimensional Morse theory and homotopy theory*, The Floer memorial volume, Progr. Math., vol. 133, Birkhäuser, 1995, pp. 297–325.

[7] J.-M. Cordier and T. Porter, *Shape theory: Categorical methods of approximation*, Ellis Horwood Series: Mathematics and Its Applications, 1989.

[8] J. Duskin, *Pro-objects (after Verdier)*, Dualité de Poincaré (Séminaire Heidelberg-Strasbourg 1966/67), IRMA Strasbourg **3** (1969), Exposé 6.

[9] W.G. Dwyer and J. Spaliński, *Homotopy theories and model categories*, Handbook of algebraic topology, North-Holland, 1995, pp. 73–126,

[10] D.A. Edwards and H.M. Hastings, *Cech and Steenrod homotopy theories with applications to geometric topology*, Lecture Notes in Mathematics, vol. 542, Springer Verlag, 1976.

[11] A.D. Elmendorf, I. Kriz, M.A. Mandell, and J.P. May, *Rings, modules, and algebras in stable homotopy theory*, Mathematical Surveys and Monographs, vol. 47, American Mathematical Society, 1997.

[12] J. Grossman, *A homotopy theory of pro-spaces*, Trans. Amer. Math. Soc. **201** (1975), 161–176.

[13] H.M. Hastings, *Addendum to: Cech and Steenrod homotopy theories with applications to geometric topology*, Colloq. Math. **41** (1979), no. 1, 161–163.

[14] P.S. Hirschhorn, *Model categories and their localizations*, Mathematical Surveys and Monographs, vol. 99, American Mathematical Society, 2003.

[15] Mark Hovey, *Model categories*, Mathematical Surveys and Monographs, vol. 63, American Mathematical Society, 1999.

[16] M. Hovey, B. Shipley, and J. Smith, *Symmetric spectra*, J. Amer. Math. Soc. **13** (2000), no. 1, 149–208.

[17] D.C. Isaksen, *A model structure on the category of pro-simplicial sets*, Trans. Amer. Math. Soc. **353** (2001), no. 7, 2805–2841.

[18] D.C. Isaksen, *Calculating limits and colimits in pro-categories*, Fund. Math. **175** (2002), no. 2, 175–194.

[19] D.C. Isaksen, *Completions of pro-spaces*, in preparation.

[20] S. Mardešić and J. Segal, *Shape theory: The inverse system approach*, North-Holland Mathematical Library, vol. 26, North-Holland, 1982.

[21] C.V. Meyer, *Approximation filtrante de diagrammes finis par Pro-C*, Ann. Math. Sci. Québec **4** (1980), no. 1, 35–57.

[22] T. Porter, *Abstract homotopy theory in procategories*, Cahiers Topologie Géom. Différentielle **17** (1976), no. 2, 113–124.

[23] T. Porter, *Coherent prohomotopical algebra*, Cahiers Topologie Géom. Différentielle **18** (1977), no. 2, 139–179.

[24] T. Porter, *On the two definitions of* Ho(pro$\mathcal{C}$), Topology Appl. **28** (1988), no. 3, 289–293.

[25] D.G. Quillen, *Homotopical algebra*, Lecture Notes in Mathematics, vol. 43, Springer Verlag, 1967.

[26] D. Sullivan, *Genetics of homotopy theory and the Adams conjecture*, Ann. of Math. (2) **100** (1974), 1–79.

Daniel C. Isaksen
Department of Mathematics
University of Notre Dame
Notre Dame, IN 46556, USA
e-mail: isaksen.1@nd.edu

Progress in Mathematics, Vol. 215
Algebraic Topology: Categorical Decomposition Techniques, 199–212

L-S Categories of Simply-connected Compact Simple Lie Groups of Low Rank

Norio Iwase and Mamoru Mimura

Abstract. We determine the L-S category of $Sp(3)$ by showing that the 5-fold reduced diagonal $\overline{\Delta}_5$ is given by ν^2, using a Toda bracket and a generalised cohomology theory h^* given by $h^*(X, A) = \{X/A, \mathcal{S}[0,2]\}$, where $\mathcal{S}[0,2]$ is the 3-stage Postnikov piece of the sphere spectrum $\mathcal{S}$. This method also yields a general result that $\mathrm{cat}(Sp(n)) \geq n+2$ for $n \geq 3$, which improves a result of Singhof [20].

1. Introduction

In this paper, we firstly discuss the L-S category of G_2 as in Theorem 1.1 to illustrate the methods to be used later in the argument for $Sp(3)$. Secondly, we prove $\mathrm{cat}(Sp(3)) = 5$ as in Theorem 1.2, although an alternative proof of it can be deduced from public sources by Lucía Fernández-Suárez, Antonio Gómez-Tato, Jeffrey Strom and Daniel Tanré [5]; the earlier version, however, appeared to the authors to contain an error. In fact, this is our starting point and motivation to write the present paper with a short and clear proof for $\mathrm{cat}(Sp(3)) = 5$. Finally we show that this argument for $Sp(3)$ partially extends to the general case.

From now on, each space is assumed to have the homotopy type of a CW complex. The (normalised) L-S category of X is the least number m such that there is a covering of X by $(m+1)$ open subsets each of which is contractible in X. Hence $\mathrm{cat}\,\{*\} = 0$. By Lusternik and Schnirelmann [13], the number of critical points of a smooth function on a manifold M is bounded below by $\mathrm{cat}\, M + 1$.

G. Whitehead showed that $\mathrm{cat}(X)$ coincides with the least number m such that the diagonal map $\Delta_{m+1} : X \to \prod^{m+1} X$ can be compressed into the 'fat wedge' $\mathrm{T}^{m+1}(X)$ (see Chapter X of [23]). Since $\prod^{m+1} X/\,\mathrm{T}^{m+1}(X)$ is the $(m+1)$-fold smash product $\wedge^{m+1}X$, we have a weaker invariant $w\mathrm{cat}\, X$, *the weak L-S category of* X, given by the least number m such that the reduced diagonal map $\overline{\Delta}_{m+1} : X \to \wedge^{m+1}X$ is trivial. Hence $w\mathrm{cat}\, X \leq \mathrm{cat}\, X$.

2000 *Mathematics Subject Classification.* Primary 55M30, Secondary 55N20, 22E20.
Key words and phrases. Lie group, Lusternik-Schnirelmann category, cohomology theory.
The authors are partially supported by the Grant-in-Aids for Scientific Research #14654016 and #12640025 from the Japan Society of Promotion of Science.

T. Ganea has also introduced a stronger invariant $\mathrm{Cat}\,X$, *the strong L-S category of* X, by the least number m such that there is a space Y homotopy equivalent to X and a covering of Y by $(m+1)$ open subsets each of which is contractible in itself. Thus $w\mathrm{cat}\,X \le \mathrm{cat}\,X \le \mathrm{Cat}\,X$. The weak and strong L-S categories usually give nice estimates of L-S category especially for manifolds. The following problems are posed by Ganea [6]:

i) (Problem 1) Determine the L-S category of a manifold.
ii) (Problem 4) Describe the L-S category of a sphere-bundle over a sphere in terms of homotopy invariants of the characteristic map of the bundle.

Problem 1 has been studied by many authors, such as Singhof [19, 20, 21], Montejano [15], Schweizer [18], Gomez-Larrañaga and Gonzalez-Acuña [7], James and Singhof [12] and Rudyak [16, 17]. In particular for compact simply-connected simple Lie groups, $\mathrm{cat}(SU(n+1)) = n$ for $n \ge 1$ by [19], $\mathrm{cat}(Sp(2)) = 3$ by [18] and $\mathrm{cat}(Sp(n)) \ge n+1$ for $n \ge 2$ by [20]. It was also announced recently that Problem 4 was solved by the first-named author [9]. The method in the present paper also provides a result for G_2, and thus we have the following result.

Theorem 1.1. *The following is the complete list of L-S categories of a simply-connected compact simple Lie group of rank* ≤ 2*:*

Lie groups	$Sp(1) = SU(2) = Spin(3)$	$SU(3)$	$Sp(2) = Spin(5)$	G_2
$w\mathrm{cat}$	1	2	3	4
cat	1	2	3	4
Cat	1	2	3	4

Although the above result is known for experts, we give a short proof for G_2. In fact, the result for G_2 has never been published and is obtained in a similar but easier manner than the following result for $Sp(3)$:

Theorem 1.2. $w\mathrm{cat}(Sp(3)) = \mathrm{cat}(Sp(3)) = \mathrm{Cat}(Sp(3)) = 5.$

Remark 1.3. *The argument given to prove Theorem* 1.2 *provides an alternative proof of Schweizer's result*

$$w\mathrm{cat}(Sp(2)) = \mathrm{cat}(Sp(2)) = \mathrm{Cat}(Sp(2)) = 3.$$

The authors know that a similar result to Theorem 1.2 is obtained by Lucía Fernández-Suárez, Antonio Gómez-Tato, Jeffrey Strom and Daniel Tanré [5]. Our method is, however, much simpler and provides the following general result:

Theorem 1.4. $n+2 \le w\mathrm{cat}(Sp(n)) \le \mathrm{cat}(Sp(n)) \le \mathrm{Cat}(Sp(n))$ *for* $n \ge 3$.

This improves Singhof's result: $\mathrm{cat}(Sp(n)) \ge n+1$ for $n \ge 2$. We propose the following conjecture.

Conjecture 1.5. *Let* G *be a simply-connected compact Lie group with* $G = \prod_{i=1}^{n} H_i$ *where* H_i *is a simple Lie group. Then* $w\mathrm{cat}(G) = \mathrm{cat}(G) = \mathrm{Cat}(G)$ *and* $\mathrm{cat}(G) = \sum_{i=1}^{n} \mathrm{cat}(H_i)$.

It might be difficult to say something about cat $Sp(n)$, but an old conjecture says the following.

Conjecture 1.6. $\operatorname{cat} Sp(n) = 2n-1$ *for all* $n \geq 1$.

The authors thank John Harper for many helpful conversations and also the referee for giving them some comments, in particular, regarding Remark 2.4.

2. Proof of Theorem 1.1

Let us recall a CW decomposition of G_2 from [14]:

$$G_2 = e^0 \cup e^3 \cup e^5 \cup e^6 \cup e^8 \cup e^9 \cup e^{11} \cup e^{14}.$$

On the other hand, we have the following cone-decomposition.

Theorem 2.1. *There is a cone-decomposition of* G_2 *as follows:*

$$G_2^{(5)} = \Sigma\mathbb{C}P^2, \quad S^5 \cup e^7 \to G_2^{(5)} \hookrightarrow G_2^{(8)},$$
$$S^8 \cup e^{10} \to G_2^{(8)} \hookrightarrow G_2^{(11)}, \quad S^{13} \to G_2^{(11)} \hookrightarrow G_2.$$

Proof. The first and the last formulae are obvious. So we show the 2nd and 3rd formulae: By taking the homotopy fibre F_1 of $G_2^{(5)} \hookrightarrow G_2$, we can easily observe using the Serre spectral sequence that the fibre has a CW structure given by $S^5 \cup e^7 \cup$(cells in dimensions ≥ 7), where the cohomology generators corresponding to S^5 and e^7 are transgressive. Thus the mapping cone of $S^5 \cup e^7 \subset F_1 \to G_2^{(5)}$ has the homotopy type of $G_2^{(8)}$. Similarly, the homotopy fibre F_2 of $G_2^{(8)} \hookrightarrow G_2$ has a CW structure given by $S^8 \cup e^{10} \cup$(cells in dimensions ≥ 10), where the cohomology generators corresponding to S^8 and e^{10} are transgressive. Thus the mapping cone of $S^8 \cup e^{10} \subset F_2 \to G_2^{(8)}$ has the homotopy type of $G_2^{(11)}$. $\square$

Corollary 2.2. $1 \geq \operatorname{Cat}(G_2^{(5)}) \geq \operatorname{Cat}(G_2^{(3)})$, $2 \geq \operatorname{Cat}(G_2^{(8)}) \geq \operatorname{Cat}(G_2^{(6)})$, $3 \geq \operatorname{Cat}(G_2^{(11)}) \geq \operatorname{Cat}(G_2^{(9)})$ *and* $4 \geq \operatorname{Cat}(G_2)$.

Let us recall the following well-known fact due to Borel.

Fact 2.3. $H^*(G_2;\mathbb{Z}/2\mathbb{Z}) \cong \mathbb{Z}/2\mathbb{Z}[x_3, x_5]/(x_3^4, x_5^2)$.

Corollary 2.4. $\operatorname{wcat}(G_2^{(5)}) \geq \operatorname{wcat}(G_2^{(3)}) \geq 1$, $\operatorname{wcat}(G_2^{(8)}) \geq \operatorname{wcat}(G_2^{(6)}) \geq 2$, $\operatorname{wcat}(G_2^{(11)}) \geq \operatorname{wcat}(G_2^{(9)}) \geq 3$ *and* $\operatorname{wcat}(G_2) \geq 4$.

Corollaries 2.2 and 2.4 yield the following.

Theorem 2.5.

Skeleta	$G_2^{(3)}$	$G_2^{(5)}$	$G_2^{(6)}$	$G_2^{(8)}$	$G_2^{(9)}$	$G_2^{(11)}$	G_2
*w*cat	1	1	2	2	3	3	4
cat	1	1	2	2	3	3	4
Cat	1	1	2	2	3	3	4

This completes the proof of Theorem 1.1.

Remark 2.6. *If we disregard the information of L-S categories of CW filtrations of G_2 and if we want only to deduce the equation* $\mathrm{wcat}(G_2) = \mathrm{cat}(G_2) = \mathrm{Cat}(G_2) = 4$, *we have an alternative short proof of it rather than the above elementary homotopy-theoretical argument: Since the manifold G_2 is 2-connected and of dimension 14, we know that* $\mathrm{cat}(G_2) \leq \frac{14}{3}$ *by James* [10]. *On the other hand, the cohomology algebra of G_2 with coefficients in $\mathbb{F}_2$ is well known by Borel as in Fact 2.3, and hence its cup-length is 4 and we get immediately that* $\mathrm{wcat}(G_2) = \mathrm{cat}(G_2) = 4$. *Concerning on the strong L-S category* $\mathrm{Cat}(G_2)$ *of a manifold G_2, we are in the range of the validity of Corollary 5.9 of Clapp and Puppe* [3] *which implies immediately that* $\mathrm{cat}(G_2) = \mathrm{Cat}(G_2)$.

3. The ring structure of $h^*(Sp(3))$

To show Theorem 1.2, we introduce a cohomology theory $h^*(-)$ such that $h^*(X, A) = \{X/A, \mathcal{S}[0, 2]\}$, where $\mathcal{S}[0, 2]$ is the spectrum obtained from the sphere spectrum $\mathcal{S}$ by killing all homotopy groups of dimensions bigger than 2. Then $\mathcal{S}[0, 2]$ is a ring spectrum with $\pi_*^S(\mathcal{S}[0, 2]) \cong \mathbb{Z}[\eta]/(\eta^3, 2\eta)$, where η is the Hopf element in $\pi_1^S(\mathcal{S}) = \pi_1^S(\mathcal{S}[0, 2])$. Thus h^* is an additive and multiplicative cohomology theory with $h^* = h^*(pt) \cong \mathbb{Z}[\varepsilon]/(\varepsilon^3, 2\varepsilon)$, $\deg \varepsilon = -1$, where $\varepsilon \in h^{-1} = \pi_0^S(\Sigma^{-1}\mathcal{S}[0, 2]) \cong \pi_1^S(\mathcal{S})$ corresponds to η.

The characteristic map of the principal $Sp(1)$-bundle

$$Sp(1) \hookrightarrow Sp(2) \to S^7$$

is given by $\omega = \langle \iota_3, \iota_3 \rangle : S^6 \to Sp(1) \approx S^3$ the Samelson product of two copies of the identity $\iota_3 : S^3 \to S^3$, which is a generator of $\pi_6(S^3) \cong \mathbb{Z}/12\mathbb{Z}$. We state the following well-known fact (see Whitehead [23]).

Fact 3.1. *Let* $\mu : S^3 \times S^3 \to S^3$ *be the multiplication of* $Sp(1) \approx S^3$. *Then we have*

$$Sp(2) \simeq S^3 \cup_{\mu\circ(1\times\omega)} S^3 \times C(S^6) = S^3 \cup_\omega C(S^6) \cup_{\hat{\mu}\circ[\iota_3, \chi_\omega]^r} C(S^9),$$

where $\hat{\mu} : S^3 \times S^3 \cup_{*\times\omega} \{*\} \times C(S^6) \to S^3 \cup_\omega C(S^6)$ *is given by* $\hat{\mu}|_{S^3\times S^3} = \mu$ *and* $\hat{\mu}|_{S^3 \cup_\omega C(S^6)} = 1$ *the identity and* $[\iota_3, \chi_\omega]^r : S^9 \to S^3 \times S^3 \cup_{*\times\omega} \{*\} \times C(S^6)$ *is the relative Whitehead product of the identity* $\iota_3 : S^3 \to S^3$ *and the characteristic map* $\chi_\omega : (C(S^6), S^6) \to (S^3 \cup e^7, S^3)$ *of the 7-cell. Thus we have* $1 \geq \mathrm{Cat}(Sp(2)^{(3)})$, $2 \geq \mathrm{Cat}(Sp(2)^{(7)})$ *and* $3 \geq \mathrm{Cat}(Sp(2))$.

Let $\nu : S^7 \to S^4$ be the Hopf element whose suspension $\nu_n = \Sigma^{n-4}\nu$ $(n \geq 4)$ gives a generator of $\pi_{n+3}(S^n) \cong \mathbb{Z}/24\mathbb{Z}$ for $n \geq 5$. Then we remark that $\omega_n = \Sigma^{n-3}\omega$ $(n \geq 3)$ satisfies the formula $\omega_n = 2\nu_n \in \pi_{n+3}(S^n)$ for $n \geq 5$. By Zabrodsky [24], there is a natural splitting

$$\Sigma(S^3 \times S^3 \cup \{*\} \times (S^3 \cup_\omega e^7)) \simeq \Sigma S^3 \vee \Sigma(S^3 \cup_\omega e^7) \vee \Sigma S^3 \wedge S^3.$$

Then by the definition of a relative Whitehead product, the composition of $[\iota_3, \omega]^r$ with the projections to S^3 and $S^3 \cup_\omega e^7$ are trivial and the composition with the projection to $S^3 \wedge S^3$ is given by $\iota_3 \wedge \omega$. Thus we have

$$\Sigma(\hat{\mu}\circ[\iota_3, \chi_\omega]^r) = H(\mu)\circ\Sigma(\iota_3\wedge\omega) = \pm\nu\circ\omega_7 = 2\nu\circ\nu_7 \neq 0$$

in $\pi_{10}(S^4) \cong \mathbb{Z}/24\mathbb{Z}\langle\nu\circ\nu_7\rangle \oplus \mathbb{Z}/2\mathbb{Z}\langle\omega_4\circ\nu_7\rangle$, and hence we have

$$\Sigma^2(\hat{\mu}\circ[\iota_3, \chi_\omega]^r) = \nu_5\circ\omega_8 = 2\nu_5^2 = 0 \in \pi_{11}(S^5) \cong \mathbb{Z}/2\mathbb{Z}$$

by Proposition 5.11 of Toda [22]. Thus we have the following well-known facts.

Fact 3.2. *We have the following homotopy equivalences:*

$$Sp(2)/S^3 \simeq (S^3 \times C(S^6))/(S^3 \times S^6) = S^3_+ \wedge \Sigma(S^6) = S^7 \vee S^{10},$$
$$\Sigma^2 Sp(2) \simeq \Sigma^2(S^3 \cup_\omega C(S^6)) \vee \Sigma^2 S^{10} = S^5 \cup_{\omega_5} C(S^8) \vee S^{12}.$$

Fact 3.3. *The* 11*-skeleton* $X^{(11)}_{3,2}$ *of* $X_{3,2} = Sp(3)/Sp(1)$ *has the homotopy type of* $S^7 \cup_{\nu_7} e^{11}$.

Restricting the principal $Sp(1)$-bundle $Sp(1) \hookrightarrow Sp(3) \xrightarrow{q} X_{3,2}$ to the subspace $X^{(11)}_{3,2} = S^7 \cup_{\nu_7} e^{11}$ of $X_{3,2}$, we obtain the subspace $q^{-1}(X^{(11)}_{3,2}) = Sp(3)^{(14)}$ of $Sp(3)$ as the total space of the principal $Sp(1)$-bundle $Sp(1) \hookrightarrow Sp(3)^{(14)} \xrightarrow{q} \Sigma(S^6 \cup_{\nu_6} e^{10})$ with a characteristic map $\phi : S^6 \cup_{\nu_6} e^{10} \to Sp(1) \approx S^3$, which is an extension of $\omega : S^6 \to S^3$.

Proposition 3.4. *We have the following homotopy equivalences:*

$$\begin{aligned}
Sp(3)^{(14)} &\simeq S^3 \cup_{\mu\circ(1\times\phi)} S^3 \times C(S^6 \cup_{\nu_6} e^{10}) \\
&= S^3 \cup_\phi C(S^6 \cup_{\nu_6} e^{10}) \cup C(S^9 \cup_{\nu_9} e^{13}), \\
Sp(3)^{(14)}/S^3 &\simeq (S^3 \times C(S^6 \cup_{\nu_6} e^{10}))/(S^3 \times (S^6 \cup_{\nu_6} e^{10})) \\
&= S^3_+ \wedge \Sigma(S^6 \cup_{\nu_6} e^{10}) = (S^7 \cup_{\nu_7} e^{11}) \vee (S^{10} \cup_{\nu_{10}} e^{14}), \\
Sp(n) &\simeq Sp(n-1) \cup Sp(n-1) \times C(S^{4n-2}),
\end{aligned}$$

where $Sp(n-1) \subset Sp(n)^{((2n+1)n-11)}$ *for* $n \geq 3$, *and hence*

$$\begin{aligned}
&Sp(n)/Sp(n)^{((2n+1)n-11)} \\
&\quad\simeq (Sp(n-1) \times C(S^{4n-2}))/(Sp(n-1) \times S^{4n-2} \\
&\qquad\qquad \cup Sp(n-1)^{((2n-1)(n-1)-11)} \times C(S^{4n-2})) \\
&\quad= (Sp(n-1)/Sp(n-1)^{((2n-1)(n-1)-11)}) \wedge \Sigma S^{4n-2} \\
&\quad= \cdots = (Sp(2)/\emptyset) \wedge \Sigma S^{10} \wedge \cdots \wedge \Sigma S^{4n-2} = (Sp(2)_+) \wedge S^{(2n+1)n-10} \\
&\quad= S^{(2n+1)n-10} \vee S^{(2n+1)n-10} \wedge Sp(2) \\
&\quad= S^{(2n+1)n-10} \vee (S^{(2n+1)n-7} \cup_{\omega_{(2n+1)n-7}} e^{(2n+1)n-3}) \vee S^{(2n+1)n}, \quad \textit{for } n \geq 3.
\end{aligned}$$

This yields the following result.

Proposition 3.5. *Let* $\hat{\mu} : S^3{\times}S^3 \cup_{*\times\phi} \{*\}{\times}(S^3 \cup_\phi C(S^6 \cup_{\nu_6} e^{10})) \to S^3 \cup_\phi C(S^6 \cup_{\nu_6} e^{10})$ *be the map given by* $\hat{\mu}|_{S^3\times S^3} = \mu$ *and* $\hat{\mu}|_{S^3\cup_\phi C(S^6\cup_{\nu_6} e^{10})} = 1$ *the identity. Then we have the following cone decomposition of* $Sp(3)$*:*

$$Sp(3) \simeq S^3 \cup_\phi C(S^6 \cup_{\nu_6} e^{10}) \cup_{\hat{\mu}\circ\hat{\phi}} C(S^9 \cup_{\nu_9} e^{13}) \cup C(S^{17}) \cup C(S^{20}).$$

Corollary 3.6. $1 \geq \mathrm{Cat}(Sp(3)^{(3)})$, $2 \geq \mathrm{Cat}(Sp(3)^{(7)})$, $3 \geq \mathrm{Cat}(Sp(3)^{(14)}) \geq \mathrm{Cat}(Sp(3)^{(11)}) \geq \mathrm{Cat}(Sp(3)^{(10)})$, $4 \geq \mathrm{Cat}(Sp(3)^{(18)})$ *and* $5 \geq \mathrm{Cat}(Sp(3))$.

To determine the ring structures of $h^*(Sp(2))$ and $h^*(Sp(3))$, we show the following lemma.

Lemma 3.7. *Let* h^* *be any multiplicative generalised cohomology theory and let* $Q = S^r \cup_f e^q$ *for a given map* $f : S^{q-1} \to S^r$ *with* $h^*(Q) \cong h^*\langle 1, x, y\rangle$*, where* x *and* y *correspond to the generators of* $h^*(S^r) \cong h^*\langle x_0\rangle$ *and* $h^*(S^q) \cong h^*\langle y_0\rangle$*. Then*

$$x^2 = \pm\bar{H}^h_1(f){\cdot}y \quad \text{in} \quad h^*(Q),$$

where $\bar{H}^h_1$ *is the composition* $\rho^h\circ\lambda_2$ *of the Boardman-Steer Hopf invariant* $\lambda_2 : \pi_{q-1}(S^r) \to \pi_q(S^{2r})$ *(see Boardman and Steer [2]) with the Hurewicz homomorphism* $\rho^h : \pi_q(S^{2r}) \to h^{2r}(S^q) \cong h^{2r-q}$ *given by* $\rho^h(g) = \Sigma_*^{-q} g^*(x_0{\otimes}x_0)$.

Remark 3.8. *By [2],* $\lambda_2(f)$ *is equal to* $\Sigma h^J_2(f)$ *the suspension of the 2nd James Hopf invariant* $h^J_2(f)$*. Hence by Remarks 2.5 and 4.3 of [8],* $\lambda_2(f) = \Sigma h^J_2(f)$ *gives the Berstein-Hilton crude Hopf invariant* $\bar{H}_1(f)$ *(see Berstein-Hilton [1] or [8]).*

Proof. By [2], $\overline{\Delta} : Q = S^r \cup_f e^q \to Q{\wedge}Q$ equals the composition $(i_Q{\wedge}i_Q)\circ\lambda_2(f)\circ q_Q$, where $q_Q : Q \to Q/S^r = S^q$ is the collapsing map and $i_Q : S^r \hookrightarrow Q$ is the bottom-cell inclusion. Then we have $i_Q^*(x) = x_0$ and $q_Q^*(y_0) = y$, and hence we obtain

$$\begin{aligned} x^2 &= \overline{\Delta}^*(x{\otimes}x) = ((i_Q{\wedge}i_Q)\circ\lambda_2(f)\circ q_Q)^*(x{\otimes}x) \\ &= q_Q^*(\lambda_2(f)^*(i_Q^*(x){\otimes}i_Q^*(x))) = q_Q^*(\lambda_2(f)^*(x_0{\otimes}x_0)) = q_Q^*(\Sigma_*^q\circ\rho^h(\lambda_2(f))). \end{aligned}$$

Since $\Sigma_*^q\circ\rho^h(\lambda_2(f))$ is $\bar{H}^h_1(f){\cdot}y_0 \in h^{2r}(S^q)$ up to sign, we proceed as

$$x^2 = q_Q^*(\pm\bar{H}^h_1(f){\cdot}y_0) = \pm\bar{H}^h_1(f){\cdot}q_Q^*(y_0) = \pm\bar{H}^h_1(f){\cdot}y.$$

This completes the proof of the lemma. □

Using cohomology long exact sequences derived from the cell structure of $Sp(3)$ and a direct calculation using Proposition 3.4 and Lemma 3.7 with the fact that $\lambda_2(\omega) = \eta_6$, we deduce the following result for the cohomology theory h^* considered at the beginning of this section.

Theorem 3.9. *The ring structures of* $h^*(Sp(2))$ *and* $h^*(Sp(3))$ *are as follows:*

$$h^*(Sp(2)) \cong h^*\{1, x_3, x_7, y_{10}\},$$
$$h^*(Sp(3)) \cong h^*\{1, x_3, x_7, x_{11}, y_{10}, y_{14}, y_{18}, z_{21}\},$$

where the suffix of each additive generator indicates its degree in the graded algebras $h^*(Sp(2))$ *and* $h^*(Sp(3))$*. Moreover we have* $x_3^2 = \varepsilon{\cdot}x_7$, $x_7^2 = 0$, $x_{11}^2 = 0$, $x_3x_7 = y_{10}$, $x_3x_{11} = y_{14}$, $x_7x_{11} = y_{18}$ *and* $x_3x_7x_{11} = z_{21}$.

Remark 3.10. *The two possible attaching maps* : $S^{10} \to S^3 \cup_\omega e^7$ *of* e^{11} *found by Lucía Fernández-Suárez, Antonio Gómez-Tato and Daniel Tanré* [4] *are homotopic in* $Sp(2)$. *So, we can not make any effective difference in the ring structure of* $h^*(Sp(3))$ *by altering, as is performed in* [5], *the attaching map of* e^{11}.

Corollary 3.11. $wcat(Sp(2)^{(3)}) \geq 1$, $wcat(Sp(2)^{(7)}) \geq 2$ *and* $wcat(Sp(2)) \geq 3$, *and also* $wcat(Sp(3)^{(3)}) \geq 1$, $wcat(Sp(3)^{(7)}) \geq 2$, $wcat(Sp(3)^{(18)}) \geq wcat(Sp(3)^{(14)}) \geq wcat(Sp(3)^{(11)}) \geq wcat(Sp(3)^{(10)}) \geq 3$ *and* $wcat(Sp(3)) \geq 4$.

Corollary 3.12.

Skeleta	$Sp(2)^{(3)}$	$Sp(2)^{(7)}$	$Sp(2)$
$wcat$	1	2	3
cat	1	2	3
Cat	1	2	3

4. Proof of Theorem 1.2

By Facts 3.1 and 3.2, the smash products $\wedge^4 Sp(3)$ and $\wedge^5 Sp(3)$ satisfy

$$(\wedge^4 Sp(3))^{(19)} \simeq S^{12} \cup_{\omega_{12}} e^{16} \vee (S^{16} \vee S^{16} \vee S^{16}) \vee (S^{19} \vee S^{19} \vee S^{19} \vee S^{19}),$$

$$(\wedge^5 Sp(3))^{(22)} \simeq S^{15} \cup_{\omega_{15}} e^{19} \vee (S^{19} \vee S^{19} \vee S^{19}) \vee (S^{22} \vee S^{22} \vee S^{22} \vee S^{22}).$$

Then we have the following two propositions.

Proposition 4.1. *The bottom-cell inclusions* $i : S^{12} \hookrightarrow \wedge^4 Sp(3)^{(18)}$ *and* $i' : S^{15} \hookrightarrow \wedge^5 Sp(3)$ *induce injective homomorphisms*

$$i_* : \pi_{18}(S^{12}) \to \pi_{18}(\wedge^4 Sp(3)^{(18)}) \quad \textit{and} \quad i'_* : \pi_{21}(S^{15}) \to \pi_{21}(\wedge^5 Sp(3)),$$

respectively.

Proof. We have the following two exact sequences

$$\pi_{18}(S^{15}) \xrightarrow{\psi} \pi_{18}(S^{12}) \xrightarrow{i_*} \pi_{18}(\wedge^4 Sp(3)^{(18)}) \to \pi_{18}(S^{16} \vee S^{16} \vee S^{16} \vee S^{16}),$$

$$\pi_{21}(S^{18}) \xrightarrow{\psi'} \pi_{21}(S^{15}) \xrightarrow{i'_*} \pi_{21}(\wedge^5 Sp(3)) \to \pi_{21}(S^{19} \vee S^{19} \vee S^{19} \vee S^{19} \vee S^{19}),$$

where $\pi_{18}(S^{12}) \cong \pi_{21}(S^{15}) \cong \mathbb{Z}/2\mathbb{Z}\nu_{15}^2$ and ψ and ψ' are induced from $\omega_{12} = 2\nu_{12}$ and $\omega_{15} = 2\nu_{15}$. Thus ψ and ψ' are trivial, and hence i_* and i'_* are injective. □

Proposition 4.2. *The collapsing maps* $q : Sp(3)^{(18)} \to Sp(3)^{(18)}/Sp(3)^{(14)} = S^{18}$ *and* $q' : Sp(3) \to Sp(3)/Sp(3)^{(18)} = S^{21}$ *induce injective homomorphisms*

$$q^* : \pi_{18}(\wedge^4 Sp(3)^{(18)}) \to [Sp(3)^{(18)}, \wedge^4 Sp(3)^{(18)}] \quad \textit{and}$$
$$q'^* : \pi_{21}(\wedge^5 Sp(3)) \to [Sp(3), \wedge^5 Sp(3)],$$

respectively.

Proof. Firstly, we show that q'^* is injective: Since we have $[Sp(3), \wedge^5 Sp(3)] = [(S^{14} \cup_{\omega_{14}} e^{18}) \vee S^{21}, \wedge^5 Sp(3)] = [S^{14} \cup_{\omega_{14}} e^{18}, \wedge^5 Sp(3)] \oplus \pi_{21}(\wedge^5 Sp(3))$ by Proposition 3.4, q'^* is clearly injective.

Secondly, we show that q^* is injective: Using a similar argument, we have $[Sp(3)^{(18)}, \wedge^4 Sp(3)^{(18)}] = [S^{14} \cup_{\omega_{14}} e^{18}, \wedge^4 Sp(3)^{(18)}]$ by Proposition 3.4. Thus it is sufficient to show that $\bar{q}^* : \pi_{18}(\wedge^4 Sp(3)^{(18)}) \to [S^{14} \cup_{\omega_{14}} e^{18}, \wedge^4 Sp(3)^{(18)}]$ is injective, where $\bar{q} : S^{14} \cup_{\omega_{14}} e^{18} \to S^{18}$ is the collapsing map. In the exact sequence

$$\pi_{15}(\wedge^4 Sp(3)^{(18)}) \overset{\omega_{15}^*}{\to} \pi_{18}(\wedge^4 Sp(3)^{(18)}) \overset{\bar{q}^*}{\to} [S^{14} \cup_{\omega_{14}} e^{18}, \wedge^4 Sp(3)^{(18)}],$$

we know that $\pi_{15}(\wedge^4 Sp(3)^{(18)}) \cong \pi_{15}(S^{12} \cup_{\omega_{12}} e^{16}) = \mathbb{Z}/2\mathbb{Z}$ is generated by the composition of ν_{12} and the bottom-cell inclusion. Since $\nu_{12} \circ \omega_{15} = 0 \in \pi_{18}(S^{12})$, the homomorphism $\omega_{15}{}^*$ is trivial, and hence $\bar{q}^*$ is injective. □

Then the following lemma implies that $\overline{\Delta}_4$ and $\overline{\Delta}_5$ are non-trivial by Propositions 4.1 and 4.2.

Lemma 4.3. *We obtain that* $\overline{\Delta}_4 = i \circ \nu_{12}^2 \circ q : Sp(3)^{(18)} \to \wedge^4 Sp(3)^{(18)}$ *and that* $\overline{\Delta}_5 = i' \circ \nu_{15}^2 \circ q' : Sp(3) \to \wedge^5 Sp(3)$.

Proof. Firstly, we show that $\overline{\Delta}_4 = i \circ \nu_{12}^2 \circ q$ implies $\overline{\Delta}_5 = i' \circ \nu_{15}^2 \circ q'$. For dimensional reasons, the image of $\overline{\Delta} : Sp(3) \to Sp(3) \wedge Sp(3)$ is in $Sp(3)^{(18)} \wedge Sp(3)^{(14)} \cup S^3 \wedge Sp(3)^{(18)}$. Since $Sp(3)^{(14)}$ is of cone-length 3 by Corollary 3.6, the restriction of the map $1 \wedge \overline{\Delta}_4$ to $Sp(3)^{(18)} \wedge Sp(3)^{(14)}$ is trivial. Thus $\overline{\Delta}_5$ is given as a composition

$$Sp(3) \to S^3 \wedge (Sp(3)^{(18)}/Sp(3)^{(14)}) \overset{1 \wedge (i \circ \nu_{12}^2)}{\to} \wedge^5 Sp(3)^{(18)} \subset \wedge^5 Sp(3),$$

since $\overline{\Delta}_4 = i \circ \nu_{12}^2 \circ q$. Thus we observe that $\overline{\Delta}_5 = i' \circ (\iota_3 \wedge \nu_{12}^2) \circ q' = i' \circ \nu_{15}^2 \circ q'$.

So, we are left to show $\overline{\Delta}_4 = i \circ \nu_{12}^2 \circ q$. For dimensional reasons, the image of $\overline{\Delta} : Sp(3)^{(18)} \to Sp(3)^{(18)} \wedge Sp(3)^{(18)}$ is in $Sp(3)^{(14)} \wedge S^3 \cup Sp(3)^{(11)} \wedge Sp(3)^{(7)} \cup Sp(3)^{(7)} \wedge Sp(3)^{(11)} \cup S^3 \wedge Sp(3)^{(14)}$. Since $S^3 \cup_\phi C(S^6 \cup_{\nu_6} e^{10})$ is of cone-length 2 by Corollary 3.6, the restriction of $\overline{\Delta}_3 : Sp(3)^{(18)} \to \wedge^3 Sp(3)^{(18)}$ to $S^3 \cup_\phi C(S^6 \cup_{\nu_6} e^{10})$ is trivial. Hence $1 \wedge \overline{\Delta}_3 : Sp(3)^{(14)} \wedge S^3 \cup Sp(3)^{(11)} \wedge Sp(3)^{(7)} \cup Sp(3)^{(7)} \wedge Sp(3)^{(11)} \cup S^3 \wedge Sp(3)^{(14)} \to \wedge^4 Sp(3)^{(18)}$ is given as a composition

$$(Sp(3) \wedge Sp(3))^{(18)} \overset{\alpha}{\to} (S^3 \cup_\omega e^7) \wedge S^{10} \cup S^3 \wedge (S^{10} \cup_{\nu_{10}} e^{14}) \overset{1 \wedge \beta}{\to} \wedge^4 (S^3 \cup_\omega e^7).$$

The map $\alpha \circ \overline{\Delta} : Sp(3)^{(18)} \to (S^3 \cup_\omega e^7) \wedge S^{10} \cup S^3 \wedge (S^{10} \cup_{\nu_{10}} e^{14})$ is given as

$$\alpha \circ \overline{\Delta} : Sp(3)^{(18)} \to S^{14} \cup_{\omega_{14}} e^{18} \to (S^3 \cup_\omega e^7) \wedge S^{10} \cup S^3 \wedge (S^{10} \cup_{\nu_{10}} e^{14}).$$

Collapsing the subspace $S^3 \wedge (S^{10} \cup_{\nu_{10}} e^{14})$ of $(S^3 \cup_\omega e^7) \wedge S^{10} \cup S^3 \wedge (S^{10} \cup_{\nu_{10}} e^{14})$, we obtain a map

$$q' \circ \alpha \circ \overline{\Delta} : Sp(3)^{(18)} \to S^7 \wedge S^{10},$$

where $q' : (S^3 \cup_\omega e^7) \wedge S^{10} \cup S^3 \wedge (S^{10} \cup_{\nu_{10}} e^{14}) \to S^3 \wedge S^{10}$ is the collapsing map. For dimensional reasons, $q' \circ \alpha \circ \overline{\Delta}$ is as follows:

$$q' \circ \alpha \circ \overline{\Delta} : Sp(3)^{(18)} \to Sp(3)^{(18)}/Sp(3)^{(14)} = S^{18} \overset{\gamma}{\to} S^7 \wedge S^{10}.$$

If γ were non-trivial, then γ would be $\eta_{17} : S^{18} \to S^{17}$, and hence we should have $x_7 y_{10} = \varepsilon \cdot y_{18} \neq 0$. However, from the ring structure of $h^*(Sp(3))$ given in Theorem 3.9, we know $x_7 y_{10} = x_3 x_7^2 = 0$, and hence we obtain $\gamma = 0$. Then the image of

$\alpha\circ\overline{\Delta}$ is in the subspace $S^3\wedge(S^{10}\cup_{\nu_{10}} e^{14})$ of $(S^3\cup_\omega e^7)\wedge S^{10}\cup S^3\wedge(S^{10}\cup_{\nu_{10}} e^{14})$, since they are 12-connected. Hence $\overline{\Delta}_4=(1\wedge\overline{\Delta}_3)\circ\overline{\Delta}$ is given as

$$\overline{\Delta}_4 : Sp(3)^{(18)} \overset{\alpha\circ\overline{\Delta}}{\to} S^3\wedge(S^{10}\cup_{\nu_{10}} e^{14}) \overset{1\wedge\beta}{\to} S^3\wedge(\wedge^3(S^3\cup_\omega e^7))^{(15)} \subset \wedge^4 Sp(3)^{(18)},$$

where $(\wedge^3(S^3\cup_\omega e^7))^{(15)}$ is

$$(S^3\cup_\omega e^7)\wedge S^3\wedge S^3 \cup S^3\wedge(S^3\cup_\omega e^7)\wedge S^3 \cup S^3\wedge S^3\wedge(S^3\cup_\omega e^7).$$

Collapsing the subspace $\wedge^3 S^3$ of $(\wedge^3(S^3\cup_\omega e^7))^{(15)}$, we obtain a map

$$q''\circ\beta : S^{10}\cup_{\nu_{10}} e^{14} \to S^7\wedge S^3\wedge S^3 \vee S^3\wedge S^7\wedge S^3 \vee S^3\wedge S^3\wedge S^7,$$

where

$$q'' : (\wedge^3(S^3\cup_\omega e^7))^{(15)} \to S^7\wedge S^3\wedge S^3 \vee S^3\wedge S^7\wedge S^3 \vee S^3\wedge S^3\wedge S^7$$

is the collapsing map. For dimensional reasons, $q''\circ\beta$ is given as

$$q''\circ\beta : S^{10}\cup_{\nu_{10}} e^{14} \to S^{14} \overset{\gamma'}{\to} S^7\wedge S^3\wedge S^3 \vee S^3\wedge S^7\wedge S^3 \vee S^3\wedge S^3\wedge S^7.$$

If γ' were non-trivial, then its projection to S^{13} would be $\eta_{13} : S^{14}\to S^{13}$, and hence we should have $x_3^2x_7 = \varepsilon\cdot y_{14} \neq 0$. However, from the ring structure of $h^*(Sp(3))$ given in Theorem 3.9, we know $x_3^2x_7=\varepsilon\cdot x_7^2=0$, and hence we obtain $\gamma'=0$. Hence the image of β lies in the subspace $\wedge^3S^3$ of $\wedge^3Sp(3)^{(18)}$.

On the other hand, for dimensional reasons, $\alpha\circ\overline{\Delta}$ is given as

$$\alpha\circ\overline{\Delta} : Sp(3)^{(18)} \to S^{14}\cup_{\omega_{14}} e^{18} \overset{\alpha'}{\to} S^3\wedge(S^{10}\cup_{\nu_{10}} e^{14}),$$

where the restriction $\alpha'|_{S^{14}}$ is given as

$$\alpha'|_{S^{14}} : S^{14} \overset{\gamma''}{\to} S^{13} \hookrightarrow S^3\wedge(S^{10}\cup_{\nu_{10}} e^{14}).$$

If it were non-trivial, then γ'' would be $\eta_{13} : S^{14}\to S^{13}$, and hence we should have $x_3y_{10}=\varepsilon\cdot y_{14}\neq 0$. However, from the ring structure of $h^*(Sp(3))$ given in Theorem 3.9, we know $x_3y_{10}=x_3^2x_7=\varepsilon\cdot x_7^2=0$, and hence $\gamma''=0$. Hence $\alpha\circ\overline{\Delta}$ is given as

$$\alpha\circ\overline{\Delta} : Sp(3)^{(18)} \overset{q}{\to} S^{18} \overset{\alpha''}{\to} S^3\wedge(S^{10}\cup_{\nu_{10}} e^{14}),$$

and hence $\overline{\Delta}_4$ is given as

$$\overline{\Delta}_4 : Sp(3)^{(18)} \overset{q}{\to} S^{18} \overset{\alpha''}{\to} S^3\wedge(S^{10}\cup_{\nu_{10}} e^{14}) \overset{1\wedge\beta}{\to} S^3\wedge(\wedge^3S^3) \overset{i}{\hookrightarrow} \wedge^4Sp(3)^{(18)}.$$

Now, we are ready to determine $\overline{\Delta}_4$: By Theorem 3.9, we know $x_3^2x_{11}=\varepsilon\cdot z_{18}$ and $x_3^2=\varepsilon\cdot x_7$, hence $\alpha'' : S^{18}\to S^{13}\cup_{\nu_{13}} e^{17}$ is a co-extension of $\eta_{16} : S^{17}\to S^{16}$ on $S^{13}\cup_{\nu_{13}} e^{17}$ and $1\wedge\beta : S^{13}\cup_{\nu_{13}} e^{17}\to S^{12}$ is an extension of $\eta_{12} : S^{13}\to S^{12}$. Thus the composition $(1\wedge\beta)\circ\alpha''$ is an element of the Toda bracket $\{\eta_{12},\nu_{13},\eta_{16}\}$ which contains a single element ν_{12}^2 by Lemma 5.12 of [22], and hence $\overline{\Delta}_4=i\circ\nu_{12}^2\circ q$. □

Corollary 4.4. *$w$cat$(Sp(3)^{(18)}) \geq 4$ and $w$cat$(Sp(3)) \geq 5$.*

This yields the following result.

Theorem 4.5.

Skeleta	$Sp(3)^{(3)}$	$Sp(3)^{(7)}$	$Sp(3)^{(10)}$	$Sp(3)^{(11)}$	$Sp(3)^{(14)}$	$Sp(3)^{(18)}$	$Sp(3)$
$w$cat	1	2	3	3	3	4	5
cat	1	2	3	3	3	4	5
Cat	1	2	3	3	3	4	5

This completes the proof of Theorem 1.2.

5. Proof of Theorem 1.4

We know that for $n \geq 4$,

$$Sp(n)^{(16)} = Sp(4)^{(15)} = Sp(3)^{(14)} \cup e^{15},$$

$$Sp(n)^{(19)} = \begin{cases} Sp(4)^{(15)} \cup (e^{18} \vee e^{18}) & n = 4, \\ Sp(4)^{(15)} \cup (e^{18} \vee e^{18}) \cup e^{19} & n \geq 5, \end{cases}$$

$$Sp(n)^{(21)} = Sp(n)^{(19)} \cup e^{21}$$

and that $w\mathrm{cat}(Sp(3)^{(14)}) = \mathrm{cat}(Sp(3)^{(14)}) = \mathrm{Cat}(Sp(3)^{(14)}) = 3$. Firstly, we show the following.

Proposition 5.1. $w\mathrm{cat}(Sp(4)^{(15)}) = 3$.

Proof. Since the pair $(Sp(4)^{(15)}, Sp(3)^{(11)})$ is 13-connected, $w\mathrm{cat}(Sp(3)^{(11)}) = 3$ implies that $\overline{\Delta}_3 : Sp(4)^{(15)} \to \wedge^3 Sp(4)^{(15)}$ is non-trivial, and hence we have $w\mathrm{cat}(Sp(4)^{(15)}) \geq 3$. Thus we are left to show $w\mathrm{cat}(Sp(4)^{(15)}) \leq 3$: For dimensional reasons, $\overline{\Delta}_4 = (\overline{\Delta}\wedge\overline{\Delta})\circ\overline{\Delta} : Sp(4)^{(15)} \to \wedge^4 Sp(4)^{(15)}$ is given as

$$\overline{\Delta}_4 : Sp(4)^{(15)} \overset{\alpha_0}{\to} Sp(4)^{(11)}\wedge Sp(4)^{(11)} \overset{\overline{\Delta}\wedge\overline{\Delta}}{\to} \wedge^4 Sp(4)^{(11)} \hookrightarrow \wedge^4 Sp(4)^{(15)},$$

for some α_0. By Fact 3.2, $\overline{\Delta} : Sp(4)^{(11)} \to \wedge^2 Sp(4)^{(11)}$ is given as

$$\overline{\Delta} : Sp(4)^{(11)} \overset{\beta_0}{\to} (S^7 \vee S^{10}) \cup e^{11} \overset{\gamma_0}{\to} \wedge^2(S^3 \cup_\omega e^7) \hookrightarrow \wedge^2 Sp(4)^{(11)},$$

for some β_0 and γ_0. Then for dimensional reasons, $(\beta_0\wedge\beta_0)\circ\alpha_0 : Sp(4)^{(15)} \to ((S^7 \vee S^{10}) \cup e^{11})\wedge((S^7 \vee S^{10}) \cup e^{11})$ and $(\gamma_0\wedge\gamma_0)|_{S^7\wedge S^7} : S^7\wedge S^7 \to \wedge^4(S^3 \cup_\omega e^7)$ are respectively equal to the compositions

$$(\beta_0\wedge\beta_0)\circ\alpha_0 : Sp(4)^{(15)} \overset{\alpha_0'}{\to} S^7\wedge S^7 \hookrightarrow ((S^7 \vee S^{10}) \cup e^{11})\wedge((S^7 \vee S^{10}) \cup e^{11}),$$

$$(\gamma_0\wedge\gamma_0)|_{S^7\wedge S^7} : S^7\wedge S^7 \overset{\gamma_0'}{\to} \wedge^4 S^3 \hookrightarrow \wedge^4(S^3 \cup_\omega e^7),$$

for some α_0' and γ_0'. Hence $\overline{\Delta}_4 : Sp(4)^{(15)} \to \wedge^4 Sp(4)^{(15)}$ is given as

$$\overline{\Delta}_4 : Sp(4)^{(15)} \overset{\alpha_0'}{\to} S^7\wedge S^7 \overset{\gamma_0'}{\to} \wedge^4 S^3 \hookrightarrow \wedge^4 Sp(4)^{(15)},$$

where $Sp(4)^{(15)} = Sp(3)^{(14)} \cup e^{15}$. By Theorem 3.9, $x_7^2 = 0$ in $h^*(Sp(3))$, and hence α_0' annihilates $Sp(3)^{(14)}$. Thus $\overline{\Delta}_4 : Sp(4)^{(15)} \to \wedge^4 Sp(4)^{(15)}$ is given as

$$\overline{\Delta}_4 : Sp(4)^{(15)} \xrightarrow{q''} S^{15} \xrightarrow{\beta_0'} S^{14} \xrightarrow{\gamma_0'} S^{12} \overset{i''}{\hookrightarrow} \wedge^4 Sp(4)^{(15)}$$

for some β_0', where $q'' : Sp(4)^{(15)} \to Sp(4)^{(15)}/Sp(4)^{(14)} = S^{15}$ is the projection and $i'' : S^{12} = S^3 \wedge S^3 \wedge S^3 \wedge S^3 \hookrightarrow \wedge^4 Sp(4)^{(15)}$ is the inclusion. Hence the non-triviality of $\overline{\Delta}_4$ implies the non-triviality of β_0' and γ_0'. Therefore $\overline{\Delta}_4$ should be $i'' \circ \eta_{12}^3 \circ q''$, if it were non-trivial. However, we also know from (5.5) of [22] that η_{12}^3 is $12\nu_{12} = 6\omega_{12}$ and that $i'' \circ \omega_{12}$ is trivial by Fact 3.1. Therefore, $\overline{\Delta}_4 : Sp(4)^{(15)} \to \wedge^4 Sp(4)^{(15)}$ is trivial, and hence $w\mathrm{cat}\, Sp(4)^{(15)} \leq 3$. This implies that $w\mathrm{cat}\, Sp(4)^{(15)} = 3$. □

Secondly, we show the following.

Proposition 5.2. *$w\mathrm{cat}(Sp(n)^{(19)}) \leq 4$ for $n \geq 4$.*

Proof. Let $n \geq 4$. Since $\overline{\Delta}_5 = ((1_{Sp(n)}) \wedge \overline{\Delta}_4) \circ \overline{\Delta} : Sp(n)^{(19)} \to \wedge^5 Sp(n)^{(19)}$, it is given as

$$\overline{\Delta}_5 : Sp(n)^{(19)} \xrightarrow{\overline{\Delta}} Sp(n)^{(16)} \wedge Sp(n)^{(16)} = Sp(4)^{(15)} \wedge Sp(4)^{(15)}$$
$$\xrightarrow{(1_{Sp(4)^{(15)}}) \wedge \overline{\Delta}_4} \wedge^5 Sp(4)^{(15)} \hookrightarrow \wedge^5 Sp(n)^{(19)},$$

which is trivial, since $\overline{\Delta}_4 : Sp(4)^{(15)} \to \wedge^4 Sp(4)^{(15)}$ is trivial by Proposition 5.1. Thus $w\mathrm{cat}(Sp(n)^{(19)}) \leq 4$ when $n \geq 4$. □

Let $p_j : Sp(n) \to X_{n,j} = Sp(n)/Sp(n-j)$ be the projection for $j \geq 1$. Then we have the following.

Proposition 5.3. *Let $q''' : Sp(n) \to Sp(n)/Sp(n)^{((2n+1)n-3)} = S^{(2n+1)n}$ be the collapsing map and $i''' : S^{(2n+1)n-6} \hookrightarrow (\wedge^5 Sp(n)) \wedge X_{n,n-3} \wedge \cdots \wedge X_{n,1}$ the inclusion. Then*

$$q'''^* \circ i'''_* : \pi_{(2n+1)n}(S^{(2n+1)n-6}) \to [Sp(n), (\wedge^5 Sp(n)) \wedge X_{n,n-3} \wedge \cdots \wedge X_{n,1}]$$

is injective.

Proof. Firstly, we have the following exact sequence

$$\pi_{(2n+1)n}(S^{(2n+1)n-3}) \xrightarrow{\psi'''} \pi_{(2n+1)n}(S^{(2n+1)n-6})$$
$$\xrightarrow{i'''_*} \pi_{(2n+1)n}((\wedge^5 Sp(n)) \wedge X_{n,n-3} \wedge \cdots \wedge X_{n,1}) \to \pi_{(2n+1)n}(\vee_5 S^{(2n+1)n-2}),$$

where $\pi_{(2n+1)n}(S^{(2n+1)n-6}) \cong \mathbb{Z}/2\mathbb{Z}\nu^2_{(2n+1)n-6}$ and ψ''' is induced from $\omega_{(2n+1)n-6} = 2\nu_{(2n+1)n-6}$. Thus ψ''' is trivial, and hence i'''_* is injective.

Secondly, since $(\wedge^5 Sp(n)) \wedge X_{n,n-3} \wedge \cdots \wedge X_{n,1})$ is $(n(2n+1)-11)$-connected and $Sp(n)/Sp(n)^{(n(2n+1)-11)} \simeq S^{(2n+1)n-10} \vee (S^{(2n+1)n-7} \cup_{\omega_{(2n+1)n-7}} e^{(2n+1)n-3})$

$\vee\ S^{(2n+1)n}$ by Proposition 3.4, we have

$$[Sp(n), (\wedge^5 Sp(n))\wedge X_{n,n-3}\wedge\cdots\wedge X_{n,1}]$$
$$= [S^{(2n+1)n-7}\cup_{\omega_{(2n+1)n-7}} e^{(2n+1)n-3}, (\wedge^5 Sp(n))\wedge X_{n,n-3}\wedge\cdots\wedge X_{n,1}]$$
$$\oplus\pi_{(2n+1)n-10}((\wedge^5 Sp(n))\wedge X_{n,n-3}\wedge\cdots\wedge X_{n,1})$$
$$\oplus\pi_{(2n+1)n}((\wedge^5 Sp(n))\wedge X_{n,n-3}\wedge\cdots\wedge X_{n,1}),$$

and hence q'''^* is injective. Thus $q'''^*\circ i'''_*$ is injective. □

Then the following lemma implies that $((1_{\wedge^5 Sp(n)})\wedge p_{n-3}\wedge\cdots\wedge p_1)\circ\overline{\Delta}_{n+2}$ is non-trivial by Proposition 5.3, and hence we obtain Theorem 1.4.

Lemma 5.4. $((1_{\wedge^5 Sp(n)})\wedge p_{n-3}\wedge p_{n-4}\wedge\cdots\wedge p_1)\circ\overline{\Delta}_{n+2} = i'''\circ\nu^2_{(2n+1)n-6}\circ q'''$.

Proof. We have

$$((1_{\wedge^5 Sp(n)})\wedge p_{n-3}\wedge\cdots\wedge p_1)\circ\overline{\Delta}_{n+2} = (\overline{\Delta}_5\wedge p_{n-3}\wedge\cdots\wedge p_1)\circ\overline{\Delta}_{n-2}$$
$$= (\overline{\Delta}_5\wedge(1_{\wedge^{n-3}Sp(n)}))\circ((1_{Sp(n)})\wedge p_{n-3}\wedge\cdots\wedge p_1)\circ\overline{\Delta}_{n-2}.$$

For dimensional reasons, the image of $((1_{Sp(n)})\wedge p_{n-3}\wedge\cdots\wedge p_1)\circ\overline{\Delta}_{n-2}$ lies in

$$Sp(n)^{(21)}\wedge S^{15}\wedge\cdots\wedge S^{4n-1}\cup Sp(n)^{(19)}\wedge X_{n,n-3}\wedge\cdots\wedge X_{n,1}.$$

From Proposition 5.2, $\overline{\Delta}_5$ annihilates $Sp(n)^{(19)}$, and hence $\overline{\Delta}_5$ is given as

$$\overline{\Delta}_5 : Sp(n)^{(21)} \to S^{21} \xrightarrow{\delta} \wedge^5 Sp(n)^{(21)}$$

for some $\delta\in\pi_{21}(\wedge^5 Sp(3))$. Using Lemma 4.3, we obtain that the following diagram except for the dotted arrow is commutative up to homotopy:

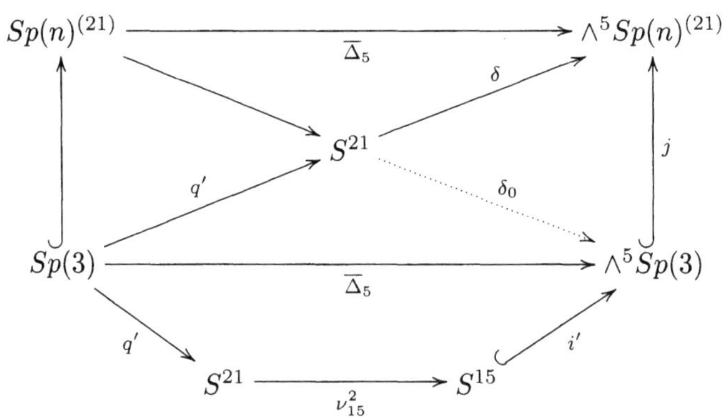

Since the pair $(\wedge^5 Sp(n), \wedge^5 Sp(3))$ is 26-connected for $n\geq 4$, we can compress δ into $\wedge^5 Sp(3)$ as $\delta\sim j\circ\delta_0$. Thus we have a homotopy relation

$$j\circ\delta_0\circ q' \sim \delta\circ q' \sim j\circ\overline{\Delta}_5 \sim j\circ i'\circ\nu^2_{15}\circ q'.$$

Now we know that $\dim Sp(3) = 21 < 26-1$, and hence we can drop j from the above homotopy relation and obtain

$$\delta_0\circ q' \sim i'\circ\nu^2_{15}\circ q'.$$

By Proposition 4.2, $q'^* : \pi_{21}(\wedge^5 Sp(3)) \to [Sp(3), \wedge^5 Sp(3)]$ is injective, and hence

$$\delta_0 \sim i' \circ \nu_{15}^2.$$

Thus $\overline{\Delta}_5$ is given as

$$\overline{\Delta}_5 : Sp(n)^{(21)} \to S^{21} \overset{\nu_{15}^2}{\to} S^{15} \hookrightarrow \wedge^5 Sp(n)^{(21)},$$

and hence $((1_{\wedge^5 Sp(n)}) \wedge p_{n-3} \wedge \cdots \wedge p_1) \circ \overline{\Delta}_{n+2}$ is given as

$$((1_{\wedge^5 Sp(n)}) \wedge p_{n-3} \wedge \cdots \wedge p_1) \circ \overline{\Delta}_{n+2} : Sp(n) \to S^{21} \wedge S^{(2n+7)(n-3)}$$
$$\overset{\nu_{(2n+1)n-6}^2}{\to} S^{15} \wedge S^{(2n+7)(n-3)} \hookrightarrow (\wedge^5 Sp(n)) \wedge X_{n,n-3} \wedge \cdots \wedge X_{n,1}.$$

This completes the proof of the lemma. □

References

[1] I. Berstein and P.J. Hilton, *Category and generalized Hopf invariants*, Illinois. J. Math. **12** (1968), 421–432.

[2] J.M. Boardman and B. Steer, *On Hopf invariants*, Comment. Math. Helv. **42** (1967), 180–221.

[3] M. Clapp and D. Puppe, *Invariants of the Lusternik-Schnirelmann type and the topology of critical sets*, Trans. Amer. Math. Soc. **298** (1986), 603–620.

[4] L. Fernández-Suárez, A. Gómez-Tato and D. Tanré, *Hopf-Ganea invariants and weak LS category*, Top. Appl. **115** (2001), 305–316.

[5] L. Fernández-Suárez, A. Gómez-Tato, J. Strom and D. Tanré, *The Lusternik-Schnirelmann category of* $Sp(3)$, http://xxx.arXiv.org/pdf/math.AT/0111263.

[6] T. Ganea, *Some problems on numerical homotopy invariants*, Symposium on Algebraic Topology, Lect. Notes in Math. **249**, Springer Verlag, Berlin (1971), 13–22.

[7] J.C. Gómez-Larrañaga and F. González-Acuña, *Lusternik-Schnirelmann category of 3-manifolds*, Topology **31** (1992), 791–800.

[8] N. Iwase, *A_∞-method in Lusternik-Schnirelmann category*, Topology **41** (2002), 695–723.

[9] N. Iwase, *L-S category of a sphere-bundle over a sphere*, Topology **42** (2003), 701–713.

[10] I.M. James, *On category, in the sense of Lusternik-Schnirelmann*, Topology **17** (1978), 331–348.

[11] I. James, *The topology of Stiefel manifolds*, Cambridge University Press, Cambridge, 1976, London Math. Soc. Lec. Notes **24**.

[12] I. James and W. Singhof, *On the category of fiber bundles, Lie groups, and Frobenius maps*, Contemp. Math. **227** (1999), 177–189.

[13] L. Lusternik and L. Schnirelmann, *Méthodes topologiques dans les Problèmes variationnels*, Hermann, Paris, 1934.

[14] M. Mimura and T. Nishimoto, *On the cellular decomposition of the exceptional Lie group* G_2, Proc. Amer. Math. Soc. **130** (2002), 2451–2459.

[15] L. Montejano, *A quick proof of Singhof's* $\mathrm{cat}(M \times S^1) = \mathrm{cat}(M) + 1$ *theorem*, Manuscripta Math. **42** (1983), 49–52.

[16] Y.B. Rudyak, *On the Ganea conjecture for manifolds*, Proc. Amer. Math. Soc. **125** (1997), 2511–2512.

[17] Y.B. Rudyak, *On category weight and its applications*, Topology **38** (1999), 37–55.

[18] P.A. Schweizer, *Secondary cohomology operations induced by the diagonal mapping*, Topology **3** (1965), 337–355.

[19] W. Singhof, *On the Lusternik-Schnirelmann category of Lie groups*, Math. Z. **145** (1975), 111–116.

[20] W. Singhof, *On the Lusternik-Schnirelmann category of Lie groups II*, Math. Z. **151** (1976), 143–148.

[21] W. Singhof, *Minimal coverings of manifolds with balls*, Manuscripta Math. **29** (1979), 385–415.

[22] H. Toda, *Composition methods in homotopy groups of spheres*, Princeton Univ. Press, Princeton, 1962, Ann. of Math. Studies, **49**.

[23] G.W. Whitehead, *Elements of homotopy theory*, Springer Verlag, Berlin, 1978, GTM series **61**.

[24] A. Zabrodsky, *Hopf spaces*, North-Holland, Amsterdam, 1976, North-Holland Math. Studies, **22**.

Norio Iwase
Faculty of Mathematics
Kyushu University
Ropponmatsu Fukuoka 810-8560, Japan
e-mail: `iwase@math.kyushu-u.ac.jp`

Mamoru Mimura
Department of Mathematics
Okayama University
Okayama 700-8530, Japan
e-mail: `mimura@math.okayama-u.ac.jp`

Progress in Mathematics, Vol. 215
Algebraic Topology: Categorical Decomposition Techniques, 213–236

The McCord Model for the Tensor Product of a Space and a Commutative Ring Spectrum

Nicholas J. Kuhn

Abstract. We begin this paper by noting that, in a 1969 paper in the Transactions, M.C. McCord introduced a construction that can be interpreted as a model for the categorical tensor product of a based space and a topological abelian group. This can be adapted to Segal's very special Γ-spaces – indeed this is roughly what Segal did – and then to a more modern situation: $K \otimes R$ where K is a based space and R is a unital, augmented, commutative, associative S-algebra.

The model comes with an easy-to-describe filtration. If one lets $K = S^n$, and then stabilizes with respect to n, one gets a filtered model for the Topological André–Quillen Homology of R. When $R = \Sigma^\infty(\Omega^\infty X)_+$, one arrives at a filtered model for the connective cover of a spectrum X, constructed from its 0th space.

Another example comes by letting K be a finite complex, and R the S-dual of a finite complex Z. Dualizing again, one arrives at G. Arone's model for the Goodwillie tower of the functor sending Z to $\Sigma^\infty \operatorname{Map}_{\mathcal{T}}(K, Z)$.

Applying cohomology with field coefficients, one gets various spectral sequences for deloopings with known E_1-terms. A few nontrivial examples are given.

In an appendix, we describe the construction for unital, commutative, associative S-algebras not necessarily augmented.

1. Introduction

The point of this paper is to tell a story that begins with a 1969 paper of M.C. McCord [McC], and ends with various disparate objects of current interest, e.g., Goodwillie towers, Topological Hochschild Homology, and Topological André–Quillen Homology. Line by line, I think most of this story is known. However, taken as a whole, I think the older work sheds some light on the newer. Moreover, in this era of tremendous activity in homotopical algebra of various sorts, it seems important to remind ourselves that the genesis of many of the most useful ideas lies way back in the literature.

2000 *Mathematics Subject Classification.* Primary 55P43; Secondary 18G55.
This research was partially supported by a grant from the National Science Foundation.

Conceptually, we feature the following categorical notion. Let $\mathcal{T}$ be the category of based topological spaces. If $\mathcal{C}$ is a category enriched over $\mathcal{T}$, there is the notion of the tensor product of a space $K \in \mathcal{T}$ with an object $X \in \mathcal{C}$: this is an object $K \otimes X \in \mathcal{C}$ satisfying

$$\mathrm{Map}_{\mathcal{C}}(K \otimes X, Y) = \mathrm{Map}_{\mathcal{T}}(K, \mathrm{Map}_{\mathcal{C}}(X, Y))$$

for all $X, Y \in \mathcal{C}$.

We give an overview of the paper.

We consider various topological categories of structured objects. Let $\mathcal{A}b$ be the category of abelian topological monoids. Let $\mathcal{T}^{\Gamma}$ be G. Segal's category of Γ-spaces [Se]: functors $X : \Gamma \to \mathcal{T}$, where Γ is the category of finite based sets. Let $\mathcal{A}lg$ be the category of commutative, associative, augmented S-algebras, where S is the sphere spectrum.

These categories are closely related. First of all, an abelian topological monoid A defines in a natural way $A^{\times} \in \mathcal{T}^{\Gamma}$. In Segal's terminology, $A^{\times}$ is an example of a 'special' object, and the abelian topological groups define 'very special objects'. A very special Γ-space is roughly the same thing as an infinite loopspace, and, in this introduction, we will tempt fate and often identify these two notions. Now we note that if X is either an abelian topological monoid or an infinite loopspace, then $\Sigma^{\infty} X_{+}$ is in the category $\mathcal{A}lg$.

With $\mathcal{C}$ equal to any of these three categories, McCord's construction yields a functor[1]

$$\mathrm{SP}^{\infty} : \mathcal{T} \times \mathcal{C} \to \mathcal{C}.$$

His construction generalizes the infinite symmetric product construction studied by Dold and Thom in the 1950's: with $\mathbb{N}$ denoting the natural numbers, $\mathrm{SP}^{\infty}(K, \mathbb{N}) = \mathrm{SP}^{\infty}(K)$.

If $\mathcal{C}$ is either $\mathcal{A}b$ or $\mathcal{A}lg$, there is an isomorphism

$$\mathrm{SP}^{\infty}(K, X) = K \otimes X. \tag{1.1}$$

See Proposition 2.2(1), and Proposition 4.8. The latter proposition (or at least its proof) seems to be new.

Equation (1.1) is almost true if $\mathcal{C} = \mathcal{T}^{\Gamma}$. S. Schwede [S], following Bousfield and Friedlander [BF], defines a model category structure on $\mathcal{T}^{\Gamma}$, having the very special Γ-spaces as the fibrant objects, and such that $ho(\mathcal{T}^{\Gamma})$ is equivalent to the homotopy category of connective spectra. With this structure, $\mathrm{SP}^{\infty}(K, \ )$ preserves fibrant objects and the natural map $K \otimes X \to \mathrm{SP}^{\infty}(K, X)$ is a weak equivalence.

McCord's interest stemmed from the following fundamental property:

(1.2) for $X \in \mathcal{A}b$, $\mathrm{SP}^{\infty}(S^1, X)$ is a classifying space for X.

Suitably interpreted, the same result is true if X is a special object of $\mathcal{T}^{\Gamma}$. For $R \in \mathcal{A}lg$, $\mathrm{SP}^{\infty}(S^1, R)$ is also of interest: $\mathrm{SP}^{\infty}(S^1, R) = S^1 \otimes R$ equals $THH(R; S)$,

[1] McCord uses the notation $B(A, K)$ where we use $\mathrm{SP}^{\infty}(K, A)$; I have borrowed my notation from [FF, Chapter 9].

the Topological Hochschild Homology of R with coefficients in the bimodule S. See Proposition 7.1; this is deduced from a similar result due to J. McClure, R. Schwänzl, and R. Vogt [MSV].

The construction has two other basic properties, both discussed by McCord when $\mathcal{C} = \mathcal{A}b$.

Firstly, there are natural isomorphisms

$$\mathrm{SP}^\infty(K \wedge L, X) = \mathrm{SP}^\infty(K, \mathrm{SP}^\infty(L, X)). \tag{1.3}$$

From this and (1.2) one quickly deduces that $\mathrm{SP}^\infty(S^n, X)$ is an n-fold delooping of X for $X \in \mathcal{A}b$, or for very special $X \in \mathcal{T}^\Gamma$. For $R \in \mathcal{A}lg$, $\mathrm{SP}^\infty(S^n, R)$ can be interpreted as 'higher' Topological Hochschild Homology of R.

Secondly, $\mathrm{SP}^\infty(K, X)$ comes with a nice increasing filtration. When $\mathcal{C} = \mathcal{A}b$, one easily sees that there is an equivalence

$$F_d\mathrm{SP}^\infty(K, X)/F_{d-1}\mathrm{SP}^\infty(K, X) \simeq K^{(d)} \wedge_{\Sigma_d} X^{\wedge d}, \tag{1.4}$$

and little variants of this hold when $\mathcal{C} = \mathcal{T}^\Gamma$ or $\mathcal{A}lg$. Here $K^{(d)}$ denotes the Σ_d-space obtained from the d-fold smash product $K^{\wedge d}$ by collapsing the fat diagonal to a point.[2]

This much of the story will be fleshed out in Sections 2, 3, and 4, with the statement about THH appearing in §7.

In [B], it is observed that $\mathcal{A}lg$ is Quillen equivalent to the category $\mathcal{A}lg'$, of commutative, nonunital S-algebras. In §5, we discuss the corresponding filtered SP^∞ construction, which again agrees with the tensor product. This reduced construction is 'smaller' than what is done in $\mathcal{A}lg$, and the category $\mathcal{E}$ of finite sets and epimorphisms replaces Γ.

The quotients of the filtration, and the use of $\mathcal{E}$, may look vaguely familiar to readers of [Ar], and this is what we explain in §6. There is a contravariant functor from $\mathcal{T}$ to $\mathcal{A}lg'$ sending a based space Z to the ring spectrum $R = D(Z)$, where D denotes the S-dual, and the multiplication on R is induced by the diagonal on Z. We note that there is a natural map in $\mathcal{A}lg'$

$$K \otimes D(Z) \to D(\mathrm{Map}_\mathcal{T}(K, Z))$$

and reinterprete the main convergence theorem of [Ar] as saying that the adjoint of this,

$$\Sigma^\infty \mathrm{Map}_\mathcal{T}(K, Z) \to D(K \otimes D(Z)),$$

is an equivalence if both Z and K are finite-dimensional complexes, and the dimension of K is less than the connectivity of Z. See Theorem 6.6. Since $K \otimes D(Z)$ has a nice increasing filtration, the S-dual is a tower of fibrations. This tower is visibly equivalent to the tower found by Arone, and is thus the Goodwillie tower associated to the functor sending a space X to the spectrum $\Sigma^\infty \mathrm{Map}_\mathcal{T}(K, Z)$.

---

[2]This notation, which the author likes, comes from [Ar].

In §7, we discuss the following construction. Given $R \in \mathcal{A}lg$, nice properties of $\mathrm{SP}^\infty(K,R)$ as a functor of K allow us to define a filtered spectrum $TAQ(R)$ by

$$TAQ(R) = \operatorname*{hocolim}_{n\to\infty} \Omega^n \mathrm{SP}^\infty(S^n, R).$$

This is one of various equivalent definitions of the Topological André–Quillen spectrum of R. When $R = \Sigma^\infty X_+$, with X an infinite loopspace, $TAQ(R)$ is the connective delooping of X. When $R = D(Z_+)$, the filtered spectrum $TAQ(R)$ is related to constructions studied by the author in [K3].

Using (1.4), one can identify the quotients of the filtration of $TAQ(R)$:

$$F_d TAQ(R)/F_{d-1}TAQ(R) \simeq \Sigma K_d \wedge_{h\Sigma_d} (R/S)^{\wedge d}, \tag{1.5}$$

where K_d is the dth partition complex which arose in the work of Arone and Mahowald on the Goodwillie tower of the identity [AM].

In §8 we note that applying ordinary cohomology with field coefficients $\mathbb{F}$ to the filtered spectra $\mathrm{SP}^\infty(S^n,R)$ and $TAQ(R)$ yield spectral sequences converging to $H^*(S^n \otimes R;\mathbb{F})$ and $H^*(TAQ(R);\mathbb{F})$, and having E_1 terms isomorphic to known functors of $H^*(R;\mathbb{F})$: see Theorem 8.1. These spectral sequences appear to be unexplored, even in the case when $R = \Sigma^\infty X_+$, with X an infinite loop space, so that, e.g., the $TAQ(R)$ spectral sequence is calculating the cohomology of a connective spectrum from knowledge of the cohomology of its 0th-space. As examples, we use results from [K3] and related papers to explain how the spectral sequence works in the cases $R = \Sigma^\infty(\mathbb{Z}/2_+)$, $\Sigma^\infty(S^1_+)$, and, most interestingly, $D(S^1_+)$.

In the appendix, we note how our models need to be slightly tweaked when one considers commutative unital S-algebras not necessarily augmented.

Very influential to me in my understanding of the older work surveyed in this paper was Chapter of 9 of the unpublished book *The actions of the classical small categories of topology* by Bill and Ed Floyd [FF]. Writing this book was Ed's project in the late 1980's, after returning to ordinary academic life after finishing a term as provost of the University of Virginia. I have also benefited from conversations with Mike Mandell, Bill Dwyer, and Greg Arone.

Versions of this work were presented at talks at the Johns Hopkins topology conference in the spring of 2000, and at the Union College topology and category theory conference of fall 2001.

2. McCord's construction

Let K be a based space with basepoint $*$, and let A be an abelian topological monoid.

Imagine using (1.1) to guide the construction $\mathrm{SP}^\infty(K,A)$. As a first experiment, suppose $A = \mathbb{N}$. In this case (1.1) tells us that, for all $A \in \mathcal{A}b$, we should have

$$\mathrm{Map}_{\mathcal{A}b}(\mathrm{SP}^\infty(K,\mathbb{N}),A) = \mathrm{Map}_{\mathcal{T}}(K, \mathrm{Map}_{\mathcal{A}b}(\mathbb{N},A)).$$

But, since $\mathbb{N}$ is the free abelian (topological) monoid on one generator, we can identify $\mathrm{Map}_{\mathcal{A}b}(\mathbb{N}, A)$ with A. Thus we are asking that $\mathrm{SP}^\infty(K, \mathbb{N})$ satisfy

$$\mathrm{Map}_{\mathcal{A}b}(\mathrm{SP}^\infty(K, \mathbb{N}), A) = \mathrm{Map}_{\mathcal{T}}(K, A).$$

In other words, $\mathrm{SP}^\infty(K, \mathbb{N})$ should be the free topological abelian monoid generated by K, a.k.a. $\mathrm{SP}^\infty(K)$.

Note that elements in $\mathrm{SP}^\infty(K)$ are words of the form $k_1^{n_1} \cdots k_d^{n_d}$, with $k_i \in K$ and $n_i \in \mathbb{N}$. This suggests a definition.

Definition 2.1. Let $\mathrm{SP}^\infty(K, A)$ be the abelian topological monoid with generators k^a with $k \in K$ and $a \in A$ subject to the relations:

(i) $*^a = *$ for all $a \in A$,
(ii) $k^0 = *$ for all $k \in K$,
(iii) $k^{a_1} k^{a_2} = k^{a_1 + a_2}$ for all $a_1, a_2 \in A$.

Viewing k^a as an element in $K \times A$, $\mathrm{SP}^\infty(K, A)$ is topologized as the evident quotient space of $\coprod_{d=0}^\infty (K \times A)^d$.

Note that the abelian topological monoid satisfying only the relations of type (i) and (ii) is $\mathrm{SP}^\infty(K \wedge A)$. If B is another abelian topological monoid, a monoid map $\mathrm{SP}^\infty(K \wedge A) \to B$ corresponds to a map of topologial spaces $\phi : K \wedge A \to B$ which itself corresponds to a map $K \to \mathrm{Map}_{\mathcal{T}}(A, B)$. This latter map takes values in $\mathrm{Map}_{\mathcal{A}b}(A, B)$ exactly when $\phi(k, a_1 + a_2) = \phi(k, a_1)\phi(k, a_2)$. Thus we see that the quotient of $\mathrm{SP}^\infty(K \wedge A)$ having type (iii) relations imposed satisfies the universal property of $K \otimes A$. (Compare with [FF, p.164].) We have checked the first part of the next proposition.

Proposition 2.2. *There are the following natural identifications in $\mathcal{A}b$.*

(1) $SP^\infty(K, A) = K \otimes A$.
(2) $SP^\infty(S^0, A) = A$.
(3) $SP^\infty(K \wedge L, A) = SP^\infty(K, SP^\infty(L, A))$.
(4) $SP^\infty(K \vee L, A) = SP^\infty(K, A) \times SP^\infty(L, A)$.

The last three parts of this proposition follow formally from statement (1). For example, (3) follows by manipulating adjunctions:

$$\begin{aligned}
\mathrm{Map}_{\mathcal{A}b}(\mathrm{SP}^\infty(K \wedge L, A), B) &= \mathrm{Map}_{\mathcal{T}}(K \wedge L, \mathrm{Map}_{\mathcal{A}b}(A, B)) \\
&= \mathrm{Map}_{\mathcal{T}}(K, \mathrm{Map}_{\mathcal{T}}(L, \mathrm{Map}_{\mathcal{A}b}(A, B))) \\
&= \mathrm{Map}_{\mathcal{T}}(K, \mathrm{Map}_{\mathcal{A}b}(\mathrm{SP}^\infty(L, A), B)) \\
&= \mathrm{Map}_{\mathcal{A}b}(\mathrm{SP}^\infty(K, \mathrm{SP}^\infty(L, A)), B).
\end{aligned}$$

For statement (4), one needs to also note that the coproduct in $\mathcal{A}b$ is the product. (In the next section, we will see that there are also reasonable direct proofs of (3) and (4).)

In the introduction to [McC], McCord comments that $\mathrm{SP}^\infty(\ , A)$ "has a tendancy to convert cofibrations ... to quasifibrations", and proves this in various cases [McC, Thm.8.8]. Note that statement (4) of the last proposition nicely illustrates his statement.

In particular, when applied to the cofibration $S^0 \hookrightarrow I \to S^1$, his observations suggest that $\mathrm{SP}^\infty(I, A) \to \mathrm{SP}^\infty(S^1, A)$ is a quasifibration with homotopy fiber $A = \mathrm{SP}^\infty(S^0, A)$. One has

Proposition 2.3. *If A has a nondegenerate base point, then $SP^\infty(S^1, A)$ is a classifying space for A.*

Combined with statement (3) of the previous proposition, this implies

Corollary 2.4. *In this case, $SP^\infty(S^n, A)$ is an n-fold classifying space of A.*

With such a mild point set hypothesis, this proposition and corollary occur as [FF, Cor. 9.16].

We end this section by noting that $\mathrm{SP}^\infty(K, A)$ is filtered by letting the dth filtration, which we denote $F_d\mathrm{SP}^\infty(K, A)$, be the evident quotient of $\coprod_{r=0}^{d}(K \times A)^r$. Let $i : \Delta(K, d) \hookrightarrow K^{\wedge d}$ denote the inclusion of the fat diagonal into the d-fold smash product. Under reasonable conditions, e.g., if K is a based C.W. complex, the inclusion i will be a Σ_d-equivariant cofibration, and we let $K^{(d)} = K^{\wedge d}/\Delta(K, d)$. Assuming the basepoint in A is also nondegenerate, the inclusion $F_{d-1}\mathrm{SP}^\infty(K, A) \hookrightarrow F_d\mathrm{SP}^\infty(K, A)$ will be a cofibration, and there is a homeomorphism

$$F_d\mathrm{SP}^\infty(K, A)/F_{d-1}\mathrm{SP}^\infty(K, A) \simeq K^{(d)} \wedge_{\Sigma_d} A^{\wedge d}. \tag{2.1}$$

Statements similar to this appear in [McC, §6].

Remark 2.5. One should note that specializing the filtration on $\mathrm{SP}^\infty(K, A)$ to $\mathrm{SP}^\infty(K, \mathbb{N}) = \mathrm{SP}^\infty(K)$ does *not* yield the standard filtration on $\mathrm{SP}^\infty(K)$. For example, in this paper an element $k^2 \in \mathrm{SP}^\infty(K)$ would be in filtration 1.

3. Γ-spaces and Segal's theorem

Our first goal in this section is to rewrite our construction $\mathrm{SP}^\infty(K, A)$ in a way allowing for generalization.

We begin by defining more precisely the category Γ.

Definition 3.1. Let Γ be the category with objects the based finite sets $\mathbf{0} = \emptyset_+$ and $\mathbf{n} = \{1, 2, \ldots, n\}_+$, $n \geq 1$, and with all based functions as morphisms. Note that $\mathbf{0}$ is both an initial and terminal object.

We remark that, unfortunately, it was the *opposite* of this category that was called Γ in [Se], and the literature is strewn with inconsistent notation.

A Γ-space is then defined to be a covariant functor $X : \Gamma \to \mathcal{T}$ that is 'based' in the sense that it sends $\mathbf{0}$ to the one point space. These are the objects of a category $\mathcal{T}^\Gamma$ having the natural transformations as morphisms. This is a category enriched over $\mathcal{T}$: the set of morphisms between two Γ-spaces, $\mathrm{Map}_\Gamma(X, X')$, has a natural topology. Similarly, a based contravariant functor $Y : \Gamma^{op} \to \mathcal{T}$ will be called a Γ^{op}-space, and these are objects in a topological category $\mathcal{T}^{\Gamma^{op}}$.

Example 3.2. If A is an abelian topological monoid, there is an associated Γ-space $A^\times$ defined as follows. Firstly, $A^\times(\mathbf{n}) = A^n$. Then, given $\alpha : \mathbf{n} \to \mathbf{m}$, the ith component of $\alpha_* : A^n \to A^m$ sends $(a_1, \ldots, a_n)$ to $\prod a_j$, with the product running over j such that $\alpha(j) = i$. This product is interpreted to be the unit of A if there are no such j.

Example 3.3. Since, for any $K \in \mathcal{T}$, $K^n = \mathrm{Map}_{\mathcal{T}}(\mathbf{n}, K)$, there is the evident Γ^{op}-space $K^\times$ with $K^\times(\mathbf{n}) = K^n$.

Note that the constructions in these last examples embed $\mathcal{A}b$ into $\mathcal{T}^\Gamma$, and $\mathcal{T}$ into $\mathcal{T}^{\Gamma^{op}}$, as full subcategories.

We now recall the coend construction. If X is a Γ-space and Y is a Γ^{op}-space, we let $Y \wedge_\Gamma X \in \mathcal{T}$ denote the quotient space

$$\bigvee_n Y(\mathbf{n}) \wedge X(\mathbf{n})/(\sim),$$

where $\alpha^*(y) \wedge x \sim y \wedge \alpha_*(x)$ generates the equivalence relation.

It is useful to observe that, because $X(\mathbf{0}) = * = Y(\mathbf{0})$, $Y \wedge_\Gamma X \in \mathcal{T} = Y \times_\Gamma X$, where $Y \times_\Gamma X$ is the quotient space

$$\coprod_n Y(\mathbf{n}) \times X(\mathbf{n})/(\sim),$$

where $\alpha^*(y) \times x \sim y \times \alpha_*(x)$ generates the equivalence relation.

By inspection, one observes

Lemma 3.4. $\mathit{SP}^\infty(K, A) = K^\times \wedge_\Gamma A^\times$.

This suggests a generalization of our construction.

Definition 3.5. Given $K \in \mathcal{T}$ and $X \in \mathcal{T}^\Gamma$, let $\mathrm{SP}^\infty(K, X) = K^\times \wedge_\Gamma X$.

A couple of remarks are now in order.

Firstly, a Yoneda's lemma type argument shows that

$$\mathrm{SP}^\infty(\mathbf{n}, X) = X(\mathbf{n}).$$

Thus, as a functor of K, $\mathrm{SP}^\infty(K, X)$ extends X to $\mathcal{T}$; more precisely, this is the left Kan extension [MacL, Chap.X].

Secondly, we can extend our construction to

$$\mathrm{SP}^\infty : \mathcal{T} \times \mathcal{T}^\Gamma \to \mathcal{T}^\Gamma$$

by first letting $X_n(\mathbf{m}) = X(\mathbf{nm})$, and then by defining

$$\mathrm{SP}^\infty(K, X)(\mathbf{n}) = \mathrm{SP}^\infty(K, X_n).$$

It is natural to wonder if $\mathrm{SP}^\infty(K,X)$ can then be interpreted as a tensor product in $\mathcal{T}^\Gamma$. Alas, this is not the case: the simple minded construction $K \wedge X$ defined by $(K \wedge X)(\mathbf{n}) = K \wedge X(\mathbf{n})$ is easily seen to play this role[3].

Since $\mathrm{SP}^\infty(K,X)$ is not the tensor product in $\mathcal{T}^\Gamma$, formal arguments used to prove the last three statements in Proposition 2.2 do not apply. However, we still can prove suitable versions of these.

First of all, the two remarks above (along with the observation that $\mathbf{1} = S^0$) combine to show that $\mathrm{SP}^\infty(S^0,X) = X$.

Less obvious are the other two. Statement (3) of Proposition 2.2 is unchanged in our greater generality.

Proposition 3.6. $SP^\infty(K \wedge L, X) = SP^\infty(K, SP^\infty(L,X))$.

To generalize statement (4), let $a : \Gamma \times \Gamma \to \Gamma$ be the functor sending $(\mathbf{m},\mathbf{n})$ to $\mathbf{m}+\mathbf{n}$. Pulling back by a defines $a^* : \mathcal{T}^\Gamma \to \mathcal{T}^{\Gamma\times\Gamma}$.

Proposition 3.7. $SP^\infty(K \vee L, X) = (K^\times \times L^\times) \times_{\Gamma\times\Gamma} a^*(X)$.

Proposition 2.2(4) follows from this, once one observes that

$$a^*(A^\times) = A^\times \times A^\times,$$

so that

$$\begin{aligned}\mathrm{SP}^\infty(K \vee L, A^\times) &= (K^\times \times L^\times) \times_{\Gamma\times\Gamma} (A^\times \times A^\times)\\ &= (K^\times \times_\Gamma A^\times) \times (L^\times \times_\Gamma A^\times)\\ &= \mathrm{SP}^\infty(K, A^\times) \times \mathrm{SP}^\infty(L, A^\times).\end{aligned}$$

Let $m : \Gamma \times \Gamma \to \Gamma$ be the functor sending $(\mathbf{m},\mathbf{n})$ to $\mathbf{mn}$[4]. We let $a_* : \mathcal{T}^{\Gamma^{op}\times\Gamma^{op}} \to \mathcal{T}^{\Gamma^{op}}$ and $m_* : \mathcal{T}^{\Gamma^{op}\times\Gamma^{op}} \to \mathcal{T}^{\Gamma^{op}}$ respectively denote the left adjoints to pulling back by a and m. We have two fundamental lemmas.

Lemma 3.8. $m_*(K^\times \times L^\times) = (K \wedge L)^\times$.

Lemma 3.9. $a_*(K^\times \times L^\times) = (K \vee L)^\times$.

[3]One can then formally deduce that, given $A, B \in \mathcal{A}b$, the natural map $K \wedge A^\times \to \mathrm{SP}^\infty(K, A^\times)$ induces a homeomorphism

$$\mathrm{Map}_\Gamma(\mathrm{SP}^\infty(K, A^\times), B^\times) = \mathrm{Map}_\Gamma(K \wedge A^\times, B^\times).$$

[4]More precisely, m is the smash product followed by the lexicographic identification of $\mathbf{m} \wedge \mathbf{n}$ with $\mathbf{mn}$.

Assuming these for the moment, Proposition 3.6 and Proposition 3.7 follow. For example, using Lemma 3.8, we have identifications

$$\begin{aligned}
\mathrm{SP}^\infty(K, \mathrm{SP}^\infty(L, X)) &= K^\times \times_\Gamma (L^\times \times_\Gamma X_*) \\
&= (K^\times \times L^\times) \times_{\Gamma\times\Gamma} m^*(X) \\
&= m_*(K^\times \times L^\times) \times_\Gamma X \\
&= (K \wedge L)^\times \times_\Gamma X \\
&= \mathrm{SP}^\infty(K \wedge L, X),
\end{aligned}$$

and Proposition 3.6 follows. The proof of Proposition 3.7 is similar.

Proof of Lemma 3.8. Given a $\Gamma^{op} \times \Gamma^{op}$-space Y, $m_*(Y)$ is explicitly given by

$$m_*(Y)(\mathbf{c}) = \operatorname*{colim}_{\mathbf{c}\downarrow m} Y$$

where $\mathbf{c} \downarrow m$ is the category with objects all triples $(\mathbf{a}, \mathbf{b}, \gamma)$ with $\gamma : \mathbf{c} \to \mathbf{ab}$, and morphisms given by pairs $(\alpha : \mathbf{a} \to \mathbf{a}', \beta : \mathbf{b} \to \mathbf{b}')$ making an appropriate diagram commute.

One such triple is $(\mathbf{c}, \mathbf{c}, \Delta)$, where $\Delta : \mathbf{c} \to \mathbf{cc}$ is the diagonal, and there is a canonical morphism from this triple to any other triple $(\mathbf{a}, \mathbf{b}, \gamma)$ given by the two components of $\gamma : \mathbf{c} \to \mathbf{ab}$. It follows that the canonical map $Y(\mathbf{c}, \mathbf{c}) \to \operatorname{colim}_{\mathbf{c}\downarrow m} Y$ is a quotient map.

Now we specialize to $Y = K^\times \times L^\times$. Both $\operatorname{colim}_{\mathbf{c}\downarrow m} K^\times \times L^\times$ and $(K \wedge L)^c$ are quotients of $K^c \times L^c$, thus we just need to verify that each maps to the other, as quotient spaces of $K^c \times L^c$.

To construct a map from the former to the latter, we observe that the smash product construction,

$$\mathrm{Map}_{\mathcal{T}}(A, K) \times \mathrm{Map}_{\mathcal{T}}(B, L) \to \mathrm{Map}_{\mathcal{T}}(A \wedge B, K \wedge L),$$

specializes to give natural maps

$$\wedge : K^a \wedge L^b \to (K \wedge L)^{ab}.$$

Thus, associated to a triple $(\mathbf{a}, \mathbf{b}, \gamma)$, there is a canonical map

$$K^a \wedge L^b \xrightarrow{\wedge} (K \wedge L)^{ab} \xrightarrow{\gamma^*} (K \wedge L)^c,$$

and these induce the needed map $\operatorname{colim}_{\mathbf{c}\downarrow m} K^\times \times L^\times \to (K \wedge L)^c$.

To construct a map in the other direction, we observe that $(K \wedge L)^c$ is the quotient of $K^c \times L^c$ obtained by collapsing to a point the subspace

$$\{(k_1, \ldots, k_c, l_1, \ldots, l_c) \mid \text{ for all } i, \text{ either } k_i = * \text{ or } l_i = *\}.$$

One checks easily that this subspace is precisely the union of the images of maps

$$\alpha^* \times \beta^* : K^a \times L^b \to K^c \times L^c$$

such that the composite $\mathbf{c} \xrightarrow{\Delta} \mathbf{cc} \xrightarrow{\alpha\beta} \mathbf{ab}$ is the constant map $\mathbf{0}$. Thus the subspace maps to the basepoint in $\operatorname{colim}_{\mathbf{c}\downarrow m} K^\times \times L^\times$, i.e., the projection $K^c \times L^c \to \operatorname{colim}_{\mathbf{c}\downarrow m} K^\times \times L^\times$ factors through $(K \wedge L)^c$. □

Sketch of proof of Lemma 3.9. This follows easily from the observation that there are canonical decompositions

$$(K \vee L)^c = \bigvee K^a \vee L^b,$$

with the wedge running over bijections $\gamma : \mathbf{c} \to (\mathbf{a}+\mathbf{b})$ which are order preserving when restricted to $\gamma^{-1}((\mathbf{a}+\mathbf{b})-\mathbf{a})$ and $\gamma^{-1}((\mathbf{a}+\mathbf{b})-\mathbf{b})$. □

Remarks 3.10. The two lemmas include the statements that wedge and smash are the left Kan extensions to $\mathcal{T} \times \mathcal{T}$ of the composites $\Gamma \times \Gamma \xrightarrow{a} \Gamma \hookrightarrow \mathcal{T}$ and $\Gamma \times \Gamma \xrightarrow{m} \Gamma \hookrightarrow \mathcal{T}$.

We suspect that Lemma 3.9 has been observed by others. Lemma 3.8 seems less familiar. (Compare our proof of Proposition 3.6 to the proof of [Se, Lemma 3.7].) We note that $\mathcal{T}^\Gamma \times \mathcal{T}^\Gamma \xrightarrow{\wedge} \mathcal{T}^{\Gamma\times\Gamma} \xrightarrow{m_*} \mathcal{T}^\Gamma$ is the smash product of [L].

The category of Γ-spaces admits products in the obvious way: if X and Y are Γ-spaces, one lets $(X \times Y)(\mathbf{n}) = X(\mathbf{n}) \times Y(\mathbf{n})$. We have

Proposition 3.11. $SP^\infty(K, X \times Y) = SP^\infty(K,X) \times SP^\infty(K,Y)$.

To prove this, we first note that $X \times Y = \Delta^*(X \times Y)$, where $\Delta^* : \mathcal{T}^{\Gamma\times\Gamma} \to \mathcal{T}^\Gamma$ is induced by the diagonal $\Delta : \Gamma \to \Gamma \times \Gamma$. Thus we have identifications

$$\begin{aligned}
\mathrm{SP}^\infty(K, X \times Y) &= K^\times \times_\Gamma \Delta^*(X \times Y) \\
&= \Delta_*(K^\times) \times_{\Gamma\times\Gamma} (X \times Y) \\
&= (K^\times \times K^\times) \times_{\Gamma\times\Gamma} (X \times Y) \\
&= (K^\times \times_\Gamma X) \times (K^\times \times_\Gamma Y) \\
&= \mathrm{SP}^\infty(K,X) \times \mathrm{SP}^\infty(K,Y),
\end{aligned}$$

where we have used the next lemma.

Lemma 3.12. $\Delta_*(K^\times) = K^\times \times K^\times$.

Proof. This can be proved in various ways. Perhaps the slickest proof is to first note that Δ is right adjoint to a. That $\Delta_* = a^*$ formally follows. Finally, it is evident that $a^*(K^\times) = K^\times \times K^\times$. □

It remains, in this section, to discuss how the SP^∞ construction interacts with the homotopy theory of Γ-spaces.

Define $\pi_*^s(X) = \mathrm{colim}_{n\to\infty} \pi_{*+n}(\mathrm{SP}^\infty(S^n, X))$. The colimit here arises from maps $S^1 \wedge \mathrm{SP}^\infty(S^{n-1}, X) \to \mathrm{SP}^\infty(S^n, X)$ which themselves are special cases ($K = S^1$ and $Y = \mathrm{SP}^\infty(S^{n-1}, X)$) of the natural transformation

$$K \wedge Y \to \mathrm{SP}^\infty(K, Y).$$

If we define *weak equivalences* to be maps $f : X \to Y$ with $\pi_*^s(f)$ an isomorphism, then Bousfield and Friedlander [BF], following Segal [Se], showed that the localized category $\mathcal{T}^\Gamma[weq^{-1}]$ is equivalent to the homotopy category of connective spectra.

Even more, this equivalence is induced by a Quillen equivalence between appropriate model categories. Schwede [S] modifies the cofibration and fibrations slightly. All these authors work with simplicial sets rather than topological spaces, but [S, Appendix B] allows for some translation into our setting.

The upshot is roughly the following. Cofibrations are maps $f : X \to Y$ where Y is obtained from X by successively attaching appropriate sorts of free Γ-spaces. Fibrant objects agree with Segal's notion of a *very special* Γ-space, where X is very special means that each map

$$X(\mathbf{a}+\mathbf{b}) \to X(\mathbf{a}) \times X(\mathbf{b}) \tag{3.1}$$

is a weak equivalence of spaces, and also

$$\text{the monoid } \pi_0(X(\mathbf{1})) \text{ is a group.} \tag{3.2}$$

Proposition 3.13. *If K is a C.W. complex, then $SP^\infty(K, \ )$ preserves cofibrations and acyclic cofibrations.*

Proposition 3.14. *If K is a C.W. complex, and X is cofibrant, then the natural map*

$$K \wedge X \to SP^\infty(K, X)$$

is a weak equivalence.

Proposition 3.15. *If K is a C.W. complex, and X is cofibrant and very special, then $SP^\infty(K, X)$ is again very special.*

Theorem 3.16. *If X is cofibrant and very special, then $SP^\infty(\ , X)$ takes cofibration sequences of C.W. complexes to a homotopy fibration sequence. In particular, there are weak equivalences of spaces*

$$X(\mathbf{1}) \xrightarrow{\sim} \Omega SP^\infty(S^1, X)(\mathbf{1}) \xrightarrow{\sim} \Omega^2 SP^\infty(S^2, X)(\mathbf{1}) \xrightarrow{\sim} \dots.$$

We briefly indicate why the propositions hold. Firstly, under the cofibrancy hypotheses, $\mathrm{SP}^\infty(K, X)$ will be nicely filtered, and satisfy

$$F_d\mathrm{SP}^\infty(K, X)/F_{d-1}\mathrm{SP}^\infty(K, X) = K^{(d)} \wedge_{\Sigma_d} (X(\mathbf{d})/X_{sing}(\mathbf{d})) \tag{3.3}$$

where $X_{sing}(\mathbf{d})$ denotes the union of all the images of maps $X(\mathbf{c}) \to X(\mathbf{d})$ with $c < d$.

It follows then that $K \wedge X(\mathbf{1}) \to \mathrm{SP}^\infty(K, X)$ is a weak equivalence through a stable range, and the first two of the propositions easily can be deduced.

For the next proposition, we note that, if (3.1) holds, then

$$X_{a+b} \to X_a \times X_b$$

is a *strict equivalence* of Γ-spaces, where a map is a strict equivalence if evaluating on any $\mathbf{n}$ yields a weak equivalence of spaces. Then we have equivalences

$$\begin{aligned}\mathrm{SP}^\infty(K, X)(\mathbf{a}+\mathbf{b}) &= \mathrm{SP}^\infty(K, X_{a+b}) \xrightarrow{\sim} \mathrm{SP}^\infty(K, X_a \times X_b) \\ &= \mathrm{SP}^\infty(K, X_a) \times \mathrm{SP}^\infty(K, X_b) \\ &= \mathrm{SP}^\infty(K, X)(\mathbf{a}) \times \mathrm{SP}^\infty(K, X)(\mathbf{b}),\end{aligned}$$

showing that $\mathrm{SP}^\infty(K, X)$ again satisfies (3.1).

For the theorem, see [Se, Prop.3.2] and [BF, Lemma 4.3]. It follows that if X is cofibrant and very special, then $X(\mathbf{1})$ is canonically weakly equivalent to an infinite loop space. Furthermore, for any C.W. complex K, there are weak equivalences

$$(3.4) \qquad F_d\mathrm{SP}^\infty(K,X)/F_{d-1}\mathrm{SP}^\infty(K,X) \simeq K^{(d)} \wedge_{\Sigma_d} X(\mathbf{1})^{\wedge d}.$$

4. Commutative ring spectra

We now show that the results of the previous sections extend nicely to the world of structured ring spectra.

We work within the category $\mathcal{S}$, the category of S-modules studied in the book [EKMM]. Given $K \in \mathcal{T}$ and $X \in \mathcal{S}$, $\Sigma^\infty K$, $K \wedge X$, and $\mathrm{Map}(K,X)$ will denote the usual S-modules[5].

Let $\mathcal{A}lg$ be the category of unital, commutative, associative, augmented S-algebras. Thus an object in $\mathcal{A}lg$ is an S-module R, together with multiplication $\mu : R \wedge R \to R$, unit $\eta : S \to R$ and counit $\epsilon : R \to S$ satisfying the usual identities. Morphisms preserve all structure.

This category is enriched over $\mathcal{T}$: given $R, Q \in \mathcal{A}lg$, the morphism space $\mathrm{Map}_{\mathcal{A}\mathrm{lg}}(R,Q)$ is based with basepoint $R \xrightarrow{\epsilon} S \xrightarrow{\eta} Q$. We also note that the coproduct in $\mathcal{A}lg$ of R and Q is $R \wedge Q$.

As observed in [B, §1], results in [EKMM] show that $\mathcal{A}lg$ has a topological model category structure in which weak equivalences are morphisms that are weak equivalences as maps of S-modules[6].

We have two important sources of examples.

Example 4.1. If $A \in \mathcal{A}b$, then $\Sigma^\infty A_+ \in \mathcal{A}lg$. More generally, if X is an E_∞-space (e.g., an infinite loop space), then $\Sigma^\infty X_+$ is naturally an object in $\mathcal{A}lg$. (See [M, Ex. IV.1.10] and [EKMM, §II.4].)

Example 4.2. Given a based space Z, let $D(Z_+)$ denote $\mathrm{Map}(Z_+, S)$. This is an object in $\mathcal{A}lg$: the unit and the counit are respectively induced by $Z_+ \to S^0$ and $S^0 \to Z_+$, and the diagonal $\Delta : Z \to Z \times Z$ induces the multiplication

$$D(Z_+) \wedge D(Z_+) \to D(Z_+ \wedge Z_+) \xrightarrow{\Delta^*} D(Z_+).$$

Given $R \in \mathcal{A}lg$, we let $R^\wedge : \Gamma \to \mathcal{S}$ denote the functor with $R^\wedge(\mathbf{n}) = R^{\wedge n}$ analogous to Example 3.2.

Definition 4.3. Given $K \in \mathcal{T}$ and $R \in \mathcal{A}lg$, let $\mathrm{SP}^\infty(K,R) = K^\times \wedge_\Gamma R^\wedge$.

We will momentarily see that $\mathrm{SP}^\infty(K,R)$ is again an object in $\mathcal{A}lg$.

Proofs from §3 extend immediately to prove the next proposition.

[5] What we are calling $\mathrm{Map}(K,X)$ is $F_S(\Sigma^\infty K, X)$ in [EKMM].

[6] In Basterra's notation, $\mathcal{A}lg$ is denoted $\mathcal{C}_{S/S}$.

Proposition 4.4. *There are the following natural identifications.*

(1) $SP^\infty(S^0, R) = R$.

(2) $SP^\infty(K \vee L, R) = SP^\infty(K, R) \wedge SP^\infty(L, R)$.

(3) $SP^\infty(K, R \wedge Q) = SP^\infty(K, R) \wedge SP^\infty(K, Q)$.

A consequence of this proposition is that $\mathrm{SP}^\infty(K, R)$ takes values in $\mathcal{A}lg$, with multiplication given by the composite

$$\mathrm{SP}^\infty(K,R) \wedge \mathrm{SP}^\infty(K,R) = \mathrm{SP}^\infty(K, R \wedge R) \xrightarrow{\mathrm{SP}^\infty(K,\mu)} \mathrm{SP}^\infty(K,R).$$

We note that this multiplication agrees with the composite

$$\mathrm{SP}^\infty(K,R) \wedge \mathrm{SP}^\infty(K,R) = \mathrm{SP}^\infty(K \vee K, R) \xrightarrow{\mathrm{SP}^\infty(\nabla,R)} \mathrm{SP}^\infty(K,R),$$

where $\nabla : K \vee K \to K$ is the fold map.

With this structure, all the identifications in the last proposition are as objects in $\mathcal{A}lg$, and we also have the next proposition, whose proof follows from the arguments of the last section.

Proposition 4.5. $SP^\infty(K \wedge L, R) = SP^\infty(K, SP^\infty(L, R))$.

Now we check that $\mathrm{SP}^\infty(K, R)$ is the categorical tensor product in $\mathcal{A}lg$. The following lemmas are easily verified, where we use the following notation: with $\mathcal{C}$ either $\mathcal{T}$ or $\mathcal{S}$, and X and Y functors from Γ to $\mathcal{C}$, $\mathrm{Map}^\Gamma_{\mathcal{C}}(X, Y)$ denotes the space of natural transformations from X to Y.

Lemma 4.6. *For all* $K, L \in \mathcal{T}$, $\mathrm{Map}^\Gamma_{\mathcal{T}}(K^\times, L^\times) = \mathrm{Map}_{\mathcal{T}}(K, L)$.

Lemma 4.7. *For all* $R, Q \in \mathcal{A}lg$, $\mathrm{Map}^\Gamma_{\mathcal{S}}(R^\wedge, Q^\wedge) = \mathrm{Map}_{\mathcal{A}\mathrm{lg}}(R, Q)$.

Proposition 4.8. *For all* $K \in \mathcal{T}$ *and* $R \in \mathcal{A}lg$, $SP^\infty(K, R)$ *is naturally isomorphic to* $K \otimes R$.

Proof. We check that $\mathrm{SP}^\infty(K, R)$ satisfies the universal property of the tensor. Given $K \in \mathcal{T}$, and $R, Q \in \mathcal{A}lg$, we have

$$\begin{aligned}
\mathrm{Map}_{\mathcal{A}\mathrm{lg}}(\mathrm{SP}^\infty(K,R), Q) &= \mathrm{Map}^\Gamma_{\mathcal{S}}(\mathrm{SP}^\infty(K,R)^\wedge, Q^\wedge) \\
&= \mathrm{Map}^\Gamma_{\mathcal{S}}(\mathrm{SP}^\infty(K,R^\wedge), Q^\wedge) \\
&= \mathrm{Map}^\Gamma_{\mathcal{S}}(K^\times \wedge_\Gamma m^*(R^\wedge), Q^\wedge) \\
&= \mathrm{Map}^\Gamma_{\mathcal{T}}(K^\times, \mathrm{Map}^\Gamma_{\mathcal{S}}(m^*(R^\wedge), Q^\wedge)) \\
&= \mathrm{Map}^\Gamma_{\mathcal{T}}(K^\times, \mathrm{Map}_{\mathcal{A}\mathrm{lg}}(R^\wedge, Q)) \\
&= \mathrm{Map}^\Gamma_{\mathcal{T}}(K^\times, \mathrm{Map}_{\mathcal{A}\mathrm{lg}}(R, Q)^\times) \\
&= \mathrm{Map}_{\mathcal{T}}(K, \mathrm{Map}_{\mathcal{A}\mathrm{lg}}(R,Q)).
\end{aligned}$$

Here $m : \Gamma \times \Gamma \to \Gamma$ is multiplication as in the last section. □

As before, $\mathrm{SP}^\infty(K,R)$ is naturally filtered. Let R/S denote the cofiber of $\eta: S \to R$. If K is a C.W. complex, and η is a cofibration, then the inclusion $F_{d-1}\mathrm{SP}^\infty(K,R) \hookrightarrow F_d\mathrm{SP}^\infty(K,R)$ will be a cofibration, and there is an isomorphism of S-modules

$$F_d\mathrm{SP}^\infty(K,R)/F_{d-1}\mathrm{SP}^\infty(K,R) \simeq K^{(d)} \wedge_{\Sigma_d} (R/S)^{\wedge d}. \tag{4.1}$$

5. The reduced model

It is sometimes useful to replace $\mathcal{A}lg$ by a slightly different category. Let $\mathcal{A}lg'$ be the category of nonunital, commutative, associative S-algebras (the category denoted $\mathcal{N}_S$ in [B]). Basterra observes that the functor $S\vee : \mathcal{A}lg' \to \mathcal{A}lg$, that wedges a unit S onto a nonunital algebra, has as right adjoint the augmentation ideal functor $J : \mathcal{A}lg \to \mathcal{A}lg'$, defined by letting $J(R)$ be the fiber of $R \xrightarrow{\epsilon} S$. She then notes that, with a natural topological model category on $\mathcal{A}lg'$, these adjoint functors form a Quillen pair, and thus induce adjoint equivalences on the associated homotopy categories.

Example 5.1. If Z is a based space, $J(D(Z_+)) = D(Z)$. The multiplication on $D(Z)$ is induced by the reduced diagonal $\Delta : Z \to Z \wedge Z$.

Our $\mathrm{SP}^\infty(K, \ )$ construction has a 'reduced' analogue in $\mathcal{A}lg'$.

Definition 5.2. Let $\mathcal{E}$ be the category with objects $\mathbf{n}$, for $n \geq 1$, and with morphisms from $\mathbf{n}$ to $\mathbf{m}$ equal to all epimorphisms from $\{1,\dots,n\}$ to $\{1,\dots,m\}$.

As observed in [Ar] (see also [AK]), a based space K defines a functor $K^\wedge : \mathcal{E}^{op} \to \mathcal{T}$ with $K^\wedge(\mathbf{n}) = K^{\wedge n}$. Also, $J \in \mathcal{A}lg'$ defines $J^\wedge : \mathcal{E} \to \mathcal{S}$ in the obvious way.

Definition 5.3. Given $K \in \mathcal{T}$ and $J \in \mathcal{A}lg'$, let $\mathrm{SP}^\infty(K,J) = K^\wedge \wedge_{\mathcal{E}} J^\wedge$.

The analogues of all the properties of $\mathrm{SP}^\infty(K,R)$ proved in the last section hold in our setting, with virtually identical proofs. In particular, $\mathrm{SP}^\infty(K,J)$ is again an object in $\mathcal{A}lg'$, and it agrees with the categorical tensor product $K \otimes J$.

From the above comments, one can formally deduce the following isomorphism in $\mathcal{A}lg$.

Proposition 5.4. *$SP^\infty(K,J) \vee S = SP^\infty(K, J \vee S)$.*

Though we will not show this here, this proposition can also be given a direct proof, and there are analogues in other contexts. Readers may wish to compare this result with observations in [P].

As usual, $\mathrm{SP}^\infty(K,J)$ is filtered: if $\mathcal{E}_d$ denote the full subcategory of $\mathcal{E}$ with objects $\mathbf{n}$ for $n \leq d$, then we let $F_d\mathrm{SP}^\infty(K,J) = K^\wedge \wedge_{\mathcal{E}_d} J^\wedge$. If K is a C.W. complex, then the inclusion $F_{d-1}\mathrm{SP}^\infty(K,J) \hookrightarrow F_d\mathrm{SP}^\infty(K,J)$ will be a cofibration, and there is an isomorphism of S-modules

$$F_d\mathrm{SP}^\infty(K,J)/F_{d-1}\mathrm{SP}^\infty(K,J) \simeq K^{(d)} \wedge_{\Sigma_d} J^{\wedge d}. \tag{5.1}$$

We note that the isomorphism of the last proposition is filtration preserving.

6. Reinterpretation of Arone's tower for $\Sigma^\infty \mathrm{Map}_{\mathcal{T}}(K, X)$

In this section, we let K be a finite C.W. complex.

In [Ar], G. Arone described a model for the Goodwillie tower of the functor sending a based space Z to the S-module $\Sigma^\infty \mathrm{Map}_{\mathcal{T}}(K, Z)$. Here we show that, if Z is a finite complex, his tower arises as the S-dual of the filtered object $K \otimes D(Z)$ of the last section.

We recall Arone's construction and some of its properties [Ar]. For more detail, see also [AK].

Definitions 6.1. Let Z be a based space.

(i) Let $P^K(Z) = \mathrm{Map}_{\mathcal{S}}^{\mathcal{E}}(K^\wedge, Z^\wedge)$, the spectrum of natural transformations from $\Sigma^\infty K^\wedge$ to $\Sigma^\infty Z^\wedge$.

(ii) Let $P^K_d(Z) = \mathrm{Map}_{\mathcal{S}}^{\mathcal{E}_d}(K^\wedge, Z^\wedge)$.

(iii) Let $\Phi(K, Z) : \Sigma^\infty \mathrm{Map}_{\mathcal{T}}(K, Z) \to P^K(Z)$ be the natural transformation that sends $f : K \to Z$ to the natural transformation with n^{th} component equal to $\Sigma^\infty f^{\wedge n} : \Sigma^\infty K^{\wedge n} \to \Sigma^\infty Z^{\wedge n}$.

The spectrum $P^K(Z)$ is the inverse limit of the tower of fibrations

$$\cdots \to P^K_{d+1}(Z) \to P^K_d(Z) \to P^K_{d-1}(Z) \to \cdots,$$

and the dth fiber is isomorphic to

$$\mathrm{Map}_{\mathcal{S}}^{\Sigma_d}(K^{(d)}, Z^{\wedge d}),$$

the spectrum of Σ_d-equivariant maps from $\Sigma^\infty K^{(d)}$ to $\Sigma^\infty Z^{\wedge d}$. Because $K^{(d)}$ is finite, and the Σ_d action on this is free away from the basepoint, this fiber is naturally homotopy equivalent to the homotopy orbit spectrum

$$(D(K^{(d)}) \wedge Z^{\wedge d})_{h\Sigma_d}.$$

From this last description one sees that the tower has the form of a Goodwillie tower, and also that the connectivity of the fibers goes up if the connectivity of Z is greater than the dimension of K. Arone then proves that this *is* the Goodwillie tower of $\Sigma^\infty \mathrm{Map}_{\mathcal{T}}(K, Z)$ by proving

Theorem 6.2. [Ar] *If the connectivity of Z is greater than the dimension of K, then $\Phi(K, Z)$ is a homotopy equivalence.*

Now we connect these constructions to $K \otimes D(Z)$. In the following definitions, and the subsequent discussion, we use $D(X)$ to denote the S-dual of an S-module X[7].

[7] Thus $D(X) = F_S(X, S)$ in the notation of [EKMM], and, if Z is a space, $D(\Sigma^\infty Z)$ is isomorphic to $D(Z)$ as S-modules.

Definitions 6.3. Let Z be a based space.

(i) Let $\tilde{\Theta}(K,Z) : K \otimes D(Z) \to D(\mathrm{Map}_{\mathcal{T}}(K,Z))$ be the map in $\mathcal{A}lg'$ adjoint to the composite

$$K \xrightarrow{\text{eval}} \mathrm{Map}_{\mathcal{T}}(\mathrm{Map}_{\mathcal{T}}(K,Z),Z) \xrightarrow{D} \mathrm{Map}_{\mathcal{A}lg'}(D(Z), D(\mathrm{Map}_{\mathcal{T}}(K,Z))).$$

(ii) Let $\Theta(K,Z) : \Sigma^\infty \mathrm{Map}_{\mathcal{T}}(K,Z) \to D(K \otimes D(Z))$ be the S-module map adjoint to $\tilde{\Theta}(K,Z)$.

(iii) Let $\alpha(K,Z) : P^K(Z) \to D(K \otimes D(Z))$ be the map of cofiltered S-modules defined as follows. Let $i : Z^\wedge \to D(D(Z)^\wedge)$ be the natural transformation adjoint to $D(Z)^\wedge \to D(Z^\wedge)$. Now let $\alpha(K,Z)$ be the composite induced by i:

$$\mathrm{Map}^{\mathcal{E}}_{\mathcal{S}}(K^\wedge, Z^\wedge) \xrightarrow{i} \mathrm{Map}^{\mathcal{E}}_{\mathcal{S}}(K^\wedge, D(D(Z)^\wedge)) = \mathrm{Map}_{\mathcal{S}}(K^\wedge \wedge_{\mathcal{E}} D(Z)^\wedge, S).$$

A check of the definitions verifies the next lemma.

Lemma 6.4. *There is a commutative diagram*

$$\begin{array}{ccc} & \Sigma^\infty \mathrm{Map}_{\mathcal{T}}(K,Z) & \\ {}^{\Phi(K,Z)}\swarrow & & \searrow^{\Theta(K,Z)} \\ P^K(Z) & \xrightarrow{\quad \alpha(K,Z) \quad} & D(K \otimes D(Z)) \end{array}$$

Lemma 6.5. *$\alpha(K,Z)$ is a homotopy equivalence if Z is a finite complex.*

Proof. If Z is finite, then $i : Z^{\wedge n} \to D(D(Z)^{\wedge n})$ is an equivalence for all n. Now the lemma follows by observing that $K^\wedge$ is a cofibrant $\mathcal{E}^{op}$-space, or more simply, note that the fibers of the towers will be equivalences, as $K^{(d)}$ is a free Σ_d-complex for all d. (This has been noted before; see, e.g., [AK, McC].) □

Summarizing, we conclude

Theorem 6.6. *If both K and Z are finite complexes, and the dimension of K is less than the connectivity of Z, then*

$$\Theta(K,Z) : \Sigma^\infty \mathrm{Map}_{\mathcal{T}}(K,Z) \to D(K \otimes D(Z))$$

is a weak equivalence, and thus the algebra map

$$\tilde{\Theta}(K,Z) : K \otimes D(Z) \to D(\mathrm{Map}_{\mathcal{T}}(K,Z))$$

can be identified as the map from a spectrum to its double dual.

We end this section by noting how the homological version of this discussion would go.

Let $\mathbb{F}$ be a field, $H\mathbb{F}$ the associated commutative S-algebra, and $\mathcal{A}lg_{\mathbb{F}}$ the category of commutative, nonunital $H\mathbb{F}$ algebras. Let $K \otimes_{\mathbb{F}} J \in \mathcal{A}lg_{\mathbb{F}}$ denote the tensor product of a based space K and an $J \in \mathcal{A}lg_{\mathbb{F}}$. As before, one learns that

$$K \otimes_{\mathbb{F}} J = K^\wedge \wedge_{\mathcal{E}} J^\wedge,$$

where smash products are taken over $H\mathbb{F}$.

Now let $D_{\mathbb{F}}(Z) = \mathrm{Map}(Z, H\mathbb{F})$, the $H\mathbb{F}$-module whose homotopy groups are the cohomology groups of Z with $\mathbb{F}$-coefficients. In this case, the natural map $i : H\mathbb{F} \wedge Z^{\wedge n} \to D_{\mathbb{F}}(D_{\mathbb{F}}(Z)^{\wedge n})$ is an equivalence for any space Z with $H_*(Z;\mathbb{F})$ of finite type. Reasoning as before, from Arone's theorem one deduces

Theorem 6.7. *If K is a finite complex of dimension less than the connectivity of Z, and $H_*(Z;\mathbb{F})$ is of finite type, then the natural map in $\mathcal{A}lg_{\mathbb{F}}$,*

$$\Theta : K \otimes_{\mathbb{F}} D_{\mathbb{F}}(Z) \to D_{\mathbb{F}}(\mathrm{Map}_{\mathcal{T}}(K, Z)),$$

is an equivalence.

Remark 6.8. It seems likely that this theorem can be deduced from older convergence results for the Anderson spectral sequence [An], and then one can run our arguments backwords, and *deduce* Arone's theorem. The novelty would then be to identify the filtration as Arone did.

7. Topological Hochschild homology and Topological André–Quillen homology

Let $THH(R;M)$ denote the Topological Hochschild homology spectrum associated to an S-algebra R and an R-bimodule M (see, e.g., [EKMM, Chap.9]). If R is commutative and augmented, then $\epsilon : R \to S$ makes S into an R-bimodule. We have

Proposition 7.1. $S^1 \otimes R = THH(R;S)$.

Proof. This is a variant of a theorem of J. McClure, R. Schwanzl, and R. Vogt [MSV]. They show that if R is a commutative S-algebra, then $THH(R;R)$ is the tensor product of R with S^1 with the tensor product in the category of commutative S-algebras. In the appendix, we note that if R is also augmented, then this would agree with $S^1_+ \otimes R \in \mathcal{A}lg$. Thus

$$THH(R;R) = S^1_+ \otimes R.$$

Applying $\otimes R$ to the pushout square in $\mathcal{T}$

$$\begin{array}{ccc} S^0 & \longrightarrow & * \\ \downarrow & & \downarrow \\ S^1_+ & \longrightarrow & S^1, \end{array}$$

yields a pushout square in $\mathcal{A}lg$

$$\begin{array}{ccc} R & \longrightarrow & S \\ \downarrow & & \downarrow \\ S^1_+ \otimes R & \longrightarrow & S^1 \otimes R, \end{array}$$

and we conclude that

$$S^1 \otimes R = (S^1_+ \otimes R) \wedge_R S = THH(R,R) \wedge_R S = THH(R;S).$$

□

Given $K \in \mathcal{T}$ and $R \in \mathcal{A}lg$, there is a natural map

$$K \wedge R \to K \otimes R,$$

and thus

$$K \wedge (L \otimes R) \to K \otimes (L \otimes R) = (K \wedge L) \otimes R.$$

This map is easily seen to be filtration preserving.

Specializing to the case when $K = S^1$, and $L = S^n$ yield filtration preserving maps

$$\Sigma(S^n \otimes R) \to S^{n+1} \otimes R,$$

or, equivalently,

$$S^n \otimes R \to \Omega(S^{n+1} \otimes R).$$

Definition 7.2. Let $TAQ(R) = \operatorname*{hocolim}_{n \to \infty} \Omega^n S^n \otimes R$.

M. Mandell has shown the author that this definition agrees with other definitions of Topological André Quillen Homology in the literature, e.g., [B]. In particular $TAQ(R)$ is homotopy equivalent to the cofiber of $J(R) \wedge J(R) \to J(R)$[8].

As the next example makes clear, $TAQ(R)$ can be viewed as an 'infinite delooping' of R.

Example 7.3. If X is a connective S-module, $TAQ(\Sigma^\infty(\Omega^\infty X)_+) \simeq X$. To see this, just recall that $S^n \otimes$ yields the $(n-1)$-connected n-fold delooping of an infinite loopspace.

Note that $TAQ(R)$ is filtered with

$$F_d TAQ(R)/F_{d+1} TAQ(R) \simeq \operatorname*{hocolim}_{n \to \infty} \Sigma^{-n} S^{n(d)} \wedge_{\Sigma_d} (R/S)^{\wedge d}.$$

As in [AM], let K_d be the unreduced suspension of the classifying space of the poset of nontrivial partitions of a set with d elements.

Lemma 7.4. [AD] *There is a Σ_d-equivariant map*

$$\operatorname*{hocolim}_{n \to \infty} \Sigma^{-n} S^{n(d)} \to \Sigma K_d$$

that is a nonequivariant equivalence.

The original short proof of this, due to Arone and Mahowald, appears in [K3, Appendix].

Corollary 7.5. *There is a homotopy equivalence*

$$F_d TAQ(R)/F_{d+1} TAQ(R) \simeq (\Sigma K_d \wedge (R/S)^{\wedge d})_{h\Sigma_d}.$$

[8]Strictly speaking, R should be replaced by a fibrant object in $\mathcal{A}lg$ and then $J(R)$ replaced by a cofibrant object in $\mathcal{A}lg'$.

8. Spectral sequences and examples

Applying homology or cohomology with $\mathbb{F}$-coefficients to our filtered models for $S^n \otimes R$ and $TAQ(R)$ yields highly structured convergent spectral sequences with E_1 terms equal to known functors of $H_*(R;\mathbb{F})$. To see why this is true, we note that there is an explicit equivariant duality map [AK]

$$F(\mathbb{R}^n, d)_+ \wedge S^{n(d)} \to S^{nd},$$

where $F(\mathbb{R}^n, d)$ is the usual configuration space of d distinct points in $\mathbb{R}^n$. Thus the homology calculations of [CLM] apply.

To be more precise, let $\{E_r^{*,*}(S^n \otimes R;\mathbb{F})\}$ and $\{E_r^{*,*}(TAQ(R);\mathbb{F})\}$ respectively denote the spectral sequences for computing $H^*(S^n \otimes R;\mathbb{F})$ and $H^*(TAQ(R);\mathbb{F})$. Let $\tilde{H}^*(R;\mathbb{F})$ denote the reduced cohomology of R, i.e., $H^*(J(R);\mathbb{F})$.

Theorem 8.1. *For $R \in \mathcal{A}lg$ with $H_*(R;\mathbb{F})$ of finite type and bounded either above or below, there are natural isomorphisms as follows.*

1. *If $\mathbb{F}$ has characteristic 0, then*

$$E_1^{*,*}(S^n \otimes R;\mathbb{F}) = S^*(\Sigma^{1-n} L(\Sigma^{-1}\tilde{H}^*(R;\mathbb{F}))),$$

$$\text{and} \qquad E_1^{*,*}(TAQ(R);\mathbb{F}) = \Sigma L(\Sigma^{-1}\tilde{H}^*(R;\mathbb{F}))).$$

2. *If $\mathbb{F}$ has characteristic p, then*

$$E_1^{*,*}(S^n \otimes R;\mathbb{F}) = S^*(\mathcal{R}_n(\Sigma^{1-n} L_r(\Sigma^{-1}\tilde{H}^*(R;\mathbb{F})))),$$

$$\text{and} \qquad E_1^{*,*}(TAQ(R);\mathbb{F}) = \mathcal{R}(\Sigma L_r(\Sigma^{-1}\tilde{H}^*(R;\mathbb{F})))).$$

In this theorem, $\Sigma^d V$ denotes the d-fold shift of a graded vector space V, L is the free Lie algebra functor, L_r is the free restricted Lie algebra functor, S^* is the free commutative algebra functor[9], and $\mathcal{R}$ and $\mathcal{R}_n$ are appropriate free Dyer–Lashof operation functors.

To deduce such calculations from [CLM], note that under the hypotheses on R, we have natural weak equivalences of $H\mathbb{F}$-modules

$$D_{\mathbb{F}}(S^{n(d)} \wedge_{\Sigma_d} (R/S)^{\wedge d}) \simeq F(\mathbb{R}^n, d)_+ \wedge_{\Sigma_d} (D_{\mathbb{F}}(\Sigma^n(R/S)))^{\wedge d}.$$

The homology of the left side for all d yields $E_1^{*,*}(S^n \otimes R;\mathbb{F})$, while computing the homology of constructions like the right side is the topic of [CLM, Part III]. Taking inverse limits over n then yields $E_1^{*,*}(TAQ(R);\mathbb{F})$.

The author plans to write more about this elsewhere, including a discussion of how Steenrod operations act if $\mathbb{F}$ has finite characteristic, a discussion of the first differential, and applications to topological nonrealization results. When $R = D(S^1_+)$, the relevant mod 2 calculations have already been studied by the author in [K3].

We end with three nontrivial examples. A more detailed discussion of the last example, which we find most interesting, will appear in [K4].

[9]One has sign conventions of the usual sort.

Example 8.2. If $R = \Sigma^\infty S^1_+$, then $TAQ(R) \simeq \Sigma H\mathbb{Z}$; this is a special case of Example 7.3. The filtration of $TAQ(R)$ will correspond to the symmetric product of spheres filtration of $H\mathbb{Z}$. One way to see this is to first note that [AD, Thm.1.14] says that the filtration quotients are correct. For a proof of this result localized at 2, one can alternatively cite [K3, Cor.A.2(2)]. The main theorems of [K1] and [KP], p-local theorems, then combine to say that having correct filtration quotients *characterizes* the filtration integrally.

In particular, localized at p, the dth associated graded spectrum is contractible unless d is a power of p,

$$F_{p^k} = \Sigma SP^{p^k}(S), \text{ and } F_{p^k}/F_{p^{k-1}} = \Sigma^{k+1}L(k).$$

Here $L(k)$ is as in [MP]. The boundary maps of the filtration yield the complex of spectra

$$\cdots \to \Sigma L(2) \to \Sigma L(1) \to \Sigma L(0) \to H\mathbb{Z}_{(p)}$$

occurring in the Whitehead conjecture [K1, KP]. The sequence is exact in homotopy, but zero in mod p homology: indeed, the spectral sequence for computing $H_*(\Sigma H\mathbb{Z}; \mathbb{F}_p) = \Sigma A/A\beta$ collapses at E_1.

Example 8.3. If $R = \Sigma^\infty \mathbb{Z}/2_+$, then $TAQ(R) \simeq H\mathbb{Z}/2$. Localized at 2, the associated graded spectra are identified in [K3, Cor. A.8.(1)]. In particular, the dth associated graded spectrum is contractible unless d is a power of 2,

$$F_{2^k} = SP^{2^k}_\Delta(S^0), \text{ and } F_{2^k}/F_{2^{k-1}} = \Sigma^k M(k).$$

Here we recall [MP] that $SP^{2^k}_\Delta(S^0)$ is defined to be the cofiber of the 'diagonal' $\Delta : SP^{2^{k-1}}(S^0) \to SP^{2^k}(S^0)$, and $M(k) = L(k) \vee L(K-1)$. As in the previous example, the boundary maps of the filtration yield the complex of spectra

$$\cdots \to M(2) \to M(1) \to M(0) \to H\mathbb{Z}/2$$

occurring in the mod 2 Whitehead conjecture [K2]. The sequence is exact in homotopy, but the spectral sequence for computing $H_*(H\mathbb{Z}/2; \mathbb{F}_2) = A$ collapses at E_1.

Example 8.4. If $R = D(S^1_+)$, then $TAQ(R) \simeq \Sigma^{-1} H\mathbb{Q}$. There are various ways to see this; in [K4], we will prove that $S^2 \otimes R \simeq \Sigma H\mathbb{Q} \vee S^0$. Localized at 2, [K3, Thm. 1.6] identifies the associated graded spectra. In particular, the dth associated graded spectrum is contractible unless d is a power of 2, and

$$F_{2^k}/F_{2^{k-1}} = \Sigma^{-1} SP^{2^k}_\Delta(S^0).$$

Thus $H^*(F_{2^k}/F_{2^{k-1}}; \mathbb{F}_2) = \Sigma^{-1} A/L_{k+1}$, where L_k is the span of all admissible sequences in the Steenrod algebra of length at least k. As will be proved in [K4], the boundary maps of the filtration yields a complex of spectra

$$\ldots \to \Sigma^{-2} SP^4_\Delta(S^0) \to \Sigma^{-1} SP^2_\Delta(S^0) \to S^0$$

that is exact in cohomology, and has the following behavior: each map sends the bottom class of the cyclic A-module to Sq^1 applied to the bottom class of the next module.

Appendix A. Augmented versus nonaugmented ring spectra

Let $\mathcal{A}lg_u$ be the category of unital commutative S-algebras, but not necessarily augmented. Thus we have forgetful maps

$$\mathcal{A}lg \to \mathcal{A}lg_u \to \mathcal{A}lg'.$$

$\mathcal{A}lg_u$ is enriched over $\mathcal{T}_u$, the category of *unbased* topological spaces, so one can look for a convenient model for $K \otimes R$ with $K \in \mathcal{T}_u$ and $R \in \mathcal{A}lg_u$. In this appendix we describe such a model, and compare this to the construction in §4.

A.1. $K \otimes R$ for unital commutative algebras

Let $\mathcal{S}_u$ be the category of S-modules under S, so an object is an S-module map $\eta : S \to X$.

Definition A.1. Given $K \in \mathcal{T}_u$ and $X \in \mathcal{S}_u$, let $K \overline{\wedge} X \in \mathcal{S}_u$ be the pushout:

$$\begin{array}{ccc} K_+ \wedge S & \longrightarrow & S \wedge S = S \\ \downarrow & & \downarrow \\ K_+ \wedge X & \longrightarrow & K \overline{\wedge} X. \end{array}$$

It is easy to check

Lemma A.2. *There is an adjunction*

$$\mathrm{Map}_{\mathcal{S}_u}(K \overline{\wedge} X, Y) = \mathrm{Map}_{\mathcal{T}_u}(K, \mathrm{Map}_{\mathcal{S}_u}(X, Y)).$$

Let $\mathcal{S}et$ be the category of finite sets. Given $K \in \mathcal{T}_u$, there is an apparent functor $K^{\times} : \mathcal{S}et^{op} \to \mathcal{T}_u$. Now note that, as it was in $\mathcal{A}lg$, $\wedge$ is the coproduct in $\mathcal{A}lg_u$. Then $R \in \mathcal{A}lg_u$ defines $R^{\wedge} : \mathcal{S}et \to \mathcal{S}_u$.

Lemma A.3. *For all* $K, L \in \mathcal{T}_u$, $\mathrm{Map}_{\mathcal{T}}^{\mathcal{S}et}(K^{\times}, L^{\times}) = \mathrm{Map}_{\mathcal{T}_u}(K, L)$.

Lemma A.4. *For all* $R, Q \in \mathcal{A}lg_u$, $\mathrm{Map}_{\mathcal{S}_u}^{\mathcal{S}et}(R^{\wedge}, Q^{\wedge}) = \mathrm{Map}_{\mathcal{A}lg_u}(R, Q)$.

Definition A.5. Given $K \in \mathcal{T}_u$ and $R \in \mathcal{A}lg_u$, let

$$\mathrm{SP}^{\infty}(K, R) = K^{\times} \overline{\wedge}_{\mathcal{S}et} R^{\wedge}.$$

As in §4, the lemmas combine to prove the analogue of Proposition 4.8.

Proposition A.6. *For all* $K \in \mathcal{T}_u$ *and* $R \in \mathcal{A}lg_u$, $SP^{\infty}(K, R)$ *is again in* $\mathcal{A}lg_u$ *and is naturally isomorphic to the categorical tensor product* $K \otimes R$.

$\mathrm{SP}^{\infty}(K, R)$ is filtered in the usual way, and one gets an isomorphism of S-modules

$$F_d\mathrm{SP}^{\infty}(K, R)/F_{d-1}\mathrm{SP}^{\infty}(K, R) \simeq (K_+)^{(d)} \wedge_{\Sigma_d} (R/S)^{\wedge d}. \tag{A.1}$$

We note that $(K_+)^{(d)}$ is just S if $d = 0$ and $K^{\times d}/(\text{fat diagonal})$ if $d > 0$.

A.2. The unbased versus the based construction

In this subsection, we denote the tensor in $\mathcal{A}lg$ by $\otimes$ and the tensor in $\mathcal{A}lg_u$ by $\otimes_u$.

Given $R \in \mathcal{A}lg_u$, the product $R \times S$ will be in $\mathcal{A}lg$, with augmentation given by projection, and unit $S \xrightarrow{\Delta} S \times S \to R \times S$. This construction is right adjoint to the forgetful functor:

Lemma A.7. *Given $Q \in \mathcal{A}lg$ and $R \in \mathcal{A}lg_u$, there is an adjunction*

$$\mathrm{Map}_{\mathcal{A}lg_u}(Q, R) = \mathrm{Map}_{\mathcal{A}lg}(Q, R \times S).$$

Note that, if $Q \in \mathcal{A}lg$ and $R \in \mathcal{A}lg_u$, then $\mathrm{Map}_{\mathcal{A}lg_u}(Q, R)$ is based with basepoint $Q \to S \to R$.

Proposition A.8. *Given $K \in \mathcal{T}$, $Q \in \mathcal{A}lg$, and $R \in \mathcal{A}lg_u$, there is an adjunction isomorphism*

$$\mathrm{Map}_{\mathcal{A}lg_u}(K \otimes Q, R) = \mathrm{Map}_{\mathcal{T}}(K, \mathrm{Map}_{\mathcal{A}lg_u}(Q, R)).$$

Proof. We have natural isomorphisms

$$\begin{aligned}\mathrm{Map}_{\mathcal{A}lg_u}(K \otimes Q, R) &= \mathrm{Map}_{\mathcal{A}lg}(K \otimes Q, R \times S) = \mathrm{Map}_{\mathcal{T}}(K, \mathrm{Map}_{\mathcal{A}lg}(Q, R \times S)) \\ &= \mathrm{Map}_{\mathcal{T}}(K, \mathrm{Map}_{\mathcal{A}lg_u}(Q, R)).\end{aligned}$$

□

Corollary A.9. *If $Q \in \mathcal{A}lg$ and $L \in \mathcal{T}_u$, then $L_+ \otimes Q = L \otimes_u Q$.*

Proof. Let $K = L_+$ in the proposition, and note that

$$\mathrm{Map}_{\mathcal{T}}(L_+, \mathrm{Map}_{\mathcal{A}lg_u}(Q, R)) = \mathrm{Map}_{\mathcal{T}_u}(L, \mathrm{Map}_{\mathcal{A}lg_u}(Q, R)) = \mathrm{Map}_{\mathcal{A}lg_u}(L \otimes_u Q, R).$$

□

References

[AK] S.T. Ahearn and N.J. Kuhn, *Product and other fine structure in polynomial resolutions of mapping spaces*, Algebraic and Geometric Topology **2**(2002), 591–647.

[An] D.W. Anderson, *A generalization of the Eilenberg–Moore spectral sequence*, Bull. A.M.S. **78**(1972), 784–788.

[Ar] G. Arone, *A generalization of Snaith-type filtration*, Trans. A.M.S. **351**(1999), 1123–1250.

[AD] G.Z. Arone and W.G. Dwyer, *Partition complexes, Tits buildings and symmetric products*, Proc. London Math. Soc. **82**(2001), 229–256.

[AM] G. Arone and M. Mahowald, *The Goodwillie tower of the identity functor and the unstable periodic homotopy of spheres*, Invent. Math. **135**(1999), 743–788.

[B] M. Basterra, *André–Quillen cohomology of commutative S-algebras*, J. Pure Applied Alg. **144**(1999), 111–143.

[BF] A.K. Bousfield and E.M. Friedlander, *Homotopy theory of Γ-spaces, spectra, and bisimplicial sets*, in Geometric Applications of Homotopy Theory II (Evanston, 1977), Springer L.N. Math. **658**(1978), 80–130.

[CLM] F.R. Cohen, T. Lada, and J.P. May, *The Homology of Iterated Loop Spaces*, Springer L.N. Math. **533**, 1976.

[EKMM] A.D. Elmendorf, I. Kriz, M.A. Mandell, J.P. May, *Rings, modules, and algebras in stable homotopy theory*, A.M.S. Math. Surveys and Monographs 47, 1997.

[FF] E.E. Floyd and W.J. Floyd, *Actions of the classical small categories of topology*, unpublished 200 page manuscript dating from 1990. Available from Bill Floyd's webpage at http://www.math.vt.edu/people/.

[K1] N.J. Kuhn, *A Kahn-Priddy sequence and a conjecture of G.W. Whitehead*, Math. Proc. Camb. Phil. Soc. **92**(1982), 467–483.

[K2] N.J. Kuhn, *Spacelike resolutions of spectra*, Proceedings of the Homotopy Theory Conference (Northwestern, 1982), A.M.S. Cont. Math. Series **19**(1983), 153–165.

[K3] N.J. Kuhn, *New relationships among loopspaces, symmetric products, and Eilenberg MacLane spaces*, in Cohomological Methods in Homotopy Theory (Barcelona, 1998), Birkhäuser Verlag Progress in Math **196**(2001), 185–216.

[K4] N.J. Kuhn, *On the S-dual of a circle*, in preparation.

[KP] N.J. Kuhn and S.B. Priddy, *The transfer and Whitehead's conjecture*, Math. Proc. Camb. Phil. Soc. **98**(1985), 459–480.

[L] M. Lydakis, *Smash products and Γ-spaces*, Math. Proc. Camb. Phil. Soc. **126**(1999), 311–328.

[MacL] S. MacLane, *Categories for the working mathematician*, Springer Graduate Texts in Math. **5**, 1971.

[M] J.P. May, *E_∞ ring spaces and E_∞ ring spectra*, Springer L.N. Math. **577**, 1977.

[McC] R. McCarthy, *On n-excisive functors of module categories*, to appear in Math. Proc. Camb. Phil. Soc.

[McC] M.C. McCord, *Classifying spaces and infinite symmetric products*, T.A.M.S. **146**(1969), 273–298.

[MSV] J.E. McClure, R. Schwänzl, and R. Vogt, *$THH(R) \simeq R \otimes S^1$ for E_∞ ring spectra*, J. Pure Appl. Algebra. **121**(1997), 137–159.

[MP] S.A. Mitchell and S.B. Priddy, *Stable splittings derived from the Steinberg module,* Topology **22**(1983), 285–298.

[P] T. Pirashvili, *Dold–Kan type theorem for Γ-groups*, Math. Ann. **318**(2000), 277–298.

[S] S. Schwede, *Stable homotopy and Γ-spaces*, Math. Proc. Camb. Phil. Soc. **126**(1999), 329–356.

[Se] G.Segal, *Categories and cohomology theories*, Topology **13**(1974), 293–312.

Nicholas J. Kuhn
Department of Mathematics
University of Virginia
Charlottesville, VA 22903, USA
e-mail: njk4x@virginia.edu

Progress in Mathematics, Vol. 215:
Algebraic Topology: Categorical Decomposition Techniques, 237–259

Spaces of Multiplicative Maps between Highly Structured Ring Spectra

A. Lazarev

Abstract. We uncover a somewhat surprising connection between spaces of multiplicative maps between A_∞-ring spectra and topological Hochschild cohomology. As a consequence we show that such spaces become infinite loop spaces after looping only once. We also prove that any multiplicative cohomology operation in complex cobordisms theory MU canonically lifts to an A_∞-map $MU \to MU$. This implies, in particular, that the Brown-Peterson spectrum BP splits off MU as an A_∞-ring spectrum.

1. Introduction

The main purpose of the present work is to provide a workable method for computing the homotopy type of spaces of A_∞-maps between A_∞-ring spectra (or S-algebras in the terminology of [7]). We make substantial use of the previous results by the author in [9] and for the reader's convenience a brief summary of these is given in Section 2. In Section 3 we collect miscellaneous technical results concerning function spectra and topological Hochschild cohomology. Some of these results are surely known to experts but never have been written down. The formula (3) (base change) deserves special mention. While easy to prove it is extremely convenient when computing with various spectral sequences.

Our first main result (Theorem 4.3) essentially states that the mapping space between two S-algebras A and B is determined after taking based loops by the spectrum of topological derivations $\mathbf{Der}(A, B)$. Therefore the problem of computing higher homotopy groups of this mapping space is a problem of stable homotopy which turns out to be quite amenable, particularly because in many cases $\mathbf{Der}(A, B)$ can be reduced to $\mathbf{THH}(A, B)$, the topological Hochschild cohomology of A with values in B.

The computation of the zeroth homotopy group of the mapping space is, of course, a completely different story. We give a simple answer in the special case when A is a connective S-algebra while B is coconnective (that is, with vanishing homotopy in positive dimensions). This is Theorem 4.8.

Even though the problem of computing homotopy classes of S-algebra maps $A \to B$ is essentially unstable it does lend itself to analysis by methods of obstruction theory developed in [9]. Our second main result (Theorem 5.4) demonstrates that any multiplicative cohomology operation in complex cobordism theory MU canonically (even uniquely in an appropriate sense) lifts to an S-algebra self-map of MU. This is used to show that for an S-algebra E belonging to a fairly large class of complex-oriented theories any multiplicative operation $MU \to E$ lifts to an S-algebra map. Another corollary is that the Brown-Peterson spectrum BP splits off MU localized at p as an S-algebra.

The paper is written in the language of S-modules of [7] and we routinely use the results and terminology of the cited reference.

Notations. In Sections 2 and 3 we work in the category of modules or algebras over a fixed q-cofibrant commutative S-algebra R, the smash product $\wedge$ and the function spectrum $F(-,-)$ are always understood as $\wedge_R$ and $F_R(-,-)$. In Section 4 we specialize to $R = S$. The free R-algebra on an R-module M is denoted by $T(M)$. The space of maps between two R-algebras A and B is denoted by $F_{R-alg}(A, B)$. If A and B are commutative R-algebras then $[A, B]_c$ denotes the set of homotopy classes of commutative algebra maps from A to B. If A and B are associative R-algebras then $[A, B]_a$ stands for homotopy classes of associative algebra maps. For an associative R-algebra A and two A-modules M and N we denote by $[M, N]_{A-mod}$ homotopy classes of A-module maps from M to N. Similarly $[M, N]_{A-bimod}$ stands for homotopy classes of bimodule maps. Finally we denote by $Mult(E, F)$ the set of multiplicative (up to homotopy) maps between ring spectra E and F. To distinguish between strict isomorphisms and weak equivalences we will use, as a rule, the symbol '$\cong$' for the former and '$\simeq$' for the latter.

2. Topological derivations and topological singular extension of S-algebras

In this short section we give an overview of some of the author's results from [9] which will be needed later on.

Let A be a q-cofibrant R-algebra and M a q-cofibrant A-bimodule. Then the R-module $A \vee M$ has the obvious structure of an R-algebra ('square-zero extension' of A). Consider the set $[A, A \vee M]_{a/A}$ of homotopy classes of R-algebra maps from A to $A \vee M$ in the category of R-algebras over A, that is the R-algebras supplied with an R-algebra map into A.

Theorem 2.1. *There exists an A-bimodule Ω_A and a natural in M isomorphism*

$$[A, A \vee M]_{a/A} \cong [\Omega_A, M]_{A-bimod}$$

where the right-hand side denotes the homotopy classes of maps in the category of A-bimodules.

Remark 2.2. *Sometimes we will need a refinement of the above theorem which is formulated as follows. Let B be an R-algebra over A, i.e., there exists a fixed R-algebra map $B \to A$. Then there is a natural isomorphism*

$$[A, B \vee M]_{a/B} \cong [A \wedge_B \Omega_B \wedge_B A, M]_{A-bimod}.$$

Furthermore an A-bimodule M can be considered as a B-bimodule and we have $[B, B \vee M]_{a/B} \cong [\Omega_B, M]_{B-bimod} \cong [A \wedge_B \Omega_B \wedge_B A, M]_{A-bimod} \cong [B, A \vee M]_{a/A}$. *The isomorphism* $[B, B \vee M]_{a/B} \cong [B, A \vee M]_{a/A}$ *will be used without explicit mention later on in this paper.*

Definition 2.3. *The topological derivations R-module of A with values in M is the function R-module* $F_{A \wedge A^{op}}(\Omega_A, M)$. *We denote it by* $\mathbf{Der}_R(A, M)$ *and its ith homotopy group by* $Der_R^{-i}(A, M)$.

The A-bimodule Ω_A is constructed as the q-cofibrant approximation of the homotopy fibre of the multiplication map $A \wedge A \to A$. There exists the following homotopy fibre sequence of R-modules:

$$\mathbf{THH}_R(A, M) \to M \to \mathbf{Der}_R(A, M). \tag{1}$$

Here $\mathbf{THH}_R(A, M)$ is the topological Hochschild cohomology spectrum of A with values in M:

$$\mathbf{THH}_R(A, M) := F_{A \wedge A^{op}}(\tilde{A}, M)$$

where $\tilde{A}$ is the q-cofibrant replacement of A as an A-bimodule.

We will also have a chance to use topological Hochschild *homology* spectrum $\mathbf{THH}^R(A, M) := A \wedge_{A \wedge A^{op}} M$. If the R-algebra A is commutative and the left and right A-module structures on M agree then both $\mathbf{THH}^R(A, M)$ and $\mathbf{THH}_R(A, M)$ are A-modules and there is a weak equivalence of A-modules

$$\mathbf{THH}_R(A, M) \cong F_A(\mathbf{THH}^R(A, M), A).$$

Furthermore in this case the sequence (1) splits giving a canonical weak equivalence $\mathbf{THH}_R(A, M) \simeq \Sigma^{-1}\mathbf{Der}_R(A, M) \vee M$.

Suppose we are given a topological derivation $d : A \to A \vee M$. Consider the following homotopy pullback diagram of R-algebras

$$\begin{array}{ccc} X & \longrightarrow & A \\ \downarrow & & \downarrow \\ A & \xrightarrow{d} & A \vee M \end{array}$$

Here the rightmost downward arrow is the canonical inclusion of a retract. Then we have the following homotopy fibre sequence of R-modules:

$$\Sigma^{-1}M \longrightarrow X \longrightarrow A\,. \tag{2}$$

Definition 2.4. *The homotopy fibre sequence* (2) *is called the topological singular extension associated with the derivation* $d : A \to A \vee M$.

Theorem 2.5. *Let* $\Sigma^{-1}M \to X \to A$ *be a singular extension of R-algebras associated with a derivation* $d : A \to A \vee M$ *and* $f : B \to A$ *a map of R-algebras. Then* f *lifts to an R-algebra map* $B \to X$ *iff a certain element in* $Der_R^0(B, M)$ *is zero. Assuming that a lifting exists the homotopy fibre of the map*

$$F_{R-alg}(B, X) \to F_{R-alg}(B, A)$$

over the point $f \in F_{R-alg}(B,A)$ *is weakly equivalent to* $\Omega^\infty \mathbf{Der}_R(B,\Sigma^{-1}M)$, *the* 0*th space of the spectrum* $\mathbf{Der}_R(B,\Sigma^{-1}M)$.

Theorem 2.6. *Assume that* R *is connective and* A *is a connective* R*-algebra. Then the Postnikov tower of* A

$$A_0 = H\pi_0 A \longleftarrow A_1 \longleftarrow \cdots \longleftarrow A_n \longleftarrow A_{n+1} \longleftarrow \cdots$$

is a tower of R*-algebras. Moreover the homotopy fibre sequences*

$$A_n \longleftarrow A_{n+1} \longleftarrow \Sigma^{n+1} H\pi_{n+1} A$$

are topological singular extensions.

3. Base change and topological derivations of supplemented R-algebras

In this section we discuss topological Hochschild cohomology and topological derivations of supplemented R-algebras and the behavior of the forgetful map l in the hypercohomology spectral sequence. This material is largely parallel to [9], Section 9 and so most of the proofs will be omitted.

We'll start with some general lemmas.

Lemma 3.1. *Let* A, B, C *be* R*-algebras and* M, N, L *be an* $A\wedge B$*-module, an* $C\wedge A^{op}$*-module and a* $C\wedge B$*-module respectively. Then there is a natural isomorphism of* R*-modules:*

$$F_{A\wedge B}(M, F_C(N,L)) \cong F_{C\wedge B}(N\wedge_A M, L).$$

Proof. Let us first check the above equivalences for $M = A\wedge B\wedge \tilde{M}$ and $N = C\wedge A^{op}\wedge \tilde{N}$. We have:

$$\begin{aligned} F_{A\wedge B}(M, F_C(N,L)) &\cong F_{A\wedge B}(A\wedge B\wedge \tilde{M}, F_C(C\wedge A^{op}\wedge \tilde{N}, L)) \\ &\cong F(\tilde{M}, F(A^{op}\wedge \tilde{N}, L)) \\ &\cong F(\tilde{M}\wedge A^{op}\wedge \tilde{N}, L). \end{aligned}$$

Likewise,

$$\begin{aligned} F_{C\wedge B}(N\wedge_A M, L) &\cong F_{C\wedge B}(C\wedge A^{op}\wedge \tilde{N}\wedge_A A\wedge B\wedge \tilde{M}, L) \\ &\cong F(A^{op}\wedge \tilde{N}\wedge \tilde{M}, L). \end{aligned}$$

Observe that the above isomorphisms are natural in M and N that is, with respect to arbitrary maps of $A\wedge B$ modules $A\wedge B\wedge \tilde{M}_1 \to A\wedge B\wedge \tilde{M}_2$ and of $C\wedge A^{op}$-modules $C\wedge A^{op}\wedge \tilde{N}_1 \to C\wedge A^{op}\wedge \tilde{N}_2$ (not only those coming from $\tilde{M}_1 \to \tilde{M}_2$ and $\tilde{N}_1 \to \tilde{N}_2$). To obtain the general case it suffices to notice that for any M and N there exist standard split coequalizers of R-modules

$$(A\wedge B)^{\wedge 2}\wedge M \rightrightarrows A\wedge B\wedge M \longrightarrow M$$

and

$$(C \wedge A^{op})^{\wedge 2} \wedge N \rightrightarrows C \wedge A^{op} \wedge N \longrightarrow N .$$

With this Lemma 3.1 is proved. □

Now let $C = N$ be an A-bimodule via an R-algebra map $f : A \to C$. Then the $C \wedge B$-module L acquires a structure of an $A \wedge B$-module via the map

$$A \wedge B \xrightarrow{f \wedge id} C \wedge B .$$

Furthermore simple diagram chase shows that the isomorphism $F_C(C, L) \cong L$ is in fact an isomorphism of $A \wedge B$-modules. This gives the following

Corollary 3.2. *There exists the following natural isomorphism of R-modules:*

$$F_{C\wedge B}(C \wedge_A M, L) \cong F_{A\wedge B}(M, F_C(C, L)) \cong F_{A\wedge B}(M, L). \tag{3}$$

We will refer to the isomorphism (3) as base change. Related formulae are found in [7], III.6.

Corollary 3.3. *If an R-algebra B is an A-bimodule via a q-cofibration of R-algebras $A \to B$, then $\mathbf{THH}_R(A, B) \simeq F_{B\wedge A^{op}}(\tilde{B}, \tilde{B})$ where $\tilde{B}$ is the q-cofibrant approximation of the $B \wedge A^{op}$-module B. In particular, $\mathbf{THH}_R(A, B)$ is an R-algebra under the composition product.*

Proof. Denoting by $\tilde{A}$ the q-cofibrant approximation of the A-bimodule A we have the following isomorphisms of R-modules:

$$\begin{aligned} \mathbf{THH}_R(A, B) &\cong F_{A\wedge A^{op}}(\tilde{A}, B) \\ &\cong F_{\tilde{A}\wedge A^{op}}(A, F_B(B, B)) \\ &\cong F_{B\wedge A^{op}}(B \wedge_A \tilde{A}, B). \end{aligned}$$

The $B \wedge A^{op}$-module $B \wedge_A \tilde{A} \cong B \wedge A^{op} \wedge_{A\wedge A^{op}} \tilde{A}$ is a q-cofibrant $B \wedge A^{op}$-module because the functor $? \to B \wedge A^{op} \wedge_{A\wedge A^{op}} ?$ preserves q-cofibrant modules. Therefore $F_{B\wedge A^{op}}(B \wedge_A \tilde{A}, B)$ represents derived function $B \wedge A^{op}$-module and is equivalent to $F_{B\wedge A^{op}}(\tilde{B}, \tilde{B})$. □

We now discuss topological Hochschild cohomology and derivations of supplemented R-algebras. Let A be a q-cofibrant R-algebra. We say that A is supplemented if it is supplied with an R-algebra morphism $\epsilon : A \to B$ which we will assume to be a q-cofibration of R-algebras. Denote by Ω_A^B the homotopy fibre of the map $B \wedge A \to B$ that determines the structure of a right A-module on B. We will assume without loss of generality that Ω_A^B is a q-cofibrant right A-module. Recall that the module of differentials Ω_A for A is defined from the homotopy fibre sequence

$$\Omega_A \to A \wedge A \to A$$

where the second arrow is the multiplication map. Smashing this fibre sequence on the left with B over A we get the fibre sequence

$$B \wedge_A \Omega_A \to B \wedge A \to B.$$

That shows that Ω_A^B is weakly equivalent to $B \wedge_A \Omega_A$ as a $B \wedge A$-module. Further, base change gives a weak equivalence

$$F_{B\wedge A^{op}}(\Omega_A^B, B) \simeq F_{B\wedge A^{op}}(B \wedge_A \Omega_A, B) \simeq F_{A\wedge A^{op}}(\Omega_A, B) \cong \mathbf{Der}_R(A, B).$$

Recall from [9] that there is a 'universal derivation' $d : A \to \Omega_A$ which is defined as the composite map

$$A \to A \vee \Omega_A \to \Omega_A$$

where the first map is the map of algebras over A adjoint to the identity map $\Omega_A \to \Omega_A$ and the second map is the projection. The universal derivation allows one to define the forgetful map $l : \mathbf{Der}_R(A, B) \to F(A, B)$ as the composite map

$$\mathbf{Der}_R(A, B) \simeq F_{A\wedge A^{op}}(\Omega_A, B) \to F(\Omega_A, B) \to F(A, B)$$

where the last map is induced by d. In terms of Ω_A^B the forgetful map l admits the following description. The fibre sequence $\Omega_A^B \to B \wedge A \to B$ splits via the map $B \cong B \wedge R \stackrel{id\wedge 1}{\longrightarrow} B \wedge A$ so there is a weak equivalence of R-modules $B \wedge A \simeq B \vee \Omega_A^B$. Denote by $\overline{d} : A \to \Omega_A^B$ the following composite map

$$A \xrightarrow{\epsilon\wedge id} B \wedge A \simeq B \vee \Omega_A^B \longrightarrow \Omega_A^B$$

the last arrow being the projection onto the wedge summand. Then l coincides with the following composition:

$$\mathbf{Der}_R(A, B) = F_{B\wedge A^{op}}(\Omega_A^B, B) \longrightarrow F(\Omega_A^B, B) \longrightarrow F(A, B)$$

the first arrow being the forgetful map and the second one is induced by $\overline{d}$.

Next we discuss the behavior of the forgetful map l in the hypercohomology spectral sequence. To do this we need to review algebraic Hochschild cohomology for supplemented algebras. The exposition will be somewhat sketchy since it is parallel to [9], Section 9.

Definition 3.4. *Let A_* be a graded algebra over a graded commutative algebra R_* supplied with an R_*-algebra map $\epsilon : A_* \to B_*$ (supplementation). Then algebraic Hochschild cohomology of A_* with coefficients in B_* is defined as*

$$HH^*_{R_*}(A_*, B_*) = Ext^*_{B_*\otimes^L_{R_*} A^{op}_*}(B_*, B_*)$$

where $\otimes^L_{R_*}$ denotes the derived tensor product

Remark 3.5. *If A_* is flat as an R_*-module, then this definition is equivalent to the standard one found in, e.g.,* [5]

We also have a generalization of the standard complex which computes Hochschild cohomology. Let $\tilde{A}_*$ be a differential graded supplemented R_*-algebra which is quasiisomorphic to A_* and R_*-projective. Denote by $\tilde{\epsilon} : \tilde{A}_* \to B_*$ its supplementation. Consider the bar-resolution of the right $\tilde{A}_*$-module B_*; here and later on $\otimes$ stands for $\otimes_{R_*}$

$$B_* \longleftarrow B_* \otimes \tilde{A}_* \longleftarrow B_* \otimes \tilde{A}_* \otimes \tilde{A}_* \longleftarrow \cdots \tag{4}$$

with the usual bar differential

$$\partial(b \otimes \tilde{a_1} \otimes \ldots \otimes \tilde{a_n}) = \pm b\tilde{\epsilon}(\tilde{a}_1) \otimes \tilde{a}_2 \ldots \otimes \tilde{a}_n + \Sigma \pm b \otimes \tilde{a}_1 \otimes \ldots \otimes \tilde{a}_i\tilde{a}_{i+1} \otimes \ldots \otimes \tilde{a}_n.$$

We do not specify the signs in this well-known formula, see, e.g., [11], Chapter X. This is actually a bicomplex since $\tilde{A}$ is a differential graded algebra. Applying the functor $Hom_{B_*\otimes\tilde{A}_*^{op}}(?, B_*)$ to (4) we get the standard Hochschild cohomology (bi)complex

$$C^{ij}(A_*, B_*) = Hom^i(\tilde{A}_*^{\otimes j}, B_*).$$

Now define the module of differentials $\Omega_{A_*}^{B_*}$ from the following short exact sequence

$$0 \longrightarrow \Omega_{A_*}^{B_*} \longrightarrow B_* \otimes \tilde{A}_* \longrightarrow B_* \longrightarrow 0\,.$$

Clearly $\Omega_{A_*}^{B_*}$ is quasiisomorphic as a complex of right $\tilde{A}_*$-modules to the truncated bar-resolution:

$$B_* \otimes \tilde{A}_*^{\otimes 2} \longleftarrow B_* \otimes \tilde{A}_*^{\otimes 3} \longleftarrow \cdots \tag{5}$$

The universal derivation $d : \tilde{A}_* \to \Omega_{A_*}^{B*}$ is induced by the map $\tilde{A}_* \to B_* \otimes \tilde{A}_*$, $\tilde{a} \to 1 \otimes \tilde{a} - \tilde{\epsilon}(\tilde{a}) \otimes 1$. If we take the complex (5) as a model for $\Omega_{A_*}^{B_*}$ then the universal derivation d is a map of complexes

$$d : \tilde{A} \longrightarrow \{B_* \otimes \tilde{A}_*^{\otimes 2} \to B_* \otimes \tilde{A}_*^{\otimes 3} \to \cdots\}$$

where $\tilde{A}$ is considered to be a complex concentrated in degree 0 and $d(\tilde{a}) = -1 \otimes \tilde{a} \otimes 1$.

Further define algebraic derivations of A_* with coefficients in B_* as

$$Der^*_{R_*}(A_*, B_*) = Ext^*_{B_*\otimes\tilde{A}_*^{op}}(\Omega_{A_*}^{B_*}, B_*).$$

Then the (truncated) standard resolution (5) provides a (bi)complex for computing $Der^*_{R_*}(A_*, R_*)$:

$$\overline{C}^*(A_*, B_*) : Hom(\tilde{A}_*, B_*) \longrightarrow Hom(\tilde{A}_*^{\otimes 2}, B_*) \longrightarrow \cdots .$$

(This is indeed a bicomplex, the additional differential being induced from the internal differential in $\tilde{A}_*$.) As in the topological case the universal derivation determines the forgetful map

$$l_{alg} : Der^*_{R_*}(A_*, B_*) = Ext^*_{B_*\otimes\tilde{A}_*^{op}}(\Omega_{A_*}^{B_*}, B_*) \longrightarrow Hom^*(\tilde{A}_*, B_*).$$

Then we have the obvious

Proposition 3.6. *The forgetful map*

$$Der^*_{R_*}(A_*, B_*) \to Hom^*(\tilde{A}_*, B_*) = Ext^*_{R_*}(A_*, B_*)$$

is induced by the projection $\overline{C}^*(A_*, B_*) \to Hom^*(\tilde{A}_*, B_*)$ *times* (-1).

Returning to our topological situation we have the following result which is analogous to Proposition 9.3 in [9]:

Proposition 3.7. *Let A, B be R-algebras, $A \to B$ is an R-algebra map. Suppose that the Kunneth spectral sequence for $\pi_*(B \wedge A^{op})$ collapses and there is a ring isomorphism*

$$\pi_*(B \wedge A^{op}) \cong B_* \otimes^L_{R_*} A^{op}_*.$$

Then there are the following spectral sequences

$$^1E_2^{**} = Der^*_{R_*}(A_*, B_*) = Ext^*_{Tor^{R_*}_*(B_*, A^{op}_*)}(\Omega^{B_*}_{A_*}, R_*) \Rightarrow Der^*(A, B);$$

$$^2E_2^{**} = Ext^*_{R_*}(A_*, B_*) \Rightarrow [A, B]^*.$$

Furthermore, the forgetful map $l : Der^*(A, B) \to [A, B]^*$ induces a map of spectral sequences $^1E_*^{**} \to {}^2E_*^{**}$ which on the level of E_2-terms gives the forgetful map $l_{alg} := Der^*_{R_*}(A_*, B_*) \to Ext^*_{R_*}(A_*, B_*)$.

4. Mappings spaces via derivations

In this section we show that for two R-algebras A and B the higher homotopy groups of the space $F_{R-alg}(A, B)$ can be reduced to the computation of certain topological derivations. This is important because in many cases topological derivations can be further reduced to topological Hochschild cohomology which is an essentially stable object, so that one could apply standard methods of homological algebra for computation. As usual, we assume that A is a q-cofibrant R-algebra.

Now consider the R-module $A \vee \Sigma^{-d}A$, $d > 0$. It can be supplied canonically with the structure of an R-algebra over A so that $\Sigma^{-d}A$ is a 'square-zero ideal'. Denote this R-algebra by A_d.

Let us also introduce the algebra $A(d) := A^{S^d}$, the cotensor of A and the d-sphere S^d. Then $A(d) \cong F(\mathbf{R}\Sigma^\infty S^d_+, A)$ as an R-module (here $\mathbf{R}$ stands for the free R-module functor. The structure of an R-algebra on $A(d)$ is induced by the R-algebra structure on A and the topological diagonal $S^d \to S^d \times S^d$. The coefficient rings of $A(d)$ and A_d are both isomorphic to the exterior algebra $\Lambda_{A_*}(y)$ where y has degree $-d$. There is also a weak equivalence of R-modules:

$$A(d) \simeq A \vee \Sigma^{-d}A \cong A_d.$$

Notice that both $A(d)$ and A_d are R-algebras over A, that is there exist maps of R-algebras $A(d) \to A$ and $A_d \to A$. (The first map is induced by choosing a base point in S^d, the second map is the canonical projection.)

Theorem 4.1. *The R-algebras $A(d)$ and A_d are weakly equivalent in the category of R-algebras.*

Proof. First consider the case $A = R$. Since R is an $R(d)$-module it makes sense to consider self-maps of R in the category of $R(d)$-modules. Notice that $R(d)$ is actually a commutative R-algebra so we need not distinguish between left and right $R(d)$-modules.

Lemma 4.2.

$$\pi_* F_{R(d)}(R, R) = R_*[[x]]$$

where the element x has degree $d-1$.

Proof. Assume first that $R = S$, the sphere spectrum. We need this special case because the connectiveness of S will be used. If R is connective this step could be skipped. Consider the spectral sequence

$$\begin{aligned} E_2^{**} &= Ext^{**}_{S(d)_*}(S_*, S_*) \\ &= Ext^{**}_{\Lambda_{S_*}(y)}(S_*, S_*) \\ &= S_*[[x]] \Longrightarrow \pi_* F_{S(d)}(S, S). \end{aligned}$$

Here the element x has degree $d-1$. This spectral sequence collapses for dimensional reasons. By Boardman's criterion [3] it converges strongly to its target which is complete with respect to the (cobar) filtration. Since this filtration coincides with the x-adic filtration on the associated graded S_*-module we conclude that

$$\pi_* F_{S(d)}(S, S) = S_*[[x]].$$

Notice that the fact that S is connective was used to show the collapse of our spectral sequence. For instance if $d = 1$ the elements x^k are located along the line of slope 1 and the whole spectral sequence E_2^{**} lies above it.

Now let R be an arbitrary commutative S-algebra. Consider the spectral sequence

$$\begin{aligned} {}'E_2^{**} &= Ext^{**}_{R(d)_*}(R_*, R_*) \\ &= Ext^{**}_{\Lambda_{R_*}(y)}(R_*, R_*) \\ &= R_*[[x]] \Longrightarrow \pi_* F_{R(d)}(R, R) \end{aligned}$$

and notice that the unit map $S \to R$ determines the map of spectral sequences $E_2^{**} \to {}'E_2^{**}$ taking x to x. It follows that ${}'E_2^{**}$ collapses proving our claim. □

Let us now return to the proof of the theorem; recall that we are still handling the special case $A = R$. Consider the set of maps $R(d) \to R_d$ in the homotopy category of R-algebras over R. By Theorem 2.1 this set is an abelian group of topological derivations of $R(d)$ with values in $\Sigma^{-d} R$. Since $R(d)$ is a commutative R-algebra there is a canonical splitting

$$THH_R^*(R(d), R) \simeq R \vee Der_R^{*-1}(R(d), R).$$

So the computation of $Der_R^*(R(d), R)$ reduces to the computation of topological Hochschild cohomology $THH_R^*(R(d), R)$. Further Corollary 3.3 provides an isomorphism

$$THH_R^*(R(d), R) \cong F_{R(d)}(R, R).$$

It follows from Lemma 4.2 that the spectral sequence

$$\begin{aligned} E_2^{ij} &= Der_{R_*}(R(d)_*, R_*) \\ &= Ext^*_{R(d)_*}(\Omega^{R_*}_{R(d)_*}, R_*) \Rightarrow Der_R^*(R(d), R) \end{aligned}$$

collapses and

$$Der_R^{*-1}(R(d),R) = R_*[[x]]/R_*.$$

Further,

$$[R(d),R]^* = R_* \oplus \Sigma^d R_* = \Lambda_{R_*}(z)$$

where the symbol z has degree d (of course, we do not claim the existence of any multiplicative structure). The element z maps the wedge summand $\Sigma^{-d}R$ of the R-module $R(d)$ isomorphically to $\Sigma^{-d}R$, and the other wedge summand maps to zero. It follows from Proposition 3.7 that the image of $x \in Der^*(R(d),R)$ in $[R(d),R]^*$ under the forgetful map is z (up to an invertible factor).

In other words we proved that there exists a topological derivation of $R(d)$ with values in R, that is a map in the homotopy category of R-algebras over R

$$R(d) \to R_d \tag{6}$$

such that the wedge summand $\Sigma^{-d}R$ of $R(d)$ maps isomorphically onto the corresponding wedge summand of R_d. Therefore the map (6) is a weak equivalence of R-algebras and our theorem is proved (in the special case $A = R$). To get the general case consider the canonical map

$$R(d) \wedge A = F(\mathbf{R}\Sigma^\infty S^d_+, R) \wedge A \to F(\mathbf{R}\Sigma^\infty S^d_+, A) = A(d). \tag{7}$$

Since A is a q-cofibrant R-algebra the point-set level smash product $F(\mathbf{R}\Sigma^\infty S^d_+, R) \wedge A$ represents the derived smash product. Further (7) is a weak equivalence since $\mathbf{R}\Sigma^\infty S^d_+$ is a finite cell R-module and diagram chase shows that this is an R-algebra map. So we have the following equivalences of R-algebras:

$$A_d \cong R_d \wedge A \simeq R(d) \wedge A \simeq A(d).$$

With this Theorem 4.1 is proved. □

Now suppose that we have another R-algebra B and a map $f : A \to B$ of R-algebras. Then the pair $(F_{R-alg}(A,B), f)$ is a pointed topological space. This space turns out to be closely related to $Der_R(A,B)$. We have the following theorem:

Theorem 4.3. *For a q-cofibrant algebra A and a map of R-algebras $f : A \to B$ the space of d-fold loops $\Omega^d(F_{R-alg}(A,B), f)$ is weakly equivalent to the space $\Omega^\infty \mathbf{Der}_R(A, \Sigma^{-d}B)$.*

Proof. For two topological spaces X and Y we will denote the space of maps between them by $\mathcal{T}(X,Y)$ ($\mathcal{T}_*(X,Y)$ in the pointed case). Then we have the following commutative diagram of spaces where both rows are homotopy fibre sequences:

$$\begin{array}{ccccc}
\mathcal{T}_*(S^d, F_{R-alg}(A,B)) & \longrightarrow & \mathcal{T}(S^d, F_{R-alg}(A,B)) & \longrightarrow & F_{R-alg}(A,B)\,. \\
\downarrow & & \downarrow & & \| \\
? & \longrightarrow & F_{R-alg}(A, B^{S^d}) & \longrightarrow & F_{R-alg}(A,B)
\end{array}$$

Here the horizontal rightmost arrows are both induced by the inclusion of the base point into S^d. Since the right and the middle vertical arrows are weak equivalences (even isomorphisms) it follows that the map $\mathcal{T}_*(S^d, Map(A,B)) \to ?$ is a weak equivalence. But Theorem 4.1 tells us that the R-algebra B^{S^d} is weakly equivalent as an R-algebra to $B \vee \Sigma^{-d}B$. In other words the term ? is weakly equivalent to the topological space of maps $A \to B \vee \Sigma^{-d}B$ which specialize to the given map f when composed with the projection $B \vee \Sigma^{-d}B \to B$. Therefore ? is weakly equivalent to $\Omega^\infty \mathbf{Der}_R(A, \Sigma^{-d}B)$ and our theorem is proved. □

Corollary 4.4. *For a q-cofibrant algebra A and a map of R-algebras $f : A \to B$ there is a bijection between sets $\pi_d(F_{R-alg}(A,B), f)$ and $Der_R^{-d}(A,B)$ for $d \geq 1$. If $d \geq 2$ then this bijection is an isomorphism of abelian groups.*

Remark 4.5. *One might wonder whether Theorem 4.3 remains true in the context of commutative S-algebras. The answer is no. The crucial point is the weak equivalence of S-algebras $S \vee S^{-1}$ and S^{S^1}. It is clear that $\pi_0 S \wedge_{S\vee S^{-1}} S$ is the divided power ring. However N. Kuhn and M. Mandell proved that $\pi_0 S \wedge_{S^{S^1}} S$ is the ring of numeric polynomials. Therefore $S \vee S^{-1}$ and S^{S^1} cannot be weakly equivalent as commutative S-algebras.*

We see, that the space $F_{R-alg}(A,B)$ when looped only once becomes an infinite loop space. This is somewhat surprising since $F_{R-alg}(A,B)$ is hardly ever an infinite loop space itself. In particular the set of connected components of $F_{R-alg}(A,B)$ does not have to be a group, let alone an abelian group. Therefore the connection between $\pi_0 F_{R-alg}(A,B)$ and $Der_R^0(A,B)$ (provided the latter is defined) may be rather weak. For instance the set of homotopy classes of A_∞ self-maps of the p-completed K-theory spectrum is the multiplicative group of p-adic integers whereas the corresponding topological derivations spectrum can be proved to be contractible. A generalization of this example is discussed in author's work [10]. However there is some evidence for the following

Conjecture 4.6. *For an R-algebra map $f : A \to B$ the connected component of f in $F_{R-alg}(A,B)$ is weakly equivalent to the connected component of $\Omega^\infty \mathbf{Der}_R(A,B)$. In particular it is an infinite loop space.*

To see why this conjecture has a chance of being true notice that the Whitehead products in the homotopy groups of $F_{R-alg}(A,B)$ determine via Theorem 4.3 various brackets in $Der_R^*(A,B)$ and, for commutative A and B - also in $THH_R^*(A,B)$. No such brackets have been recorded so far and it seems likely that they should all vanish. This suggests that the connected component of f in $F_{R-alg}(A,B)$ is an H-space.

There is another interesting question raised by Theorem 4.3. In a recent work [12] J. McClure and J. Smith introduced the Gerstenhaber bracket on $\mathbf{THH}_R(A,A)$. Their work probably implies the existence of the bracket on $\mathbf{Der}_R(A,A)$. This is surely the case if A is commutative since then $\mathbf{Der}_R(A,A)$ splits off $\mathbf{THH}_R(A,A)$ as a wedge summand. Then via Theorem 4.3 a Poisson bracket is defined on $\pi_* F_{R-alg}(A,A)$ for $* > 0$.

Conjecture 4.7. *The bracket described above agrees with the Whitehead product on $BF_{R-alg}(A, A)$, the classifying space of the monoid $F_{R-alg}(A, A)$.*

We see that the problem of computing $\pi_0 F_{R-alg}(A, B)$ differs sharply from computing higher homotopy groups. This problem is usually much harder, being essentially nonabelian. However there is one case when it is possible to give a complete general answer.

Theorem 4.8. *Assuming that R is connective let A be a connective q-cofibrant R-algebra, and B a coconnective R-algebra (i.e., $\pi_i B = 0$ for $i > 0$). Then any $\pi_0 R$-algebra map $\pi_0 A \to \pi_0 B$ lifts to a unique R-algebra map $A \to B$ so that the forgetful map $[A, B]_a \to Hom_{\pi_0 R-alg}(\pi_0 A, \pi_0 B)$ is bijective. Moreover the topological space $F_{R-alg}(A, B)$ is homotopically discrete, i.e., $\pi_i F_{R-alg}(A, B) = 0$ for $i > 0$.*

Similarly if A and B are both commutative R-algebras, where B is coconnective and A is q-cofibrant and connective then the forgetful map $[A, B]_c \to Hom_{\pi_0 R-alg}(\pi_0 A, \pi_0 B)$ is bijective and the space of commutative R-algebra maps from A to B is homotopically discrete.

Proof. We will deal only with the associative case, the commutative one being completely analogous. Picking a system of generators and relations for the $\pi_0 R$-algebra $\pi_0 A$ we construct the following pushout diagram in the category of R-algebras:

$$\begin{array}{ccc} T(\coprod_J R) & \longrightarrow & R \\ \downarrow & & \downarrow \\ T(\coprod_I R) & \longrightarrow & A^0 \end{array} \qquad (8)$$

Here the sets I and J run respectively through the systems of generators and relations in $\pi_0 A$. There is a canonical R-algebra map from A^0 to A that induces an isomorphism on zeroth homotopy group. The R-algebra A^0 is the zeroth skeleton of A in the category of R-algebras and (the CW-approximation of) A is obtained from A^0 by attaching R-algebra cells in higher dimensions. Then induction up the CW-filtration of A shows that the map $A^0 \to A$ induces a weak equivalence $F_{R-alg}(A, B) \simeq F_{R-alg}(A^0, B)$.

Further applying the functor $F_{R-alg}(?, B)$ to the diagram (8) we get the following homotopy pullback of topological spaces:

$$\begin{array}{ccc} \prod_J \pi_0 B & \longrightarrow & pt \\ \downarrow & & \downarrow \\ \prod_I \pi_0 B & \longrightarrow & F_{R-alg}(A^0, B) \end{array}$$

It follows that the space $F_{R-alg}(A^0, B)$ is homotopically discrete with the set of connected components being equal to $Hom_{\pi_0 R-alg}(\pi_0 A, \pi_0 B)$. $\square$

Now let k be an associative ring. Recall that according to [7], Proposition IV.3.1 the Eilenberg-MacLane spectrum Hk admits a structure of an S-algebra or a commutative S-algebra if k is commutative. Theorem 4.8 shows that this structure is unique up to a weak equivalence of S-algebras or commutative S-algebras. We also have the following evident corollary which will be used in the next section.

Corollary 4.9. *Let A be either a connective q-cofibrant S-algebra or a connective commutative S-algebra. Then the topological space of S-algebra maps (or commutative S-algebra maps) from A to $H\pi_0 A$ is homotopically discrete and*

$$\pi_0 F_{S-alg}(A, H\pi_0 A) = End_{rings}(\pi_0 A).$$

5. Spaces of multiplicative self-maps of MU

In this section we study the homotopy groups of A_∞-maps from the complex cobordism spectrum MU into itself. Our main result here is that any homotopy multiplicative operation $MU \to MU$ lifts canonically to an S-algebra map. We also calculate completely higher homotopy groups of S-algebra maps out of MU into an arbitrary MU-algebra E. In this section we work with various homotopy categories and so smash products and function spectra are understood in the derived sense.

Before we state our main theorem we need to introduce the notion of $\mathbb{Q}$-commutative S-algebras and $\mathbb{Q}$-preferred S-algebra maps.

Definition 5.1. *Let A be an S-algebra and denote by $A_\mathbb{Q}$ its rationalization. We say that A is $\mathbb{Q}$-commutative if the $A_\mathbb{Q}$ is weakly equivalent as an S-algebra to a commutative S-algebra.*

Remark 5.2. *Later on all $\mathbb{Q}$-commutative S-algebras which we encounter will in fact be commutative. Notice, however, that it is not always the case. Denote by $S[x_i]$ the free S-algebra on the S-module S_S^{2i}, the cell approximation of the $2i$-dimensional sphere. Then clearly $S_\mathbb{Q}[x_i]$ is weakly equivalent to the free commutative S-algebra on $S_\mathbb{Q}^{2i}$. Therefore $S[x_i]$ is a $\mathbb{Q}$-commutative S-algebra which is not commutative unless $i = 0$.*

Consider two $\mathbb{Q}$-commutative S-algebras $A_\mathbb{Q}$ and $B_\mathbb{Q}$. We have the following maps:

$$k : [A_\mathbb{Q}, B_\mathbb{Q}]_c \longrightarrow [A_\mathbb{Q}, B_\mathbb{Q}]_a \longleftarrow [A, B]_a : q\ .$$

Here k is the forgetful map and q is induced by rationalization.

Definition 5.3. *A map $f \in [A, B]_a$ is called $\mathbb{Q}$-preferred if $q(f) \in [A_\mathbb{Q}, B_\mathbb{Q}]_a$ is in the image of k. Similarly for an A-bimodule M which is $\mathbb{Q}$-symmetric (that is, the square-zero extension $A \vee M$ is $\mathbb{Q}$-commutative) an S-algebra derivation $d : A \to A \vee M$ is called $\mathbb{Q}$-preferred if d is $\mathbb{Q}$-preferred as an S-algebra map.*

In other words a map of S-algebras (or a topological derivation) is $\mathbb{Q}$-preferred if it lifts to a map (to a derivation) of commutative S-algebras after rationalization.

Theorem 5.4. *The forgetful map of monoids*

$$[MU, MU]_a \longrightarrow Mult(MU, MU)$$

admits a unique section whose image consists of $\mathbb{Q}$-preferred S-algebra maps.

The proof will be given below after a succession of lemmas.

Remark 5.5. *The set $Mult(MU, MU)$ is relatively well understood. One can describe it for example as the set of all MU_*-algebra maps*

$$MU_*MU = MU_*[t_1, t_2, \ldots] \longrightarrow MU_* \ .$$

Our next result is the computation of topological Hochschild cohomology of MU with coefficients in an MU-algebra E. Since there is a canonical splitting of spectra

$$\mathbf{THH}_S(MU, E) \simeq E \vee \Sigma^{-1}\mathbf{Der}_S(MU, E)$$

the combination of this result with Corollary 4.4 gives a complete calculation of higher homotopy groups of the based space $F_{S-alg}(MU, E)$.

Proposition 5.6. *For an MU-algebra E considered as an MU-bimodule the following isomorphism holds*

$$THH_S^*(MU, E) \cong \hat{\Lambda}_{E_*}(y_1, y_2, \ldots)$$

where the hat denotes the completed exterior algebra and the exterior generator y_i has cohomological degree $2i - 1$.

Proof. Consider the topological Hochschild homology S-module of MU with coefficients in E,

$$\mathbf{THH}^S(MU, E) := MU \wedge_{MU \wedge MU} E \cong MU \wedge_{MU \wedge MU} MU \wedge_{MU} E.$$

We have the spectral sequence of MU_*-algebras

$$\begin{aligned} E^2_{**} &= Tor^{MU_*MU}_{**}(MU_*, MU_*) \\ &= MU_* \otimes \Lambda(\tilde{y}_1, \tilde{y}_2, \ldots) \Rightarrow \pi_*\mathbf{THH}^S(MU, MU). \end{aligned}$$

Since the differentials applied to the exterior generators $\tilde{y}_i$ are trivial for dimensional reasons we conclude that it collapses. It follows that

$$\begin{aligned} \pi_*\mathbf{THH}^S(MU, E) &= \pi_*\mathbf{THH}^S(MU, MU) \otimes_{MU_*} E_* \\ &= E_* \otimes \Lambda(\tilde{y}_1, \tilde{y}_2, \ldots). \end{aligned}$$

Now the result for topological Hochschild cohomology follows by virtue of the universal coefficients formula and the isomorphism

$$\mathbf{THH}_S(MU, E) \cong F_E(\mathbf{THH}^S(MU, E), E).$$

Proposition (5.6) is proved. □

Recall that we are using the notation $S_{\mathbb{Q}}[x_i]$ for the free commutative S-algebra on the S-module $S^{2i}_{\mathbb{Q}}$, the rationalized $2i$-sphere S-module. The coefficient ring of $S_{\mathbb{Q}}[x_i]$ is isomorphic to $\mathbb{Q}[x_i]$ where the polynomial generator x_i has degree $2i$. Further denote the infinite smash power $S_{\mathbb{Q}}[x_1]^{\wedge\infty}$ by $S_{\mathbb{Q}}[x_1, x_2, \ldots]$.

Lemma 5.7. *There is a weak equivalence of commutative S-algebras*

$$S_{\mathbb{Q}}[x_1, x_2, \ldots] \longrightarrow MU_{\mathbb{Q}} .$$

Proof. The polynomial generators x_i of the ring $MU_{\mathbb{Q}*} = \mathbb{Q}[x_1, x_2, \ldots]$ determine a collection of maps $S^{2i}_{\mathbb{Q}} \to MU_{\mathbb{Q}}$ and therefore a map of commutative algebras $S_{\mathbb{Q}}[x_1]^{\wedge\infty} \to MU_{\mathbb{Q}}$ which is clearly a weak equivalence. □

Definition 5.8. *Let E be a ring spectrum (in the traditional up to homotopy sense) with multiplication $m : E \wedge E \to E$ and M an E-bimodule spectrum with the left action $m_l : E \wedge M \to M$ and the right action $m_r : M \wedge E \to M$. We say that a map $f : E \to M$ is a primitive operation if $f \circ m$ and $m_r \circ (f \wedge id) + m_l \circ (id \wedge f)$ are homotopic as maps from $E \wedge E$ to M. The set of all primitive operation from E to M is denoted by $Prim(E, M)$*

Remark 5.9. *Perhaps it is more natural to use the term 'derivation' instead of 'primitive operation' but this term is already overworked in this paper.*

The next lemma provides a description of topological derivations of MU with coefficients in $H\mathbb{Z}$, the integral Eilenberg-MacLane spectrum.

Lemma 5.10. *There is the following isomorphism of graded abelian groups:*

$$Der^*_S(MU, H\mathbb{Z}) \cong \Lambda^{*-1}(y_1, y_2 \ldots)/\mathbb{Z}.$$

Under the forgetful map

$$l : Der^*_S(MU, H\mathbb{Z}) \longrightarrow [MU, H\mathbb{Z}]^* = Hom(\mathbb{Z}[t_1, t_2, \ldots], \mathbb{Z})$$

the elements $y_i \in Der^{2i-2}_S(MU, H\mathbb{Z})$ correspond to the derivations ∂_{t_i} evaluated at 0. Moreover the elements y_i are $\mathbb{Q}$-preferred topological derivations.

Proof. We have the spectral sequence

$$Der^*(H\mathbb{Z}_* MU, \mathbb{Z}) = \Lambda^{*-1}(y_1, y_2 \ldots)/\mathbb{Z}$$
$$\Rightarrow Der^*_{H\mathbb{Z}}(H\mathbb{Z} \wedge MU, H\mathbb{Z}) = Der^*_S(MU, H\mathbb{Z}).$$

This spectral sequence clearly collapses. Next using Proposition 3.7 we see that the image of the element y_i under the forgetful map l in the group

$$[H\mathbb{Z} \wedge MU, H\mathbb{Z}]^*_{H\mathbb{Z}-mod} = [MU, H\mathbb{Z}]^* = Hom^*(\mathbb{Z}[t_1, t_2, \ldots], \mathbb{Z})$$

is precisely the algebraic derivation ∂_{t_i} evaluated at 0 (up to elements of higher filtration). Since this image is contained in the subgroup of primitive operations $MU \to H\mathbb{Z}$ none of these elements of higher filtration are present.

To see that y_i are $\mathbb{Q}$-preferred derivations let us introduce the notation $CDer^*(MU_{\mathbb{Q}}, H\mathbb{Q})$ to denote topological derivations of $MU_{\mathbb{Q}}$ with values in $H\mathbb{Q} \simeq$

$S_{\mathbb{Q}}$ in the category of commutative S-algebras. (These commutative derivations are also known as topological André-Quillen cohomology, cf. [2].) Then since $MU_{\mathbb{Q}}$ is a free commutative S-algebra we see immediately that

$$CDer^*(MU_{\mathbb{Q}}, H\mathbb{Q}) = Der^*(MU_{\mathbb{Q}*}, \mathbb{Q}) = \mathbb{Q}\left\langle \bar{\partial}_{x_1}, \bar{\partial}_{x_2}, \ldots \right\rangle$$

the right-hand side being the set of derivations (in the usual algebraic sense) of the algebra $MU_{\mathbb{Q}*}$ with values in the rational numbers. Here we denoted by $\bar{\partial}_{x_i}$ the standard derivation ∂_{x_i} of the ring $MU_{\mathbb{Q}} = \mathbb{Q}[x_1, x_2, \ldots]$ composed with evaluation at zero.

On the other hand $Der^*_S(MU_{\mathbb{Q}}, H\mathbb{Q}) \cong \Lambda^{*-1}_{\mathbb{Q}}(y_1, y_2 \ldots)/\mathbb{Q}$. We need to prove therefore that the forgetful map

$$CDer^*(MU_{\mathbb{Q}}, H\mathbb{Q}) \longrightarrow Der^*_S(MU_{\mathbb{Q}}, H\mathbb{Q})$$

sends the elements $\bar{\partial}_{x_i}$ to y_i.

Since the commutative S-algebra $H\mathbb{Q}[x_i] \simeq S_{\mathbb{Q}}[x_i]$ is free as a commutative S-algebra as well as an (associative) S-algebra it follows that

$$CDer^*(H_{\mathbb{Q}}[x_i], H\mathbb{Q}) = Der^*(\mathbb{Q}[x_i], \mathbb{Q}) = \mathbb{Q}\left\langle \bar{\partial}_{x_i} \right\rangle.$$

There is a unique map of commutative S-algebras $MU_{\mathbb{Q}} \to H\mathbb{Q}[x_i]$ which corresponds to quotienting out the ideal $(x_1, x_2, \ldots, x_{i-1}, \hat{x}_i, x_{i+1}, \ldots)$ in the coefficient ring of $MU_{\mathbb{Q}}$. We have the following commutative diagram:

$$\begin{array}{ccc} CDer^*(H\mathbb{Q}[x_i], H\mathbb{Q}) & \xrightarrow{\cong} & Der^*(\mathbb{Q}[x_i], \mathbb{Q})) \\ \downarrow & & \downarrow \\ CDer^*(MU_{\mathbb{Q}}, H\mathbb{Q}) & \longrightarrow & Der^*_S(MU_{\mathbb{Q}}, H\mathbb{Q}) \end{array}$$

from which it is clear that the image of $\bar{\partial}_{x_i}$ in $Der^*(MU_{\mathbb{Q}}, H\mathbb{Q})$ is y_i and the lemma is proved. □

Corollary 5.11. *The set of $\mathbb{Q}$-preferred derivations of MU with values in $H\mathbb{Z}$ maps bijectively onto the set of primitive cohomology operations $MU \to H\mathbb{Z}$ under the forgetful map $Der^*_S(MU, H\mathbb{Z}) \to [MU, H\mathbb{Z}]^*$.*

Lemma 5.12. *Let A be an S-algebra and B a commutative S-algebra. Suppose that B has a structure of an A-bimodule via a map of S-algebras $A \to B$. Then $\mathbf{THH}_S(A, B)$ has a structure of a B-bimodule and there is a canonical splitting of B-bimodules $\mathbf{THH}_S(A, B) \simeq B \vee \Sigma^{-1}\mathbf{Der}_S(A, B)$.*

Proof. Consider the following sequence of S-algebra maps:

$$B \longrightarrow \mathbf{THH}_S(B, B) \cong F_{B\wedge B^{op}}(B, B) \longrightarrow F_{B\wedge A^{op}}(B, B) \cong \mathbf{THH}_S(A, B).$$

The first map exists because B is commutative, the middle map is induced by the S-algebra map $A \to B$ and the last equivalence is Corollary 3.3. The composite map $B \to \mathbf{THH}_S(A, B)$ supplies $\mathbf{THH}_S(A, B)$ with a structure of a B-bimodule and splits the canonical map $\mathbf{THH}_S(A, B) \to B$. □

Let us introduce the notation MU_n for the nth Postnikov stage of MU. Then MU_n is an S-algebra (even a commutative S-algebra).

Lemma 5.13. *There is the following weak equivalence of MU_{n*}-modules:*

$$THH^*_S(MU, MU_n) \cong \Lambda_{MU_{n*}}(y_1, y_2, \ldots)$$

where MU_n is considered as an MU-bimodule via any (not necessarily central) map of S-algebras $f : MU \to MU_n$.

Proof. Since MU_n is a commutative S-algebra the multiplication map

$$MU_n \wedge MU_n \xrightarrow{m} MU_n$$

is an S-algebra map. Therefore the composition

$$MU \wedge MU_n \xrightarrow{f\wedge id} MU_n \wedge MU_n \xrightarrow{m} MU_n$$

is also an S-algebra map. This gives the following weak equivalence of S-modules:

$$\begin{aligned}\mathbf{THH}_S(MU, MU_n) &\cong F_{MU\wedge MU_n}(MU_n, MU_n)\\ &\cong F_{MU_n}(MU_n \wedge_{MU\wedge MU_n} MU_n, MU_n).\end{aligned}$$

Therefore it is enough to show that

$$\pi_* MU_n \wedge_{MU\wedge MU_n} MU_n = \Lambda_{MU_{n*}}(\tilde{y}_1, \tilde{y}_2 \ldots).$$

(The exterior generators $y_i \in THH^*_S(MU, MU_n)$ will be dual to $\tilde{y}_i$.)

Consider the spectral sequence

$$Tor^{MU_*MU_n}_{**}(MU_{n*}, MU_{n*}) = \Lambda_{MU_{n*}}(\tilde{y}_1, \tilde{y}_2, \ldots) \Rightarrow \pi_* MU_n \wedge_{MU\wedge MU_n} MU_n \quad (9)$$

This spectral sequence is *not* multiplicative since the map $f : MU \to MU_n$ may not be central. However it is a spectral sequence of MU_{n*}-modules.

Let us introduce another spectral sequence

$$Tor^{MU_*MU}_{**}(MU_*, MU_*) = \Lambda_{MU_*}(\tilde{y}_1, \tilde{y}_2, \ldots) \Rightarrow \pi_* MU \wedge_{MU\wedge MU} MU. \quad (10)$$

Then the map $f : MU \to MU_n$ induces a map of spectral sequences (10)$\to$(9). Further the spectral sequence (10) is multiplicative and collapses for that reason. Therefore in (9) all elements of the form $\tilde{y_{i_1}} \wedge \tilde{y_{i_2}} \wedge \ldots \tilde{y_{i_k}}$ are permanent cycles and it follows that (9) collapses. Lemma 5.13 is proved. □

Suppose as before that we have an MU-bimodule structure on MU_{n+1} via some S-algebra map $f : MU \to MU_{n+1}$. Composing f with the canonical map in the Postnikov tower $p_n : MU_{n+1} \to MU_n$ we get an MU-bimodule structure on MU_n also. Then we have the following

Corollary 5.14. *The induced map $Der^*_S(MU, MU_{n+1}) \longrightarrow Der^*_S(MU, MU_n)$ is onto.*

Proof. Indeed, the map

$$\begin{aligned}THH_S^*(MU, MU_{n+1}) &= \Lambda_{MU_{n+1*}}(y_1, y_2, \dots) \\ &\to \Lambda_{MU_{n*}}(y_1, y_2, \dots) = THH_S^*(MU, MU_n)\end{aligned}$$

is clearly onto and our claim follows from Lemma 5.12. □

Proof of Theorem 5.4. We start by outlining the strategy of the proof. Take a multiplicative operation $f \in Mult(MU, MU)$. Define $f_n : MU \to MU_n$ as the composition

$$MU \xrightarrow{f} MU \xrightarrow{p_n} MU_n .$$

(Recall that we denoted by p_n the canonical projection onto the nth Postnikov stage.) Then by Corollary 4.9 the map $f_0 : MU \to MU_0 = H\mathbb{Z}$ is homotopic to a unique S-algebra map which we will denote by $\tilde{f}_0$. Proceeding by induction assume that there exists a unique $\mathbb{Q}$-preferred S-algebra map $\tilde{f}_n : MU \to MU_n$ which is homotopic to f_n when considered as a map of S-modules. We will see that

- $\tilde{f}_n$ admits a $\mathbb{Q}$-preferred lifting to an S-algebra map $MU \to MU_{n+1}$ and
- there is a one-to-one correspondence between such liftings and the set of liftings of $\tilde{f}_n$ to a *homotopy* multiplicative map $MU \to MU_{n+1}$.

In particular f_{n+1} being one of such liftings can be realized as a $\mathbb{Q}$-preferred S-algebra lifting in a unique fashion.

We now proceed to realize the above program in detail. The first thing is to show that there exists a lifting of $\tilde{f}_n$ in the category of S-algebras. The homotopy fibre sequence

$$\Sigma^{n+1} H\pi_{n+1}MU \longrightarrow MU_{n+1} \longrightarrow MU_n \tag{11}$$

is a topological singular extension by Theorem 2.6. Then Theorem 2.5 tells us that the obstruction to an S-algebra lifting $\tilde{f}_n$ to MU_{n+1} is a certain element $\sigma \in Der_S^0(MU, \Sigma H\pi_{n+1}MU)$. More precisely, the extension (11) is associated with a derivation

$$d : MU_n \longrightarrow MU_n \vee \Sigma^{n+2} H\pi_{n+1}MU$$

and $\sigma : MU \to MU_n \vee \Sigma^{n+2} H\pi_{n+1}MU$ is the composition of d with $p_n : MU \to MU_n$.

Furthermore, notice that the set of S-algebra maps $MU_{\mathbb{Q}} \to MU_{\mathbb{Q}n+1}$ is in bijective correspondence with the set of ring maps $MU_{\mathbb{Q}*} \to MU_{\mathbb{Q}n+1*}$. Since $MU_{\mathbb{Q}*}$ is a polynomial algebra we see that a lift of $\tilde{f}_n$ does exist after rationalization. Therefore the image of σ in $Der_S^0(MU, \Sigma^{n+2} H\pi_{n+1}MU_{\mathbb{Q}})$ is zero. But the spectrum $H\pi_{n+1}MU$ is a wedge of suspensions of $H\mathbb{Z}$ and according to Lemma 5.10 the abelian group $Der_S^0(MU, H\pi_{n+1}MU)$ has no torsion. It follows that $\sigma = 0$ as an element in the group $Der_S^0(MU, \Sigma H\pi_{n+1}MU)$ and a lift of $\tilde{f}_n$ exists integrally (though not necessarily $\mathbb{Q}$-preferred). By Theorem 2.5 the homotopy fibre of the map

$$F_{S-alg}(MU, MU_{n+1}) \longrightarrow F_{S-alg}(MU, MU_n)$$

taken over the point $\tilde{f}_n \in F_{S-alg}(MU, MU_n)$ is weakly equivalent to the zeroth space of the spectrum $\mathbf{Der}_S(MU, \Sigma^{n+1} H\pi_{n+1}MU)$.

Therefore denoting by $[MU, MU_{n+1}]_a^{lift} \subset [MU, MU_{n+1}]_a$ the set of homotopy classes of S-algebra maps $MU \to MU_{n+1}$ lifting $\tilde{f}_n$ we have the following long exact sequence:

$$\begin{aligned} \ldots \longrightarrow Der_S^{-1}(MU, \Sigma^{n+1} H\pi_{n+1}MU) &\longrightarrow \pi_1 F_{S-alg}(MU, MU_{n+1}) \\ \longrightarrow \pi_1 F_{S-alg}(MU, MU_n) &\longrightarrow Der_S^0(MU, \Sigma^{n+1} H\pi_{n+1}MU) \\ \longrightarrow [MU, MU_{n+1}]_a^{lift} &\longrightarrow pt \end{aligned}$$

which is the same (by Theorem 4.3) as the long exact sequence

$$\begin{aligned} \ldots \longrightarrow Der_S^{-1}(MU, \Sigma^{n+1} H\pi_{n+1}MU) &\longrightarrow Der_S^{-1}(MU, MU_{n+1}) \\ \longrightarrow Der_S^{-1}(MU, MU_n) &\longrightarrow Der_S^0(MU, \Sigma^{n+1} H\pi_{n+1}MU) \\ \longrightarrow [MU, MU_{n+1}]_a^{lift} &\longrightarrow pt. \end{aligned}$$

By Corollary 5.14 the map

$$Der_S^{-1}(MU, MU_{n+1}) \longrightarrow Der_S^{-1}(MU, MU_n)$$

is onto and we conclude that the map

$$Der^0(MU, \Sigma^{n+1} H\pi_{n+1}MU) \longrightarrow [MU, MU_{n+1}]_a^{lift}$$

is bijective. So the set of all possible lifts of $\tilde{f}_k$ is in one-to-one correspondence with elements in the group $Der_S^0(MU, \Sigma^{n+1} H\pi_{n+1}MU)$. Clearly the set of all $\mathbb{Q}$-preferred lifts corresponds under this isomorphism to the set of $\mathbb{Q}$-preferred topological derivations of MU with values in $\Sigma^{n+1} H\pi_{n+1}MU$. By Corollary 5.11 these $\mathbb{Q}$-preferred derivations are identified with the set of primitive operations from MU to $\Sigma^{n+1} H\pi_{n+1}MU$. So we established a one-to-one correspondence between the set of $\mathbb{Q}$-preferred lifts of $\tilde{f}_n$ and $Prim(MU, \Sigma^{n+1} H\pi_{n+1}MU)$.

Now we examine the question of lifting the map $\tilde{f}_n$ up to homotopy to a homotopy multiplicative map $MU \to MU_{n+1}$. Clearly the homotopy class of any map of S-modules $MU \to MU_k$ is determined by its rationalization, i.e., the rationalization map $[MU, MU_k] \to [MU_{\mathbb{Q}}, MU_{\mathbb{Q}k}]$ is injective. It follows that the map

$$Mult(MU, MU_k) \longrightarrow Mult(MU_{\mathbb{Q}}, MU_{\mathbb{Q}k})$$

is also injective.

Further we have the following bijection (for any k)

$$Mult(MU_{\mathbb{Q}}, MU_{k\mathbb{Q}}) \cong Hom_{rings}(MU_{\mathbb{Q}*}, MU_{\mathbb{Q}k*}).$$

Therefore there is a short exact sequence

$$\begin{aligned} 0 \longrightarrow\ & Prim(MU_{\mathbb{Q}}, \Sigma^{n+1} H\pi_{n+1} MU_{\mathbb{Q}}) = Der(MU_{\mathbb{Q}*}, \pi_{n+1} MU_{\mathbb{Q}}) \\ \longrightarrow\ & Mult(MU_{\mathbb{Q}}, MU_{\mathbb{Q}n+1}) \longrightarrow Mult(MU_{\mathbb{Q}}, MU_{\mathbb{Q}n}) \longrightarrow pt. \end{aligned} \tag{12}$$

Of course the last three terms are only sets. The exactness here means that $Mult(MU_{\mathbb{Q}}, MU_{\mathbb{Q}n+1})$ has a faithful action of $Prim(MU_{\mathbb{Q}}, \Sigma^{n+1} H\pi_{n+1} MU_{\mathbb{Q}})$ so that the quotient is isomorphic to $Mult(MU_{\mathbb{Q}}, MU_{\mathbb{Q}n})$.

Consider the diagram of fibre sequences

$$\begin{array}{ccccc} \Sigma^{n+1} H\pi_{n+1} MU & \longrightarrow & MU_{n+1} & \longrightarrow & MU_n \\ \downarrow & & \downarrow & & \downarrow \\ \Sigma^{n+1} H\pi_{n+1} MU_{\mathbb{Q}n+1} & \longrightarrow & MU_{\mathbb{Q}n+1} & \longrightarrow & MU_{\mathbb{Q}n} \end{array}.$$

Taking into account the fact that $MU^*_{k+1}MU$ surjects onto $MU^*_k MU$ for any k we obtain a map of short exact sequences

$$\begin{array}{ccccc} 0 \to [MU, \Sigma^{n+1} H\pi_{n+1} MU] & \longrightarrow & [MU, MU_{n+1}] & \longrightarrow & [MU, MU_n] \to 0 \\ \downarrow & & \downarrow & & \downarrow \\ 0 \to [MU_{\mathbb{Q}}, \Sigma^{n+1} H\pi_{n+1} MU_{\mathbb{Q}}] & \longrightarrow & [MU_{\mathbb{Q}}, MU_{\mathbb{Q}n+1}] & \longrightarrow & [MU_{\mathbb{Q}}, MU_{\mathbb{Q}n}] \to 0 \end{array}.$$

Notice that all downward maps are injections. Combining this with (12) we find that there is a short exact sequence

$$\begin{aligned} 0 \longrightarrow\ & Prim(MU, \Sigma^{n+1} H\pi_{n+1} MU) \\ \longrightarrow\ & Mult(MU, MU_{n+1}) \longrightarrow Mult(MU, MU_n) \longrightarrow 0. \end{aligned}$$

That is the indeterminacy in lifting the map $\tilde{f}_n : MU \to MU_n$ is precisely the set of primitive cohomology operations $Prim(MU, \Sigma^{n+1} H\pi_{n+1} MU)$. We see that the set of lifts of the map $\tilde{f}_n$ to a homotopy multiplicative map is in one-to-one correspondence with $\mathbb{Q}$-preferred S-algebra lifts. This completes the inductive step and shows that the original homotopy multiplicative map $f : MU \to MU$ can be improved in a unique way to a $\mathbb{Q}$-preferred S-algebra map.

So we succeeded in finding a section $i : Mult(MU, MU) \to [MU, MU]_a$ of the forgetful map $j : [MU, MU]_a \to Mult(MU, MU)$ so that the image of i consists of $\mathbb{Q}$-preferred S-algebra self-maps of MU. To see that i respects composition notice that for $f, g \in Mult(MU, MU)$ the S-algebra map $i(f) \circ i(g)$ is $\mathbb{Q}$-preferred and $j(i(f) \circ i(g)) = f \circ g$. Since there is a unique $\mathbb{Q}$-preferred S-algebra self-map whose image under j is $f \circ g$ we conclude that $i(f) \circ i(g) = i(f \circ g)$. With this the proof of Theorem 5.4 is completed.

Remark 5.15. *Using the Bousfield-Kan mapping space spectral sequence (cf. [4]) it is possible to calculate the set of all S-algebra self-maps of MU. However this approach leads to the identification of $[MU, MU]_a$ only as a set, not as a monoid. It seems that the monoid structure on $[MU, MU]_a$ should be related to the Gerstenhaber bracket on $THH_S^*(MU, MU)$*

Now consider an S-algebra E with a fixed map of S-algebras $f : MU \to E$ Suppose that E satisfies the following condition:

(S) The unit map $f : MU_* \to E_*$ is surjective.

Remark 5.16. *In [6] it was proved that a rather broad class of C-oriented spectra (namely those which are obtained by killing any regular ideal in the ring MU) can be supplied with MU-algebra structures. For this class of spectra the condition (S) is obviously satisfied.*

Corollary 5.17. *For an S-algebra E satisfying the condition (S) any multiplicative operation $MU \to E$ can be lifted (perhaps in a non-unique way) to an S-algebra map.*

Proof. The condition (S) guarantees that the map

$$Mult(MU, MU) \to Mult(MU, E)$$

induced by the given map $f : MU \to E$ is surjective. In other words any multiplicative operation $g : MU \to E$ can be represented as a composition $h \circ f$ where $h \in Mult(MU, MU)$. Since h can be lifted to an S-algebra self-map of MU our claim follows. □

As another consequence of Theorem 5.4 we will show that the p-local Brown-Peterson spectrum BP is an A_∞-retract of $MU_{(p)}$, the spectrum MU localized at p. Recall from e.g., [13], 4.1 that there exists a multiplicative cohomology operation $g : MU_{(p)} \to MU_{(p)}$ which is idempotent and whose image is the p-local spectrum BP.

Theorem 5.18. *There exists an S-algebra map $f : MU_{(p)} \to BP$ which has a right inverse S-algebra map $h : BP \to MU_{(p)}$.*

Proof. According to Theorem 5.4 the multiplicative operation $g : MU_{(p)} \to MU_{(p)}$ determines a map of S-algebras which we will denote by the same letter. Without loss of generality we can assume g to be a q-cofibration of S-algebras. Consider the diagram in the category of S-algebras:

$$\begin{array}{ccccc} MU_{(p)} & \xrightarrow{g} & MU_{(p)} & \xrightarrow{g} & \cdots . \\ \downarrow{\scriptstyle g} & & \downarrow{\scriptstyle g} & & \\ MU_{(p)} & \xrightarrow{id} & MU_{(p)} & \xrightarrow{id} & \cdots \end{array}$$

Each square in this diagram is commutative since the operation g is idempotent. The colimit of the upper row taken in the category of S-algebras coincides with the colimit taken in the category of spectra by Cofibration Hypothesis ([7], VII.4) and both are equivalent to BP. (That shows that BP is an S-algebra.) Now the map $f : MU_{(p)} \to BP$ is just the canonical map to the colimit. Next the colimit of the lower row is obviously $MU_{(p)}$ and therefore there exists an S-algebra map $h : BP \to MU_{(p)}$. It follows that $f \circ h : BP \to BP$ is homotopic to the identity and Theorem 5.18 is proved. □

Remark 5.19. *It can be shown (cf.* [9], [1], [6]*) that BP actually supports a structure of an MU-algebra.*

Remark 5.20. *It seems natural to conjecture that any $\mathbb{Q}$-preferred S-algebra self-map of MU lifts to a commutative S-algebra self-map. This conjecture, if true, would imply the existence of a canonical E_∞ ring structure on BP, a long-standing problem posed by P. May. The first (to author's knowledge) serious attack on this problem was undertaken by I. Kriz in his* 1993 *preprint* [8]. *This paper inspired much activity in the area, however it is still regarded as a program for further work rather than a definitive solution.*

Even though we do not know whether BP is a commutative S-algebra we can use Theorem 5.18 to compute homotopy classes of A_∞-maps out of BP.

Corollary 5.21. *For an S-algebra E satisfying the condition (S) every multiplicative operation $BP \to E$ lifts to an A_∞ ring map $BP \to E$ (perhaps in a non-unique way).*

Proof. The composition of the multiplicative operation $BP \to E$ with the canonical projection $MU \to BP$ determines a multiplicative operation $MU \to E$. This operation lifts to an S-algebra map. Composing this S-algebra map with the splitting map $BP \to MU$ (which we know is an S-algebra map by Theorem 5.18) we find the desired S-algebra map $BP \to E$. □

Acknowledgement

The author would like to thank P. Goerss and N. Kuhn for many stimulating discussions.

References

[1] A. Baker A. Jeanneret, *Brave new Hopf algebroids and extensions of MU-algebras,* Homology, Homotopy and Applications, to appear.

[2] M. Basterra, *Andre-Quillen cohomology of commutative S-algebras.* J. Pure Appl. Algebra 144 (1999), no. 2, 111–143.

[3] M. Boardman, *Conditionally convergent spectral sequences.* Contemp. Math. 239.

[4] A.K. Bousfield, *Homotopy spectral sequences and obstructions.* Israel J. Math. 66 (1989), no. 1–3, 54–104.

[5] H. Cartan, S. Eilenberg, *Homological algebra.* Princeton University Press, Princeton, N. J., 195.

[6] P. Goerss, *Associative MU-algebras*, preprint 2001.

[7] A.D. Elmendorf, I. Kriz, M.A. Mandell, J.P. May, *Rings, Modules and Algebras in Stable Homotopy Theory.* Mathematical Surveys and Monographs. Vol. 47. AMS, 1996.

[8] I. Kriz, *Towers of E_∞-ring spectra with applications to BP*, preprint, 1993.

[9] A. Lazarev, *Homotopy theory of A_∞ ring spectra and applications to $M\mathrm{U}$-modules.* K-Theory 24 (2001), no. 3, 243–281.

[10] A. Lazarev, *Towers of MU-algebras and a generalized Hopkins-Miller theorem*, to appear in the Proc. of the London Math. Soc.

[11] S. MacLane, *Homology.* Berlin, Göttingen-Heidelberg: Springer 1963.

[12] J. McClure, J. Smith, *A solution of Deligne's conjecture*, Recent progress in homotopy theory (Baltimore, MD, 2000), 153–193, Contemp. Math., 293, Amer. Math. Soc., Providence, RI, 2002.

[13] D. Ravenel, *Complex cobordism and stable homotopy groups of spheres*, Academic Press, 1986.

A. Lazarev
Mathematics Department
Bristol University
Bristol, BS8 1TW, England
e-mail: a.lazarev@bristol.ac.uk

Progress in Mathematics, Vol. 215
Algebraic Topology: Categorical Decomposition Techniques, 261–291

Colimits, Stanley-Reisner Algebras, and Loop Spaces

Taras Panov, Nigel Ray, and Rainer Vogt

Abstract. We study diagrams associated with a finite simplicial complex K, in various algebraic and topological categories. We relate their colimits to familiar structures in algebra, combinatorics, geometry and topology. These include: right-angled Artin and Coxeter groups (and their complex analogues, which we call *circulation groups*); Stanley-Reisner algebras and coalgebras; Davis and Januszkiewicz's spaces $DJ(K)$ associated with toric manifolds and their generalisations; and coordinate subspace arrangements. When K is a flag complex, we extend well-known results on Artin and Coxeter groups by confirming that the relevant circulation group is homotopy equivalent to the space of loops $\Omega DJ(K)$. We define homotopy colimits for diagrams of topological monoids and topological groups, and show they commute with the formation of classifying spaces in a suitably generalised sense. We deduce that the homotopy colimit of the appropriate diagram of topological groups is a model for $\Omega DJ(K)$ for an arbitrary complex K, and that the natural projection onto the original colimit is a homotopy equivalence when K is flag. In this case, the two models are compatible.

1. Introduction

In this work we study diagrams associated with a finite simplicial complex K, in various algebraic and topological categories. We are particularly interested in colimits and homotopy colimits of such diagrams.

We are motivated by Davis and Januszkiewicz's investigation [12] of toric manifolds, in which K arises from the boundary of the quotient polytope. In the course of their cohomological computations, Davis and Januszkiewicz construct real and complex versions of a space whose cohomology ring is isomorphic to the Stanley-Reisner algebra (otherwise known as the face ring [33]) of K, over $\mathbb{Z}/2$ and $\mathbb{Z}$ respectively. We denote spaces of this homotopy type by $DJ(K)$, and follow Buchstaber and Panov [7] by describing them as colimits of diagrams of classifying

Key words and phrases. Colimit, flag complex, homotopy colimit, loop space, right-angled Artin group, right-angled Coxeter group, Stanley-Reisner algebra, topological monoid.
The first author was supported by a Royal Society/NATO Postdoctoral Fellowship at the University of Manchester, and by the Russian Foundation for Basic Research, grant 01-01-00546.

spaces. In this context, an exterior version arises naturally as an alternative. Suggestively, the cohomology algebras and homology coalgebras of the $DJ(K)$ may be expressed as the limits and colimits of analogous diagrams in the corresponding algebraic category.

When colimits of similar diagrams are taken in a category of discrete groups, they yield right-angled Coxeter and Artin groups. These are more usually described by a complementary construction involving only the 1-skeleton $K^{(1)}$ of K. Whenever K is determined entirely by $K^{(1)}$ it is known as a flag complex, and results such as those of [12] and [22] may be interpreted as showing that the associated Coxeter and Artin groups are homotopy equivalent to the loop spaces $\Omega DJ(K)$, in the real and exterior cases respectively. In other words, the groups are discrete models for the loop spaces. These observations raise the possibility of modelling $\Omega DJ(K)$ in the complex case, and for arbitrary K, by colimits of diagrams in a suitably defined category of topological monoids. Our primary aim is to carry out this programme.

Before we begin, we must therefore confirm that our categories are sufficiently cocomplete for the proposed colimits to exist. We show that this is indeed the case (as predicted by folklore), and explain how the complex version of $\Omega DJ(K)$ is modelled by the colimit of a diagram of tori whenever K is flag. We refer to the colimit as a *circulation group*, and consider it as the complex analogue of the corresponding right-angled Coxeter and Artin groups. Of course, it is also determined by $K^{(1)}$. On the other hand, there are simple examples of non-flag complexes for which the colimit groups cannot possibly model $\Omega DJ(K)$ in any of the real, exterior, or complex cases. More subtle constructions are required.

Since we are engaged with homotopy theoretic properties of colimits, it is no great surprise that the appropriate model for arbitrary complexes K is a homotopy colimit. Considerable care has to be taken in formulating the construction for topological monoids, but the outcome clarifies the status of the original colimits when K is flag; flag complexes are precisely those for which the colimit and the homotopy colimit coincide. Our main result is therefore that $\Omega DJ(K)$ is modelled by the homotopy colimit of the relevant diagram of topological groups, in all three cases and for arbitrary K. When K is flag, the natural projection onto the original colimit is a homotopy equivalence, and is compatible with the two model maps. Our proof revolves around the fact that homotopy colimits commute with the classifying space functor, in a context which is considerably more general than is needed here.

For particular complexes K, our constructions have interesting implications for traditional homotopy theoretic invariants such as Whitehead products, Samelson products, and their higher analogues and iterates. We hope to deal with these issues in subsequent work [27].

We now summarise the contents of each section.

It is particularly convenient to use the language of enriched category theory, so we devote Section 2 to establishing the notation, conventions and results that we need. These include a brief discussion of simplicial objects and their realisations, and verification of the cocompleteness of our category of topological monoids in the enriched setting. Readers who are familiar with this material, or willing to refer back to Section 2 as necessary, may proceed directly to Section 3, where we introduce the relevant categories and diagrams associated with a simplicial complex K. They include algebraic and topological examples, amongst which are the exponential diagrams G^K; here G denotes one of the cyclic groups C_2 or C, or the circle group T, in the real, exterior, and complex cases respectively.

We devote Section 4 to describing the limits and colimits of these diagrams. Some are identified with standard constructions such as the Stanley-Reisner algebra of K and the Davis-Januszkiewicz spaces $DJ(K)$, whereas the G^K yield right-angled Coxeter and Artin groups, or circulation groups respectively. In Section 5 we study aspects of the diagrams involving associated fibrations and homotopy colimits. We note connections with coordinate subspace arrangements.

We introduce the model map $f_K\colon \operatorname{colim}^{\text{TMG}} G^K \to \Omega DJ(K)$ in Section 6, and determine the connectivity of its homotopy fibre in terms of combinatorial properties of K. The results confirm that f_K is a homotopy equivalence whenever K is flag, and quantify its failure for general K. In our final Section 7 we consider suitably well-behaved diagrams D of topological monoids, and prove that the homotopy colimit of the induced diagram of classifying spaces is homotopy equivalent to the classifying space of the homotopy colimit of D, taken in the category of topological monoids. By application to the exponential diagrams G^K, we deduce that our generalised model map $h_K\colon \operatorname{hocolim}^{\text{TMG}} G^K \to \Omega DJ(K)$ is a homotopy equivalence for all complexes K. We note that the two models are compatible, and homotopy equivalent, when K is flag.

We take the category TOP of k-spaces X and continuous functions $f\colon X \to Y$ as our underlying topological framework, following [35]. Every function space Y^X is endowed with the corresponding k-topology. Many of the spaces we consider have a distinguished basepoint $*$, and we write TOP_+ for the category of pairs $(X, *)$ and basepoint preserving maps; the forgetful functor $\text{TOP}_+ \to \text{TOP}$ is faithful. For any object X of TOP, we may add a disjoint basepoint to obtain a based space X_+. The k-function space $(Y,*)^{(X,*)}$ has the trivial map $X \to *$ as basepoint. In some circumstances we need $(X, *)$ to be *well pointed*, in the sense that the inclusion of the basepoint is a closed cofibration, and we emphasise this requirement as it arises.

Several other useful categories are related to TOP_+. These include TMONH, consisting of topological monoids and homotopy homomorphisms [5] (essentially equivalent to Sugawara's strongly homotopy multiplicative maps [34]), and its subcategory TMON, in which the homorphisms are strict. Again, the forgetful functor $\text{TMON} \to \text{TOP}_+$ is faithful. Limiting the objects to topological groups defines a further subcategory TGRP, which is full in TMON. In all three cases the identity element e is the basepoint, and we may sometimes have to insist that objects are well pointed. The Moore loop space ΩX is a typical object in TMONH for any pair $(X, *)$, and the canonical inclusion $M \to \Omega BM$ is a homotopy homomorphism for any well-pointed topological monoid M.

For each $m \geq 0$ we consider the small categories $\mathrm{ID}(m)$, which consist of m objects and their identity morphisms; in particular, we use the based versions $\mathrm{ID}_{\varnothing}(m)$, which result from adjoining an initial object $\varnothing$. Given a topological monoid M, the associated topological category $\mathrm{C}(M)$ consists of one object, and one morphism for each element of M. Segal's [32] classifying space $B\mathrm{C}(M)$ then coincides with the standard classifying space BM.

Given objects X_0 and X_n of any category C, we denote the set of n-composable morphisms

$$X_0 \xrightarrow{f_1} X_1 \xrightarrow{f_2} \cdots \xrightarrow{f_n} X_n$$

by $\mathrm{C}_n(X_0, X_n)$, for all $n \geq 0$. Thus $\mathrm{C}_1(X, Y)$ is the morphism set $\mathrm{C}(X, Y)$ for all objects X and Y, and $\mathrm{C}_0(X, X)$ consists solely of the identity morphism on X.

In order to distinguish between them, we write T for the multiplicative topological group of unimodular complex numbers, and S^1 for the circle. Similarly, we discriminate between the cyclic group C_2 and the ring of residue classes $\mathbb{Z}/2$, and between the infinite cyclic group C and the ring of integers $\mathbb{Z}$. We reserve the symbol G exclusively for one of the groups C_2, C, or T, in contrast to an arbitrary topological group Γ.

The first and second authors benefitted greatly from illuminating discussions with Bill Dwyer at the International Conference on Algebraic Topology held on the Island of Skye during June 2001. They are particularly grateful to the organisers for providing the opportunity to work in such magnificent surroundings.

2. Categorical prerequisites

We refer to the books of Kelly [21] and Borceux [3] for notation and terminology associated with the theory of enriched categories, and to Barr and Wells [1] for background on the theory of monads (otherwise known as triples). For more specific results, we cite [14] and [18]. Unless otherwise stated, we assume that all our categories are enriched in one of the topological senses described below, and that functors are continuous. In many cases the morphism sets are finite, and therefore invested with the discrete topology.

Given an arbitrary category R, we refer to a covariant functor $D\colon \mathrm{A} \to \mathrm{R}$ as an A-*diagram* in R, for any small category A. Such diagrams are the objects of a category $[\mathrm{A},\mathrm{R}]$, whose morphisms are natural transformations of functors. We may interpret any object X of R as a constant diagram, which maps each object of A to X and every morphism to the identity.

Examples 2.1. *Let Δ be the category whose objects are the sets $(n) = \{0, 1, \ldots, n\}$, where $n \geq 0$, and whose morphisms are the nondecreasing functions; then Δ^{op}- and Δ-diagrams are simplicial and cosimplicial objects of R respectively. In particular, $\varDelta\colon \Delta \to \mathrm{TOP}$ is the cosimplicial space which assigns the standard n-simplex $\varDelta(n)$ to each object (n). Its pointed analogue $\varDelta_+$ is given by $\varDelta_+(n) = \varDelta(n)_+$.*

If M is a topological monoid, then $\mathrm{C}(M)$- and $\mathrm{C}(M)^{op}$-diagrams in TOP are left and right M-spaces respectively.

We recall that $(\mathrm{S}, \Box, \Phi)$ is a *symmetric monoidal* category if the bifunctor $\Box\colon \mathrm{S} \times \mathrm{S} \to \mathrm{S}$ is coherently associative and commutative, and Φ is a coherent unit object. Such an S is *closed* if there is a bifunctor $\mathrm{S} \times \mathrm{S}^{op} \to \mathrm{S}$, denoted by $(Z, Y) \mapsto [Y, Z]$, which satisfies the adjunction

$$\mathrm{S}(X \Box Y, Z) \cong \mathrm{S}(X, [Y, Z])$$

for all objects X, Y, and Z of S. A category R is S-*enriched* when its morphism sets are identified with objects of S, and composition factors naturally through $\Box$. A closed symmetric monoidal category is canonically self-enriched, by identifying $\mathrm{S}(X, Y)$ with $[X, Y]$. Henceforth, S denotes such a category.

Example 2.2. *A small* S*-enriched category* A *determines a diagram* $A\colon \mathrm{A} \times \mathrm{A}^{op} \to \mathrm{S}$, *whose value at* (a, b) *is the morphism object* $\mathrm{A}(b, a)$.

An S-*functor* $\mathrm{Q} \to \mathrm{R}$ of S-enriched categories acts on morphism sets as a morphism of S. The category $[\mathrm{Q},\mathrm{R}]$ of such functors has morphisms consisting of natural transformations, and is also S-enriched. The S-functors $F\colon \mathrm{Q} \to \mathrm{R}$ and $U\colon \mathrm{R} \to \mathrm{Q}$ are S-*adjoint* if there is a natural isomorphism

$$\mathrm{R}(F(X), Y) \cong \mathrm{Q}(X, U(Y))$$

in S, for all objects X of Q and Y of R.

Examples 2.3. *The categories* TOP *and* TOP_+ *are symmetric monoidal under cartesian product* $\times$ *and smash product* $\wedge$ *respectively, with unit objects the one-point space* $*$ *and the zero-sphere* $*_+$. *Both are closed, and therefore self-enriched, by identifying* $[X, Y]$ *with* Y^X *and* $(Y, *)^{(X,*)}$ *respectively.*

Since $(Y, *)^{(X,*)}$ inherits the subspace topology from Y^X, the induced TOP-enrichment of TOP_+ is compatible with its self-enrichment. Both TMON and TGRP are TOP_+-enriched by restriction.

In certain situations it is helpful to reserve the notation T for either or both of the self-enriched categories TOP and TOP_+. Similarly, we reserve TMG for either or both of the TOP_+-enriched categories TMON and TGRP.

It is well known that TOP and TOP_+ are complete and cocomplete, in the standard sense that every small diagram has a limit and colimit. Completeness is equivalent to the existence of products and equalizers, and cocompleteness to the existence of coproducts and coequalizers. Both TOP and TOP_+ actually admit *indexed limits* and *indexed colimits* [21], involving topologically parametrized diagrams in the enriched setting; in other words, T is T-*complete* and T-*cocomplete.* A summary of the details for TOP can be found in [26].

Amongst indexed limits and colimits, the enriched analogues of products and coproducts are particularly important.

Definitions 2.4. An S-enriched category R is *tensored* and *cotensored* over S if there exist bifunctors $\mathrm{R} \times \mathrm{S} \to \mathrm{R}$ and $\mathrm{R} \times \mathrm{S}^{op} \to \mathrm{R}$ respectively, denoted by

$$(X, Y) \longmapsto X \otimes Y \quad \text{and} \quad (X, Y) \longmapsto X^Y,$$

together with natural isomorphisms

$$\text{(2.5)} \qquad \mathrm{R}(X \otimes Y, Z) \cong \mathrm{S}(Y, \mathrm{R}(X, Z)) \cong \mathrm{R}(X, Z^Y)$$

in S, for all objects X, Z of R and Y of S. □

For any such R, there are therefore natural isomorphisms

$$\text{(2.6)} \qquad X \otimes \Phi \cong X \cong X^{\Phi} \quad \text{and} \quad X \otimes (Y \,\square\, W) \cong (X \otimes Y) \otimes W.$$

Every S is tensored over itself by □, and cotensored by [,].

Examples 2.7. *The categories* T *are tensored and cotensored over themselves; so* $X \otimes Y$ *and* X^Y *are given by* $X \times Y$ *and* X^Y *in* TOP, *and by* $X \wedge Y$ *and* $(Y, *)^{(X,*)}$ *in* TOP_+.

The rôle of tensors and cotensors is clarified by the following results of Kelly [21, (3.69)–(3.73)].

Theorem 2.8. *An* S*-enriched category is* S*-complete if and only if it is complete, and cotensored over* S*; it is* S*-cocomplete if and only if it is cocomplete, and tensored over* S.

Theorem 2.8 asserts that standard limits and colimits may themselves be enriched in the presence of tensors and cotensors, since they are special cases of indexed limits and colimits. Given an A-diagram D in R, where A is also S-enriched, we deduce that the natural bijections

$$\text{(2.9)} \quad \mathrm{R}(X, \lim D) \longleftrightarrow [\mathrm{A},\mathrm{R}](X, D) \qquad \text{and} \qquad \mathrm{R}(\operatorname{colim} D, Y) \longleftrightarrow [\mathrm{A},\mathrm{R}](D, Y)$$

are isomorphisms in S, for any objects X and Y of R. Henceforth, we assume that S is complete and cocomplete in the standard sense.

It is convenient to formulate several properties of TMON and TGRP by observing that both categories are TOP_+-complete and -cocomplete. We appeal to the monad associated with the forgetful functor $U\colon \mathrm{TMG} \to \mathrm{TOP}_+$; in both cases it has a left TOP_+-adjoint, given by the free monoid or free group functor F. The composition $U \cdot F$ defines a TOP_+-monad $L\colon \mathrm{TOP}_+ \to \mathrm{TOP}_+$, whose category TOP_+^L of algebras is precisely TMG.

Proposition 2.10. *The categories* TMON *and* TGRP *are* TOP_+*-complete and* TOP_+*-cocomplete.*

Proof. We consider the forgetful functor $\mathrm{TOP}_+^L \to \mathrm{TOP}_+$, noting that TOP_+ is TOP_+-complete by Theorem 2.8.

Part (i) of [14, VII, Proposition 2.10] asserts that the forgetful functor creates all indexed limits, confirming that TMG is TOP_+-complete. Part (ii) (whose origins lie in work of Hopkins) asserts that TOP_+^L is TOP_+-cocomplete if L preserves reflexive coequalizers, which need only be verified for U because F preserves colimits. The result follows for an arbitrary reflexive pair (f, g) in TMG by using the right inverse to show that the coequalizer of $(U(f), U(g))$ in TOP_+ is itself in the image of U, and lifts to the coequalizer of (f, g). □

Pioneering results on the completeness and cocompleteness of categories of topological monoids and topological groups may be found in [6]. Our main deduction from Proposition 2.10 is that TMON and TGRP are tensored over TOP_+. By studying the isomorphisms (2.5), we may construct the tensors explicitly; they are described as pushouts in [30, 2.2].

Construction 2.11. For any objects M of TMON and Y of TOP_+, the *tensored monoid* $M \circledast Y$ is the quotient of the free topological monoid on $U(M) \wedge Y$ by the relations

$$(m, y)(m', y) = (mm', y) \quad \text{for all } m,\ m' \in M \text{ and } y \in Y.$$

For any object Γ of TGRP, the *tensored group* $\Gamma \circledast Y$ is the topological group $V(\Gamma) \circledast Y$, where V denotes the forgetful functor TGRP $\to$ TMON.

The *cotensored monoid* M^Y and *cotensored group* Γ^Y are the function spaces $\text{TOP}_+(Y, M)$ and $\text{TOP}_+(Y, \Gamma)$ respectively, under pointwise multiplication. □

Lemma 2.12. *The forgetful functor* $V\colon$ TGRP $\to$ TMON *preserves indexed limits and colimits.*

Proof. Since V is right TOP_+-adjoint to the universal group functor, it preserves indexed limits. Construction 2.11 confirms that V preserves tensors, so we need only show that it preserves coequalizers, by the results of [21]. But the coequalizer of topological groups $\Gamma_1 \rightrightarrows \Gamma_2$ in TMON is also a group; and inversion is continuous, being induced by the continuous isomorphism $\gamma \mapsto \gamma^{-1}$ on Γ_2. □

The constructions of Section 7 involve indexed colimits in TMG, and Lemma 2.12 ensures that these may be formed in TMON, even when working exclusively with topological groups.

Given a category R which is tensored and cotensored over S, we may now describe several categorical constructions. They are straightforward variations on [18, 2.3], and initially involve three diagrams. The first is $D\colon \mathrm{B}^{op} \to \mathrm{R}$, the second $E\colon \mathrm{B} \to \mathrm{S}$, and the third $F\colon \mathrm{B} \to \mathrm{R}$.

Definitions 2.13. The *tensor product* $D \otimes_{\mathrm{B}} E$ is the coequalizer of

$$\coprod_{g: b_0 \to b_1} D(b_1) \otimes E(b_0) \underset{\beta}{\overset{\alpha}{\rightrightarrows}} \coprod_b D(b) \otimes E(b)$$

in R, where g ranges over the morphisms of B, and $\alpha|_g = D(g) \otimes 1$ and $\beta|_g = 1 \otimes E(g)$. The *homset* $\mathrm{Hom}_{\mathrm{B}}(E, F)$ is the equalizer of

$$\prod_b F(b)^{E(b)} \underset{\beta}{\overset{\alpha}{\rightrightarrows}} \prod_{g: b_0 \to b_1} F(b_1)^{E(b_0)}$$

in R, where $\alpha = \prod_g \cdot E(g)$ and $\beta = \prod_g F(g) \cdot$. □

We may interpret the elements of $\mathrm{Hom}_{\mathrm{B}}(E, F)$ as mappings from the diagram E to the diagram F, using the cotensor pairing.

Examples 2.14. *Consider the case* $\mathrm{R} = \mathrm{S} = \mathrm{TOP}$ *or* TOP_+*, with* $\mathrm{B} = \Delta$*. Given simplicial spaces* $X_\bullet\colon \Delta^{op} \to \mathrm{TOP}$ *and* $Y_\bullet\colon \Delta^{op} \to \mathrm{TOP}_+$*, the tensor products*

$$|X_\bullet| = X_\bullet \times_\Delta \varDelta \qquad \text{and} \qquad |Y_\bullet| = Y_\bullet \wedge_\Delta \varDelta_+$$

represent their topological realisation [24] *in* TOP *and* TOP_+ *respectively. If we choose* $\mathrm{R} = \mathrm{TMG}$ *and* $\mathrm{S} = \mathrm{TOP}_+$*, a simplicial object* $M_\bullet\colon \Delta^{op} \to \mathrm{TMG}$ *has internal and topological realisations*

$$|M_\bullet|_{\mathrm{TMG}} = M_\bullet \circledast_\Delta \varDelta_+ \qquad \text{and} \qquad |M_\bullet| = U(M)_\bullet \wedge_\Delta \varDelta_+$$

in TMG *and* TOP_+ *respectively. Since* $|\ \ |$ *preserves products,* $|M_\bullet|$ *actually lies in* TMG*.*

If $\mathrm{R} = \mathrm{S}$*, then* $D \otimes_{\mathrm{B}} \varPhi$ *is* $\operatorname{colim} D$*, where* $\varPhi$ *is the trivial* B*-diagram. Also,* $\mathrm{Hom}_{\mathrm{B}}(E, F)$ *is the morphism set* $[\mathrm{B},\mathrm{R}](E, F)$*, consisting of the natural transformations* $E \to F$*.*

For $Y_\bullet$ in Examples 2.14, its TOP- and TOP_+-realisations are homeomorphic because basepoints of the Y_n represent degenerate simplices for $n > 0$. We identify $|M_\bullet|_{\mathrm{TMG}}$ with $|M_\bullet|$ in Section 7.

We need certain generalisations of Definitions 2.13, in which analogies with homological algebra become apparent. We extend the first and second diagrams to $D\colon \mathrm{A} \times \mathrm{B}^{op} \to \mathrm{R}$ and $E\colon \mathrm{B} \times \mathrm{C}^{op} \to \mathrm{S}$, and replace the third by $F\colon \mathrm{C} \times \mathrm{D}^{op} \to \mathrm{S}$ or $G\colon \mathrm{A} \times \mathrm{C}^{op} \to \mathrm{R}$. Then $D \otimes_{\mathrm{B}} E$ becomes an $(\mathrm{A} \times \mathrm{C}^{op})$-diagram in R, and $\mathrm{Hom}_{\mathrm{C}^{op}}(E, G)$ becomes an $(\mathrm{A} \times \mathrm{B}^{op})$-diagram in R. The extended diagrams reduce to the originals by judicious substitution, such as $\mathrm{A} = \mathrm{C} = \mathrm{ID}$ in D and E.

Example 2.15. *Consider the case* $\mathrm{R} = \mathrm{S} = \mathrm{TOP}_+$*, with* $\mathrm{A} = \mathrm{C} = \mathrm{ID}$ *and* $\mathrm{B} = \Delta$*. Given* $E = \varDelta_+$ *as before, and* G *a constant diagram* $Z\colon \mathrm{ID} \to \mathrm{TOP}_+$*, then* $\mathrm{Hom}_{\mathrm{C}^{op}}(E, G)$ *coincides with the total singular complex* $Sin(Z)$ *as an object of* $[\Delta^{op},\mathrm{TOP}_+]$*. If* $\mathrm{R} = \mathrm{TMG}$ *and* $N\colon \mathrm{ID} \to \mathrm{TMG}$ *is a constant diagram, then* $Sin(N)$ *is an object of* $[\Delta^{op},\mathrm{TMG}]$*.*

Important properties of tensor products are described by the natural equivalences

$$D \otimes_{\mathrm{B}} B \cong D \qquad \text{and} \qquad (D \otimes_{\mathrm{B}} E) \otimes_{\mathrm{C}} F \cong D \otimes_{\mathrm{B}} (E \otimes_{\mathrm{C}} F) \tag{2.16}$$

of $(\mathrm{A} \times \mathrm{B}^{op})$- and $(\mathrm{A} \times \mathrm{D}^{op})$-diagrams respectively, in R. The first equivalence applies Example 2.2 with $\mathrm{A} = \mathrm{B}$, and the second uses the isomorphism of (2.6). The adjoint relationship between $\otimes$ and Hom is expressed by the equivalences

$$\begin{aligned}[\mathrm{A}\times\mathrm{C}^{op},\mathrm{R}](D \otimes_{\mathrm{B}} E, G) &\cong [\mathrm{B}\times\mathrm{C}^{op},\mathrm{S}](E, [\mathrm{A},\mathrm{R}](D, G)) \\ &\cong [\mathrm{A}\times\mathrm{B}^{op},\mathrm{R}](D, \mathrm{Hom}_{\mathrm{C}^{op}}(E, G)),\end{aligned} \tag{2.17}$$

which extend the tensor-cotensor relations (2.5), and are a consequence of the constructions.

Examples 2.18. *Consider the data of Example* 2.15, *and suppose that D is a simplicial pointed space $Y_\bullet\colon \Delta^{op} \to \mathrm{TOP}_+$. Then the adjoint relation* (2.17) *provides a homeomorphism*

$$\mathrm{TOP}_+(|Y_\bullet|, Z) \cong {}_{[\Delta^{op},\mathrm{TOP}_+]}(Y_\bullet, Sin(Z)).$$

If $\mathrm{R} = \mathrm{TMG}$ *and* $\mathrm{S} = \mathrm{TOP}_+$, *and $M_\bullet$ is a simplicial object in* TMG, *we obtain a homeomorphism*

$$\mathrm{TMG}(|M_\bullet|_{\mathrm{TMG}}, N) \cong {}_{[\Delta^{op},\mathrm{TMG}]}(M_\bullet, Sin(N))$$

for any object N of TMG.

If $\mathrm{R} = \mathrm{S}$ *and* $E = \Phi$, *the relations* (2.17) *reduce to the second isomorphism* (2.9).

The first two examples extend the classic adjoint relationship between $|\ |$ and *Sin*.

We now assume $\mathrm{R} = \mathrm{S} = \mathrm{TOP}$. We let D be an $(\mathrm{A} \times \mathrm{B}^{op})$-diagram as above, and define $B_\bullet(*, \mathrm{A}, D)$ to be a degenerate form of the 2-*sided bar construction.* It is a B^{op}-diagram of simplicial spaces, given as a $\mathrm{B}^{op} \times \Delta^{op}$-diagram in TOP by

$$(2.19) \qquad (b, (n)) \longmapsto \coprod_{a_0, a_n} D(a_0, b) \times \mathrm{A}_n(a_0, a_n)$$

for each object b of B; the face and degeneracy maps are described as in [18] by composition (or evaluation) of morphisms and by the insertion of identities respectively. The topological realisation $B(*, \mathrm{A}, D)$ is a B^{op}-diagram in TOP. These definitions ensure the existence of natural equivalences

$$(2.20) \qquad \begin{aligned} B_\bullet(*, \mathrm{A}, D) \times_{\mathrm{B}} E &\cong B_\bullet(*, \mathrm{A}, D \times_{\mathrm{B}} E) \\ \text{and} \quad B(*, \mathrm{A}, D) \times_{\mathrm{B}} E &\cong B(*, \mathrm{A}, D \times_{\mathrm{B}} E) \end{aligned}$$

of C^{op}-diagrams in $[\Delta^{op},\mathrm{TOP}]$ and TOP respectively.

Examples 2.21. *If* $\mathrm{B} = \mathrm{ID}$, *the homotopy colimit* [4] *of a diagram* $D\colon \mathrm{A} \to \mathrm{TOP}$ *is given by*

$$\operatorname{hocolim} D = B(*, \mathrm{A}, D),$$

as explained in [18]; *using* (2.16) *and* (2.20), *it is homeomorphic to both of*

$$B(*, \mathrm{A}, A) \times_{\mathrm{A}} D \cong D \times_{\mathrm{A}^{op}} B(*, \mathrm{A}, A).$$

In particular, $B_\bullet(*, \mathrm{A}, *)$ *is the nerve* [32] $B_\bullet\mathrm{A}$ *of* A, *whose realisation is the classifying space* $B\mathrm{A}$ *of* A. *The natural projection* $\operatorname{hocolim} D \to \operatorname{colim} D$ *is given by the map*

$$D \times_{\mathrm{A}^{op}} B(*, \mathrm{A}, A) \longrightarrow D \times_{\mathrm{A}^{op}} *,$$

induced by collapsing $B(*, \mathrm{A}, A)$ *onto* $*$.

If $\mathrm{A} = \mathrm{C}(M)$, *where M is an arbitrary topological monoid, then D is a left M-space and* $B(*, \mathrm{C}(M), C(M))$ *is a universal contractible right M-space EM* [13]. *So*

$$\operatorname{hocolim} D = B(*, \mathrm{C}(M), C(M)) \times_{\mathrm{C}(M)} D$$

is a model for the Borel construction $EM \times_M D$.

3. Basic constructions

We choose a universal set V of vertices $v_1,\dots,v_m$, and let K denote a simplicial complex with faces $\sigma \subseteq V$. The integer $|\sigma|-1$ is the *dimension* of σ, and the greatest such integer is the dimension of K. For each $1 \leq j \leq m$, the faces of dimension less than or equal to j form a subcomplex $K^{(j)}$, known as the *j-skeleton* of K; in particular, the 1-skeleton $K^{(1)}$ is a graph. We abuse notation by writing V for the zero-skeleton of K, more properly described as $\{\{v_j\} : 1 \leq j \leq m\}$. At the other extreme we have the $(m-1)$-simplex, which is the complex containing *all* subsets of V; it is denoted by 2^V in the abstract setting and by $\Delta(V)$ when emphasising its geometrical realisation. Any simplicial complex K therefore lies in a chain

$$V \longrightarrow K \longrightarrow 2^V \tag{3.1}$$

of subcomplexes. Every face σ may also be interpreted as a subcomplex of K, and so masquerades as a $(|\sigma|-1)$-simplex.

A subset $W \subseteq V$ is a *missing face* of K if every proper subset lies in K, yet W itself does not; its dimension is $|W|-1$. We refer to K as a *flag complex*, or write that K *is flag*, when every missing face has two vertices. The boundary of a planar m-gon is therefore flag whenever $m \geq 4$, as is the barycentric subdivision K' of an arbitrary complex K. The *flagification* $Fl(K)$ of K is the minimal flag complex containing K as a subcomplex, and is obtained from K by adjoining every missing face containing three or more vertices.

Example 3.2. *For any $n > 2$, the simplest non-flag complex on n vertices is the boundary of an $(n-1)$-simplex, denoted by $\partial(n)$; then $Fl(\partial(n))$ is $\Delta(n-1)$ itself.*

Given a subcomplex $K \subseteq L$ on vertices V, it is useful to define $W \subseteq V$ as a missing face of the pair (L, K) whenever W fails to lie in K, yet every proper subset lies in L.

Every finite simplicial complex K gives rise to a finite category $\mathrm{CAT}(K)$, whose objects are the faces σ and morphisms the inclusions $\sigma \subseteq \tau$. The empty face $\varnothing$ is an initial object. For any subcomplex $K \subseteq L$, the category $\mathrm{CAT}(K)$ is a full subcategory of $\mathrm{CAT}(L)$; in particular, (3.1) determines a chain of subcategories

$$\mathrm{ID}_{\varnothing}(m) \longrightarrow \mathrm{CAT}(K) \longrightarrow \mathrm{CAT}(2^V). \tag{3.3}$$

For each face σ, we define the *undercategory* $\sigma \downarrow \mathrm{CAT}(K)$ by restricting attention to those objects τ for which $\sigma \subseteq \tau$; thus σ is an initial object. Insisting that the inclusion $\sigma \subset \tau$ be strict yields the subcategory $\sigma \Downarrow \mathrm{CAT}(K)$, obtained by deleting σ. The *overcategories* $\mathrm{CAT}(K) \downarrow \sigma$ and $\mathrm{CAT}(K) \Downarrow \sigma$ are defined likewise, and may be rewritten as $\mathrm{CAT}(\sigma)$ and $\mathrm{CAT}(\partial(\sigma))$ respectively.

A complex K also determines a simplicial set $S(K)$, whose nondegenerate simplices are exactly the faces of K [24]. So the nerve $B_\bullet\mathrm{CAT}(K)$ coincides with the simplicial set $S(Con(K'))$, where $Con(K')$ denotes the cone on the barycentric subdivision of K, and the cone point corresponds to $\varnothing$. More generally, $B(\sigma \downarrow \mathrm{CAT}(K))$ is the cone on $B(\sigma \Downarrow \mathrm{CAT}(K))$.

Examples 3.4. *If $K = V$, then $B\mathrm{ID}_{\varnothing}(m)$ is the cone on m disjoint points. If $K = 2^V$, then $B\mathrm{CAT}(2^V)$ is homeomorphic to the unit cube $I^V \subset \mathbb{R}^V$, and defines its canonical simplicial subdivision; the homeomorphism maps each vertex $\sigma \subseteq V$ to its characteristic function χ_σ, and extends by linearity. If K is the subcomplex $\partial(m)$, then $B\mathrm{CAT}(\partial(m))$ is obtained from the boundary ∂I^m by deleting all faces which contain the maximal vertex $(1, \ldots, 1)$.*

The undercategories define a $\mathrm{CAT}(K)^{op}$-diagram $\downarrow \mathrm{CAT}(K)$ in the category of small categories. It takes the value $\sigma \downarrow \mathrm{CAT}(K)$ on each face σ, and the inclusion functor $\tau \downarrow \mathrm{CAT}(K) \subseteq \sigma \downarrow \mathrm{CAT}(K)$ on each reverse inclusion $\tau \supseteq \sigma$. The formation of classifying spaces yields a $\mathrm{CAT}(K)^{op}$-diagram $B(\downarrow \mathrm{CAT}(K))$ in TOP_+, which consists of cones and their inclusions. It takes the value $B(\sigma \downarrow \mathrm{CAT}(K))$ on σ and $B(\tau \downarrow \mathrm{CAT}(K)) \subseteq B(\sigma \downarrow \mathrm{CAT}(K)$ on $\tau \supseteq \sigma$, and its colimit is the final space $B\mathrm{CAT}(K)$. Following [18], we note the isomorphism

$$B(\downarrow \mathrm{CAT}(K)) \cong B(*, \mathrm{CAT}(K), CAT(K)) \tag{3.5}$$

of $\mathrm{CAT}(K)^{op}$-diagrams in TOP_+ (using the notation of Example 2.2).

We refer to the cones $B(\sigma \downarrow \mathrm{CAT}(K))$ as *faces* of $B\mathrm{CAT}(K)$, amongst which we distinguish the *facets* $B(v \downarrow \mathrm{CAT}(K))$, defined by the vertices v. The facets determine the faces, according to the expression

$$B(\sigma \downarrow \mathrm{CAT}(K)) = \bigcap_{v \in \sigma} B(v \downarrow \mathrm{CAT}(K))$$

for each $\sigma \in K$, and form a panel structure on $B\mathrm{CAT}(K)$ as described by Davis [11]. This terminology is motivated by our next example, which lies at the heart of recent developments in the theory of toric manifolds.

Example 3.6. *The boundary of a simplicial polytope P is a simplicial complex K_P, with faces σ. The polar P^* of P is a simple polytope of the same dimension, whose faces F_σ are dual to those of P (it is convenient to consider $F_\varnothing$ as P^* itself). There is a homeomorphism $B\mathrm{CAT}(K_P) \to P^*$, which maps each vertex σ to the barycentre of F_σ, and transforms each face $B(\sigma \downarrow \mathrm{CAT}(K))$ PL-homeomorphically onto F_σ.*

Classifying the categories and functors of (3.3) yields the chain of subspaces

$$Con(V) \longrightarrow B\mathrm{CAT}(K) \longrightarrow I^m. \tag{3.7}$$

So $B\mathrm{CAT}(K)$ contains the unit intervals along the coordinate axes, and is a subcomplex of I^m. It is therefore endowed with the induced cubical structure, as are all subspaces $B(\sigma \downarrow \mathrm{CAT}(K))$. In particular, the simple polytope P^* of Example 3.6 admits a natural cubical decomposition.

In our algebraic context, we utilise the category GRP of discrete groups and homomorphisms. Many constructions in GRP may be obtained by restriction from those we describe in TMON, and we leave readers to provide the details. In particular, GRP is a full subcategory of TMG, and is complete and cocomplete.

Given a commutative ring Q (usually the integers, or their reduction mod 2), we consider the category ${}_Q\text{MOD}$ of left Q-modules and Q-linear maps, which is symmetric monoidal with respect to the tensor product $\otimes_Q$ and closed under $(Z, Y) \mapsto {}_Q\text{MOD}(Y, Z)$. We usually work in the related category G_QMOD of connected graded modules of finite type, or more particularly in the categories G_QCALG and G_QCOCOA, which are dual; the former consists of augmented commutative Q-algebras and their homomorphisms, and the latter of supplemented cocommutative Q-coalgebras and their coalgebra maps.

As an object of ${}_Q\text{MOD}$, the polynomial algebra $Q[V]$ on V has a basis of monomials $v_W = \prod_W v_j$, for each multiset W on V. Henceforth, we assign a common dimension $d(v_j) > 0$ to the vertices v_j for all $1 \leq j \leq m$, and interpret $Q[V]$ as an object of G_QCALG; products are invested with appropriate signs if $d(v_j)$ is odd and $2Q \neq 0$. Then the quotient map

$$Q[V] \longrightarrow Q[V]/(v_\lambda : \lambda \subseteq V \text{ and } \lambda \notin K)$$

is a morphism in G_QCALG, whose target is known as the graded *Stanley-Reisner Q-algebra* of the simplicial complex K, and written $SR_Q(K)$. This ring is a fascinating invariant of K, and reflects many of its combinatorial and geometrical properties, as explained in [33]. Its Q-dual is a graded incidence coalgebra [20], which we denote by $SR^Q(K)$.

We define a $\text{CAT}(K)^{op}$-diagram D_K in TOP_+ as follows. The value of D_K on each face σ is the discrete space σ_+, obtained by adjoining $+$ to the vertices, and the value on $\tau \supseteq \sigma$ is the projection $\tau_+ \to \sigma_+$, which fixes the vertices of σ and maps the vertices of $\tau \setminus \sigma$ to $+$.

Definition 3.8. Given objects $(X, *)$ of TOP_+ and M of TMG, the *exponential diagrams* X^K and M^K are the cotensor homsets $\text{Hom}_{\text{ID}}(D_K, X)$ and $\text{Hom}_{\text{ID}}(D_K, M)$ respectively; they are $\text{CAT}(K)$-diagrams in TOP_+ and TMG. Alternatively, they are the respective compositions of the exponentiation functors $X^{(\ )}\colon \text{TOP}_+^{op} \to \text{TOP}_+$ and $M^{(\ )}\colon \text{TOP}_+^{op} \to \text{TMG}$ with D_K^{op}. □

So the value of X^K on each face σ is the product space X^σ, whose elements are functions $f\colon \sigma \to X$, and the value of X^K on $\sigma \subseteq \tau$ is the inclusion $X^\sigma \subseteq X^\tau$ obtained by extending f over τ by the constant map $*$. The space $X^\varnothing$ consists only of $*$. In the case of M^K, each M^σ is invested with pointwise multiplication, so H^K takes values in GRP for a discrete group H.

In G_QCALG, we define a $\text{CAT}(K)^{op}$-diagram $Q[K]$ by analogy. Its value on σ is the graded polynomial algebra $Q[\sigma]$, and on $\tau \supseteq \sigma$ is the projection $Q[\tau] \to Q[\sigma]$. We denote the dual $\text{CAT}(K)$-diagram $\text{Hom}_{\text{ID}}(Q[K], Q)$ by $Q\langle K\rangle$, and note that it lies in G_QCOCOA. Its value on σ is the free Q-module $Q\langle S(\sigma)\rangle$ generated by simplices z in the simplicial set $S(\sigma)$, and on $\sigma \subseteq \tau$ is the corresponding inclusion of coalgebras. The coproduct is given by $\delta(z) = \sum z_1 \otimes z_2$, where the sum ranges over all partitions of z into subsimplices z_1 and z_2.

When $Q = \mathbb{Z}/2$ we let the vertices have dimension 1. Every monomial v_U therefore has dimension $|U|$ in the graded algebra $\mathbb{Z}/2[\sigma]$, and every j-simplex in

$S(\sigma)$ has dimension $j+1$ in $\mathbb{Z}/2\langle S(\sigma)\rangle$. We refer to this as the *real case*. When $Q = \mathbb{Z}$ we consider two possibilities. First is the *complex case*, in which the vertices have dimension 2, so that the additive generators of $\mathbb{Z}[\sigma]$ and $\mathbb{Z}\langle S(\sigma)\rangle$ have twice the dimension of their real counterparts. Second is the *exterior case*, in which the dimension of the vertices reverts to 1. Every squarefree monomial v_U then has dimension $|U|$ in $\mathbb{Z}[\sigma]$, and anticommutativity ensures that every monomial containing a square is zero; every j-face of σ has dimension $j+1$ in $\mathbb{Z}\langle S(\sigma)\rangle$, and every degenerate j-simplex z represents zero. To distinguish between the complex and exterior cases, we write Q as $\mathbb{Z}$ and $\wedge$ respectively.

In the real and complex cases, Davis and Januszkiewicz [12] introduce homotopy types $DJ_{\mathbb{R}}(K)$ and $DJ_{\mathbb{C}}(K)$. The cohomology rings $H^*(DJ_{\mathbb{R}}(K);\mathbb{Z}/2)$ and $H^*(DJ_{\mathbb{C}}(K);\mathbb{Z})$ are isomorphic to the graded Stanley-Reisner algebras $SR_{\mathbb{Z}/2}(K)$ and $SR_{\mathbb{Z}}(K)$ respectively. We shall deal with the exterior case below, and discuss alternative constructions for all three cases. We write $DJ(K)$ as a generic symbol for Davis and Januszkiewicz's homotopy types, and refer to them as *Davis-Januszkiewicz* spaces for K. They are represented by objects in TOP_+.

4. Colimits

In this section we introduce the colimits which form our main topic of discussion, appealing to the completeness and cocompleteness of T and TMG as described in Section 2. We consider colimits of the diagrams X^K, M^K, G^K, and $Q\langle K\rangle$ in the appropriate categories, and label them $\operatorname{colim}^+ X^K$, $\operatorname{colim}^{\text{TMG}} M^K$, $\operatorname{colim}^{\text{TMG}} G^K$, and $\operatorname{colim} Q\langle K\rangle$ respectively. Similarly, we write the limit of $Q[K]$ as $\lim Q[K]$. As we shall see, these limits and colimits coincide with familiar constructions in several special cases.

As an exercise in acclimatisation, we begin with the diagrams associated to (3.3). Exponentiating with respect to $(X,*)$ and taking colimits provides the chain of subspaces

$$\bigvee_{j=1}^{m} X_j \longrightarrow \operatorname{colim}^+ X^K \longrightarrow X^m, \tag{4.1}$$

thereby sandwiching $\operatorname{colim}^+ X^K$ between the axes and the cartesian power. On the other hand, using an object M of TMG yields the chain of epimorphisms

$$\mathop{*}_{j=1}^{m} M_j \longrightarrow \operatorname{colim}^{\text{TMG}} M^K \longrightarrow M^m, \tag{4.2}$$

giving a presentation of $\operatorname{colim}^{\text{TMG}} M^K$ which lies between the m-fold free product of M and the cartesian power.

The following example emphasises the influence of the underlying category on the formation of colimits, and is important later.

Example 4.3. *If K is the non-flag complex $\partial(m)$ of Example 3.2 (where $m > 2$), then* $\operatorname{colim}^{+} X^{K}$ *is the* fat wedge *subspace* $\{(x_1, \ldots, x_m) : \textit{some } x_j = *\}$*; on the other hand,* $\operatorname{colim}^{\mathrm{TMG}} M^{K}$ *is isomorphic to* M^m *itself.*

By construction, $\operatorname{colim}^{\mathrm{TMG}} C_2^{K}$ in GRP enjoys the presentation

$$\langle a_1, \ldots, a_m : a_j^2 = 1, (a_i a_j)^2 = 1 \text{ for all } \{v_i, v_j\} \text{ in } K \rangle$$

and is isomorphic to the *right-angled Coxeter group* $Cox(K^{(1)})$ determined by the 1-skeleton of K. Readers should not confuse $K^{(1)}$ with the more familiar Coxeter graph of the group, which is almost its complement!

Similarly, $\operatorname{colim}^{\mathrm{TMG}} C^{K}$ has the presentation

$$\langle b_1, \ldots, b_m : [b_i, b_j] = 1 \text{ for all } \{v_i, v_j\} \text{ in } K \rangle$$

(where $[b_i, b_j]$ denotes the commutator $b_i b_j b_i^{-1} b_j^{-1}$), and so is isomorphic to the *right-angled Artin group* $Art(K^{(1)})$. Such groups are sometimes called *graph groups*, and are special examples of *graph products* [10]. As explained to us by Dave Benson, neither should be confused with the *graphs of groups* described in [31].

In the continuous case, we refer to $\operatorname{colim}^{\mathrm{TMG}} T^{K}$ as the *circulation group* $Cir(K^{(1)})$ in TMG. Every element of $Cir(K^{(1)})$ may therefore be represented as a word

$$t_{i_1}(1) \cdots t_{i_k}(k), \tag{4.4}$$

where $t_{i_j}(j)$ lies in the i_jth factor T_{i_j} for each $1 \leq j \leq k$. Two elements $t_r \in T_r$ and $t_s \in T_s$ commute whenever $\{r, s\}$ is an edge of K.

Following (4.2), we abbreviate the generating subgroups $G^{v_j} < \operatorname{colim} G^{K}$ to G_j, where $1 \leq j \leq m$, and call them the *vertex groups*. Since $\operatorname{colim}^{\mathrm{TMG}} G^{K}$ is presented as a quotient of the free product $*_{j=1}^{m} G_j$, its elements g may be assigned a wordlength $l(g)$. In addition, the arguments of [8] apply to decompose every g *from the right* as

$$g = \prod_{j=1}^{n} s_j(g) \tag{4.5}$$

for some $n \leq l(g)$, where each subword $s_j(g)$ contains the maximum possible number of mutually commuting letters, and is unique.

Given any subset $W \subseteq V$ of vertices, we write K_W for the complex obtained by restricting K to W. The following Lemma is a simple restatement of the basic properties of $\operatorname{colim}^{\mathrm{TMG}} G^{K}$.

Lemma 4.6. *We have that*

1. *the subgroup* $\operatorname{colim}^{\mathrm{TMG}} G^{K_W} \leq \operatorname{colim}^{\mathrm{TMG}} G^{K}$ *is abelian if and only if* $K_W^{(1)}$ *is a complete graph, in which case it is isomorphic to* G^{W}*;*
2. *when K is flag, each subword $s_j(g)$ of (4.5) lies in a subgroup G^{σ_j} for some face σ_j of K.*

Other algebraic examples of our colimits relate to the Stanley-Reisner algebras and coalgebras of K. By construction, there are algebra isomorphisms

$$\lim \mathbb{Z}/2[K] \cong SR_{\mathbb{Z}/2}(K),\ \lim \mathbb{Z}[K] \cong SR_{\mathbb{Z}}(K),\ \text{and}\ \lim \wedge[K] \cong SR_{\wedge}(K), \tag{4.7}$$

where the limits are taken in $\mathrm{G}_{\mathbb{Z}}\mathrm{CALG}$. Dually, there are coalgebra isomorphisms

$$\operatorname{colim} \mathbb{Z}/2\langle K\rangle \cong SR^{\mathbb{Z}/2}(K), \quad \operatorname{colim} \mathbb{Z}\langle K\rangle \cong SR^{\mathbb{Z}}(K), \quad \text{and} \quad \operatorname{colim} \wedge\langle K\rangle \cong SR^{\wedge}(K) \tag{4.8}$$

in $\mathrm{G}_{\mathbb{Z}}\mathrm{COCOA}$. The analogues of 4.1 display these limits and colimits as

$$\bigoplus_{j=1}^{m} Q[v_j] \longleftarrow \lim Q[K] \longleftarrow Q[V] \quad \text{and} \quad \bigoplus_{j=1}^{m} DP^Q(v_j) \longrightarrow \operatorname{colim} Q\langle K\rangle \longrightarrow DP^Q(V) \tag{4.9}$$

respectively; here $DP^Q(W)$ denotes the divided power Q-coalgebra of multisets on $W \subseteq V$, graded by dimension.

If we let $(X,*)$ be one of the pairs $(BC_2,*)$, $(BT,*)$, or $(BC,*)$, then simple arguments with cellular chain complexes show that the cohomology rings $H^*(\operatorname{colim}^+(BC_2)^K;\mathbb{Z}/2)$, $H^*(\operatorname{colim}^+(BT)^K;\mathbb{Z})$, and $H^*(\operatorname{colim}^+(BC)^K;\mathbb{Z})$ are isomorphic to the limits (4.7) respectively. Similarly, the homology coalgebras are isomorphic to the dual coalgebras (4.8). In cohomology, these observations are due to Buchstaber and Panov [7] in the real and complex cases, and to Kim and Roush [22] in the exterior case (at least when K is 1-dimensional). In homology, they may be made in the context of incidence coalgebras, following [29]. In both cases, the maps of (4.1) induce the homomorphisms (4.9).

Such calculations do not identify $\operatorname{colim}^+(BC_2)^K$ and $\operatorname{colim}^+(BT)^K$ with Davis and Januszkiewicz's constructions. Nevertheless, Buchstaber and Panov provide homotopy equivalences $\operatorname{colim}^+(BC_2)^K \simeq DJ_{\mathbb{R}}(K)$ and $\operatorname{colim}^+(BT)^K \simeq DJ_{\mathbb{C}}(K)$, which also follow from Corollary 5.4 below; the Corollary yields a corresponding equivalence in the exterior case. Of course, $\operatorname{colim}^+(BC)^K$ is a subcomplex of the m-dimensional torus $(S^1)^m$, and is therefore finite.

In due course, we shall use these remarks to interpret the following proposition in terms of Davis-Januszkiewicz spaces. The proof for $G = C_2$ is implicit in [12], and for $G = C$ is due to Kim and Roush [22].

Proposition 4.10. *When $G = C_2$ or C, there is a homotopy equivalence*

$$\operatorname{colim}^+(BG)^K \simeq B\operatorname{colim}^{\mathrm{TMG}} G^K$$

for any flag complex K.

Since both cases are discrete, $B\operatorname{colim}^{\mathrm{TMG}} G^K$ is, of course, an Eilenberg-Mac Lane space; Charney and Davis [9] discuss the identification of good models for BA, given *any* Artin group A. Proposition 4.10 fails for arbitrary complexes K, as our next examples show.

Examples 4.11. *Proposition* 4.10 *applies when* $K = V$*, because the discrete complex is flag; then* $\operatorname{colim}^{\text{TMG}} G^K$ *is isomorphic to the free product of* m *copies of* G*, whose classifying space is the* m*-fold wedge* $\bigvee_{j=1}^{m} BG_j$ *(by* [6]*, for example). On the other hand, when* K *is the non-flag complex* $\partial(m)$*, Example* 4.3 *confirms that* $B\operatorname{colim}^{\text{TMG}} G^K$ *is* BG^m*, whereas* $\operatorname{colim}^+(BG)^K$ *is the fat wedge subspace.*

These examples apply unchanged to the case $G = T$, and serve to motivate our extension of Proposition 4.10 to the complex case in Proposition 6.1 below. So far as C_2 and C are concerned, the proposition asserts that certain homotopy homomorphisms

$$h_K\colon \Omega\operatorname{colim}^+(BG)^K \longrightarrow \operatorname{colim}^{\text{TMG}} G^K \tag{4.12}$$

are homotopy equivalences when K is flag. We therefore view the h_K as modelling the loop spaces; in the complex case, they express $\Omega\operatorname{colim}^+(BT)^K$ in terms of the circulation groups $\operatorname{colim}^{\text{TMG}} T^K$. In Section 7 we will use homotopy colimits to describe analogues of h_K for all complexes K.

Our interest in the loop spaces $\Omega\operatorname{colim}^+(BG)^K$ has been stimulated by several ongoing programmes in combinatorial algebra. For example, Herzog, Reiner, and Welker [17] discuss combinatorial issues associated with calculating the k-vector spaces $\operatorname{Tor}^{SR_k(K)}(k,k)$ over an arbitrary ground field k, and refer to [16] for historical background. Such calculations have applications to diagonal subspace arrangements, as explained by Peeva, Reiner and Welker [28]. Since these Tor spaces also represent the E_2-term of the Eilenberg-Moore spectral sequence for $H^*(\Omega DJ(K);k)$, it seems well worth pursuing geometrical connections. We consider the algebraic implications elsewhere [27].

5. Fibrations and homotopy colimits

In this section we apply the theory of homotopy colimits to study various relevant fibrations and their geometrical interpretations. Some of the results appear in [7], but we believe that our approach offers an attractive and efficient alternative, and eases generalisation. We refer to [18] and [36] for the notation and fundamental properties of homotopy colimits. Several of the results we use are also summarised in [37], together with additional information on combinatorial applications.

We begin with a general construction, based on a well-pointed topological group Γ and a diagram $H\colon \text{A} \to \text{TMG}$ of closed subgroups and their inclusions. We assume that the maps of the classifying diagram $BH\colon \text{A} \to \text{TOP}_+$ are cofibrations, and that the Projection Lemma [37] applies to the natural projection $\operatorname{hocolim}^+ BH \to \operatorname{colim}^+ BH$, which is therefore a homotopy equivalence. The cofibrations $BH(a) \to B\Gamma$ correspond to the canonical map $f_H\colon \operatorname{colim}^+ BH \to B\Gamma$ under the homeomorphism (2.9).

By Examples 2.1 the coset spaces $\Gamma/H(a)$ define an $\text{A} \times \text{C}(\Gamma)$ diagram Γ/H in TOP, and by Examples 2.21 the cofibration $BH(a) \to B\Gamma$ is equivalent to the fibration

$$B(*, \mathrm{C}(\Gamma), C(\Gamma) \times_{\mathrm{C}(\Gamma)} \Gamma/H(a)) \longrightarrow B\mathrm{C}(\Gamma)$$

for each object a of A. So f_H is equivalent to

$$\operatorname{hocolim}^+ B(*, \mathrm{C}(\Gamma), C(\Gamma) \times_{\mathrm{C}(\Gamma)} \Gamma/H) \longrightarrow B\Gamma$$

in the homotopy category of spaces over $B\Gamma$, where the homotopy colimit is taken over A.

Proposition 5.1. *The homotopy fibre of f_H is the homotopy colimit* $\operatorname{hocolim}^+ \Gamma/H$.

Proof. We wish to identify the homotopy fibre of the projection

$$B(*, \mathrm{A}, B(*, \mathrm{C}(\Gamma), C(\Gamma) \times_{\mathrm{C}(\Gamma)} \Gamma/H)) \longrightarrow B\Gamma.$$

But we may rewrite the total space as $B(*, \mathrm{A}, \Gamma/H) \times_{\mathrm{C}(\Gamma)^{op}} B(*, \mathrm{C}(\Gamma), C(\Gamma))$, and therefore as $B(*, \mathrm{C}(\Gamma), C(\Gamma)) \times_{\mathrm{C}(\Gamma)} B(*, \mathrm{A}, \Gamma/H)$, using (2.20) and Examples 2.21. So the homotopy fibre is $B(*, \mathrm{A}, \Gamma/H)$, as required. □

Given a pair of simplicial complexes (L, K) on vertices V, we let $\mathrm{A} = \mathrm{CAT}(K)$, and choose $\Gamma = \operatorname{colim}^{\mathrm{TMG}} G^L$ and $H = G^K$; we also abbreviate the diagram Γ/H to L/K. Then f_H is the induced map

$$f_{K,L} \colon \operatorname{colim}^+ (BG)^K \longrightarrow B \operatorname{colim}^{\mathrm{TMG}} G^L, \tag{5.2}$$

and the Projection Lemma applies to $(BG)^K$ because the maps $\operatorname{colim}^+ (BG)^{\partial(\sigma)} \to BG^\sigma$ are closed cofibrations for each face σ. In particular, the canonical projection

$$\operatorname{hocolim}^+ (BG)^K \longrightarrow \operatorname{colim}^+ (BG)^K = DJ(K) \tag{5.3}$$

is a homotopy equivalence. We may also deduce the following corollary to Proposition 5.1.

Corollary 5.4. *The homotopy fibre of $f_{K,L}$ is the homotopy colimit* $\operatorname{hocolim} L/K$, *and is homeomorphic to the identification space*

$$\big(B\mathrm{CAT}(K) \times \operatorname{colim}^{\mathrm{TMG}} G^L\big) / \sim, \tag{5.5}$$

where $(p, gh) \sim (p, g)$ whenever $h \in G^\sigma$ and p lies in the face $B(\sigma \downarrow \mathrm{CAT}(K))$.

Proof. By (3.5), the homotopy colimit $B(*, \mathrm{CAT}(K), L/K)$ may be expressed as

$$B(\downarrow \mathrm{CAT}(K)) \times_{\mathrm{CAT}(K)} L/K,$$

and the inclusions $B(\sigma \downarrow \mathrm{CAT}(K)) \subseteq B\mathrm{CAT}(K)$ induce a homeomorphism with (5.5). □

We write the canonical action of $\operatorname{colim}^{\mathrm{TMG}} G^L$ on $B(*, \mathrm{CAT}(K), L/K)$ as μ for future use.

We note that $f_{K,L}$ coincides with the right-hand map of (4.1) when $L = 2^V$ and $X = BG$; the cases in which $K = L$ (abbreviated to f_K) and $L = Fl(K)$ also feature below. The space $\operatorname{hocolim} 2^V/K$ plays a significant rôle in [12], where it is described as the identification space of Corollary 5.4 and denoted by $\mathcal{Z}_P$ (with P the dual of K, in the sense of Example 3.6). To emphasise this connection, we write hocolim L/K as $\mathcal{Z}_G(K,L)$, which we abbreviate to $\mathcal{Z}_G(K)$ when $K = L$. It appears repeatedly below, by virtue of Proposition 5.1. Our examples assume that $L = 2^V$, and continue the theme of Examples 4.11.

Examples 5.6. *If $K = V$ then $\mathcal{Z}_G(K, 2^V)$ is the homotopy fibre of $\bigvee_{j=1}^m BG_j \to BG^m$, the inclusion of the axes; it has been of interest to homotopy theorists for many years. If K is the non-flag complex $\partial(m)$, then $\mathcal{Z}_G(K, 2^V)$ is homotopy equivalent to S^{m-1} for $G = \mathbb{Z}/2$, and S^{2m-1} for $G = T$.*

The second of these examples may be understood by noting that the inclusion of the fat wedge in BG^m has the Thom complex of the external product ζ^m of Hopf bundles as its cofibre.

Davis and Januszkiewicz [12] prove that the mod 2 cohomology ring of $EC_2^m \times_{C_2^m} \mathcal{Z}_{C_2}(K, 2^V)$ and the integral cohomology ring of $ET^m \times_{T^m} \mathcal{Z}_T(K, 2^V)$ are isomorphic to the Stanley-Reisner algebras $SR^*_{\mathbb{Z}/2}(K)$ and $SR^*_{\mathbb{Z}}(K)$ respectively. In view of Corollary 5.4 (in the case $L = 2^V$), we regard the spaces $\operatorname{colim}^+(BG)^K$ and the Davis-Januszkiewicz homotopy types as interchangeable from this point on.

The canonical projection $\mathcal{Z}_G(K,L) \to B\text{CAT}(K)$ is obtained by factoring out the action μ of $\operatorname{colim}^{\text{TMG}} G^L$ on hocolim L/K. The cubical structure (3.7) of the quotient lifts to an associated decomposition of $\mathcal{Z}_G(K,L)$; when $G = T$ and $L = 2^V$, for example, we recover the description of [7] in terms of polydiscs and tori.

The action μ has other important properties.

Proposition 5.7. *The isotropy subgroups of the action μ are given by the conjugates $wG^\sigma w^{-1} < \operatorname{colim}^{\text{TMG}} G^L$, as σ ranges over the faces of K.*

Proof. It suffices to note from Corollary 5.4 that each point $[x, wG^\sigma]$ is fixed by $wG^\sigma w^{-1} < \operatorname{colim}^{\text{TMG}} G^L$, for any $x \in B(\sigma \downarrow \text{CAT}(K))$. □

Corollary 5.8. *The commutator subgroup of* $\operatorname{colim}^{\text{TMG}} G^L$ *acts freely on $\mathcal{Z}_G(K,L)$ under μ.*

Proof. The isotropy subgroups are abelian, and so have trivial intersection with the commutator subgroup. □

When $K = L$ and $G = C_2$, Proposition 5.7 strikes a familiar chord. The *parabolic* subgroups of a Coxeter group H are the conjugates $w\Gamma w^{-1}$ of certain subgroups Γ, generated by subsets of the defining Coxeter system; when H is right-angled, and therefore takes the form $Cox(K^{(1)})$, such subgroups are abelian. When $L = 2^V$, each subgroup $wG^\sigma w^{-1}$ reduces to G^σ. In this case, Proposition 5.7 implies that the isotropy subgroups form an exponential $\text{CAT}^{op}(K)$-diagram in TGRP, which assigns G^σ to the face σ and the quotient homomorphism $G^\tau \to G^\sigma$ to the reverse inclusion $\tau \supseteq \sigma$.

As detailed in [7], the homotopy fibre $\mathcal{Z}_G(K, 2^V)$ is closely related to the theory of subspace arrangements and their auxiliary spaces. These spaces are defined in each of the real, complex, and exterior cases, and will feature below; we introduce them here as homotopy colimits.

Given a pointed space $(Y, 0)$, we let $Y_\times$ denote $Y \setminus 0$. For any subset $W \subseteq V$, we write $Y_W \subseteq Y^V$ for the *coordinate subspace* of functions $f\colon V \to Y$ for which $f(W) = 0$. The set of subspaces

$$\mathcal{A}_Y(K) = \{Y_W : W \notin K\}$$

is the associated *arrangement* of K, whose *complement* $U_Y(K)$ is given by the equivalent formulae

$$Y^V \setminus \bigcup_{W \notin K} Y_W \;=\; \{f : f^{-1}(0) \in K\}. \tag{5.9}$$

The $\mathrm{CAT}(K)$-diagram $Y(K)$ associates the function space $Y(\sigma) = \{f : f^{-1}(0) \subseteq \sigma\}$ to each face σ, and the inclusion $Y(\sigma) \subseteq Y(\tau)$ to each morphism $\sigma \subseteq \tau$. It follows that $Y(\sigma)$ is homeomorphic to $Y^\sigma \times (Y_\times^{V\setminus\sigma})$, and that $U_Y(K)$ is $\operatorname{colim} Y(K)$.

The exponential $\mathrm{CAT}(K)$-diagram $Y_\times^{V\setminus K}$ associates $Y_\times^{V\setminus\sigma}$ to σ; when Y is contractible, we may therefore follow Proposition 5.1 by combining the Projection Lemma and Homotopy Lemma of [37] to obtain a homotopy equivalence

$$\operatorname{hocolim} Y_\times^{V\setminus K} \;\simeq\; U_Y(K). \tag{5.10}$$

Now let us write $\mathbb{F}$ for one of the fields $\mathbb{R}$ or $\mathbb{C}$. The study of the *coordinate subspace arrangements* $\mathcal{A}_{\mathbb{F}}(K)$, together with their complements, is a special case of a well-developed theory whose history is rich and colourful (see [2], for example). In the exterior case, we replace $\mathbb{F}$ by the union of a countably infinite collection of 1-dimensional cones in $\mathbb{R}^2$, which we call a 1-*star* and write as $\mathbb{E}$. So $\mathbb{E}^V$ is an m-star; it is homeomorphic to the union of countably many m-dimensional cones in $(\mathbb{R}^2)^V$, obtained by taking products.

As G ranges over C_2, T and C, we let $\mathbb{F}$ denote $\mathbb{R}$, $\mathbb{C}$ and $\mathbb{E}$ respectively. In all three cases, the natural inclusion of G into $\mathbb{F}_\times$ is a cofibration, and $\mathbb{F}_\times$ retracts onto its image. So (5.10) applies, and may be replaced by the corresponding equivalence

$$\operatorname{hocolim} G^{V\setminus K} \;\simeq\; U_{\mathbb{F}}(K). \tag{5.11}$$

Proposition 5.12. *The space $\mathcal{Z}_G(K, 2^V)$ is homotopy equivalent to $U_{\mathbb{F}}(K)$, for any complex K.*

Proof. Substitute $L = 2^V$ in Corollary 5.4 and apply (5.11). □

By specialising results of [37] and [38], we may also describe $\left(\bigcup_{W\notin K} \mathbb{F}_W\right) \setminus 0$ as a homotopy colimit. This space is dual to $U_{\mathbb{F}}(K)$, and appears to have a more manageable homotopy type in many relevant cases. For $G = C_2$ and T, a version of Proposition 5.12 features prominently in [7].

The following examples illustrate Proposition 5.12, in the light of Examples 5.6.

Examples 5.13. *For $m > 2$ and $G = T$, the subspace arrangements of the discrete complex V and the non-flag complex $\partial(m)$ are given by*

$$\big\{\{z : z_j = z_k = 0\} : 1 \le j < k \le m\big\} \quad \textit{and} \quad \{0\}$$

respectively; the corresponding complements are

$$\{z : z_j = 0 \Rightarrow z_k \neq 0 \text{ for all } k \neq j\} \quad \text{and} \quad \mathbb{C}^m \setminus 0.$$

We conjecture that the former is homotopy equivalent to a wedge of spheres; the latter, of course, is equivalent to S^{2m-1}.

6. Flag complexes and connectivity

In this section, we examine the homotopy fibre $\mathcal{Z}_G(K, L)$ more closely. The results form the basis of our model for $\Omega DJ(K)$ when K is flag, and enable us to measure the extent of its failure for general K.

We consider a flag complex K, and substitute $K = L$ into Corollary 5.4 to deduce that $\mathcal{Z}_G(K)$ is the homotopy fibre of the cofibration $f_K\colon DJ(K) \to B\operatorname{colim}^{\mathrm{TMG}} G^K$. It is helpful to abbreviate $B(\sigma\downarrow\mathrm{CAT}(K))$ to $B(\sigma)$ throughout the following argument.

Proposition 6.1. *The cofibration f_K is a homotopy equivalence whenever K is flag.*

Proof. We prove that $\mathcal{Z}_G(K)$ is contractible.

For any face $\sigma \in K$, the space $(\operatorname{colim}^{\mathrm{TMG}} G^K)/G^\sigma$ inherits an increasing filtration by subspaces $(\operatorname{colim}^{\mathrm{TMG}} G^K)_i/G^\sigma$, consisting of those cosets wG^σ for which a representing element satisfies $l(w) \leq i$. We may therefore define a $\mathrm{CAT}(K)$-diagram K_i/K, which assigns $(\operatorname{colim}^{\mathrm{TMG}} G^K)_i/G^\sigma$ to each face σ and the corresponding inclusion to each inclusion $\sigma \subseteq \tau$. By construction, $\mathcal{Z}_G(K)$ is filtered by the subspaces $\operatorname{hocolim} K_i/K$ and each inclusion $\operatorname{hocolim} K_{i-1}/K \subset \operatorname{hocolim} K_i/K$ is a cofibration. We proceed by induction on i.

For the base case $i = 0$, we observe that $(\operatorname{colim}^{\mathrm{TMG}} G^K)_0/G^\sigma$ is the single point eG^σ for all values of σ. Thus $\operatorname{hocolim} K_0/K$ is homeomorphic to $B(\varnothing)$, and is contractible. To make the inductive step, we assume that $\operatorname{hocolim} K_i/K$ is contractible for all $i < n$, and write the quotient space $(\operatorname{hocolim} K_n/K)/(\operatorname{hocolim} K_{n-1}/K)$ as Q_n. It then suffices to prove that Q_n is contractible.

Every point of Q_n has the form (x, wG^σ), for some $x \in B(\sigma)$ and some w of length n. If the final letter of w lies in G^σ, then (x, wG^σ) is the basepoint of Q_n. Otherwise, we rewrite w as $w's$ by (4.5), where s contains the maximum possible number of mutually commuting letters. These determine a subset $\chi \subseteq V$, and Lemma 4.6 confirms that $K^{(1)}$ contains the complete graph on vertices χ. Since K is flag, we deduce that $2^\chi \in K$, and therefore that $(x, w'G^\chi)$ is the basepoint of Q_n. To describe a contraction of Q_n, we may find a canonical path p in $\mathrm{CAT}_\varnothing(K)$, starting at x and finishing at some x' in $B(\chi)$; of course p must vary continuously with (x, wG^σ), and lift to a corresponding path in Q_n. If x is a vertex of $B(\sigma)$, we choose p to run at constant speed along the edge from x to the cone point $\varnothing$, and again from $\varnothing$ to the vertex $\chi \in B(\chi)$. If x is an interior point of $B(\sigma)$, we extend the construction by linearity. Then p lifts to the path through $(p(t), w)$ for all $0 < t < 1$, as required. □

We expect Proposition 6.1 to hold for more general topological groups Γ.

The Proposition also leads to the study of $f_{K,L}\colon DJ(K) \to B\operatorname{colim}^{\mathrm{TMG}} G^L$ for any subcomplex $K \subseteq L$. We consider the missing faces of K with three or more vertices and write $c(K) \geq 2$ for their minimal dimension. We let $d(K)$ denote $c(K) - 1$ when $G = C_2$ or C, and $2c(K)$ when $G = T$; thus K is flag if and only if $c(K)$ (and therefore $d(K)$) is infinite. Finally, we define

$$c(K, L) = \begin{cases} c(K) & \text{if } L \subseteq Fl(K) \\ 1 & \text{otherwise,} \end{cases}$$

and let $d(K, L)$ be given by $c(K, L) - 1$ or $2c(K, L)$ as before.

Theorem 6.2. *For any subcomplex $K \subseteq L$, the cofibration $f_{K,L}$ is an equivalence in dimensions $\leq d(K, L)$.*

Proof. We may factorise $f_{K,L}$ as

$$DJ(K) \longrightarrow DJ(Fl(K)) \longrightarrow DJ(Fl(L)) \longrightarrow B\operatorname{colim}^{\mathrm{TMG}} G^{Fl(L)}.$$

The first map is induced by flagification, and is a $d(K)$-equivalence by construction. The second is the identity if $L \subseteq Fl(K)$; otherwise, it is 0-connected when $G = C_2$ or C, and 2-connected when $G = T$. The third map is $f_{Fl(L)}$, and an equivalence by Proposition 6.1. □

Theorem 6.2 suggests our first model for $\Omega DJ(K)$.

Proposition 6.3. *For any complex K, there is a homotopy homomorphism and $(d(K)-1)$-equivalence $h_K\colon \Omega DJ(K) \to \operatorname{colim}^{\mathrm{TMG}} G^K$; in particular, it is an equivalence when K is flag.*

Proof. We deduce that $\Omega f_K\colon \Omega DJ(K) \to \Omega B\operatorname{colim}^{\mathrm{TMG}} G^K$ is a $(d(K)-1)$-equivalence by applying Theorem 6.2 with $K = L$. The result follows by composing with the canonical homotopy homomorphism $\Omega BH \to H$, which exists for any topological group H. □

When $L = 2^V$, the missing faces of $(2^V, K)$ are precisely the non-faces of K. In this case only, we write their minimal dimension as $c'(K)$.

It is instructive to consider the homotopy commutative diagram

$$\text{(6.4)} \qquad \begin{array}{ccccc}
\mathcal{Z}_G(K,L) & \xrightarrow{id} & \mathcal{Z}_G(K,L) & \longrightarrow & * \\
\downarrow{\scriptstyle p} & & \downarrow & & \downarrow \\
\mathcal{Z}_G(K,2^V) & \longrightarrow & DJ(K) & \xrightarrow{f_{K,2^V}} & BG^m \\
\downarrow{\scriptstyle \gamma} & & \downarrow{\scriptstyle f_{K,L}} & & \downarrow{\scriptstyle id} \\
B[G,L] & \longrightarrow & B\operatorname{colim}^{\mathrm{TMG}} G^L & \xrightarrow{Ba} & BG^m
\end{array}$$

of fibrations, where a is the abelianisation homomorphism and $[G, L]$ denotes the commutator subgroup of $\operatorname{colim}^{\mathrm{TMG}} G^L$. By Theorem 6.2, $\mathcal{Z}_G(K, L)$ and $\mathcal{Z}_G(K, 2^V)$ are $(d(K, L) - 1)$- and $(d'(K) - 1)$-connected respectively, where $d(K, L) \geq d'(K)$ by definition. In fact $\mathcal{Z}_G(K, 2^V)$ is $d'(K)$-connected, by considering the homotopy exact sequence of $f_{K,2^V}$.

Corollary 5.8 confirms that

$$[G, L] \longrightarrow \mathcal{Z}_G(K, L) \xrightarrow{p} \mathcal{Z}_G(K, 2^V) \tag{6.5}$$

is a principal $[G, L]$-bundle, classified by γ. This bundle encodes a wealth of geometrical information on the pair (L, K). Its total space measures the failure of $f_{K,L}$ to be a homotopy equivalence, and its base space is the complement of the coordinate subspace arrangement $\mathcal{A}_{\mathbb{F}}(K)$ by Corollary 5.12. Moreover, Theorem 6.2 implies that γ is also a $d(K, L)$-equivalence, and so sheds some light on the homotopy type of $U_{\mathbb{F}}(K)$.

Looping (6.4) gives a homotopy commutative diagram of fibrations

$$\begin{array}{ccccc}
\Omega\mathcal{Z}_G(K, L) & \xrightarrow{id} & \Omega\mathcal{Z}_G(K, L) & \longrightarrow & 1 \\
\big\downarrow{\scriptstyle \Omega p} & & \big\downarrow & & \big\downarrow \\
\Omega U_{\mathbb{F}}(K) & \xrightarrow{i} & \Omega DJ(K) & \xrightarrow{\Omega f_{K,2^V}} & G^m \\
\big\downarrow{\scriptstyle \Omega\gamma} & & \big\downarrow{\scriptstyle \Omega f_{K,L}} & & \big\downarrow{\scriptstyle id} \\
[G, L] & \longrightarrow & \operatorname{colim}^{\mathrm{TMG}} G^L & \xrightarrow{a} & G^m
\end{array} \tag{6.6}$$

in TMONH, which offers an alternative perspective on $\Omega DJ(K)$.

Lemma 6.7. *The loop space $\Omega DJ(K)$ splits as $G^m \times \Omega U_{\mathbb{F}}(K)$ for any simplicial complex K; the splitting is not multiplicative.*

Proof. The vertex groups G_j embed in $\Omega DJ(K)$ via homotopy homomorphisms, whose product $j\colon G^m \to \Omega DJ(K)$ is left inverse to $\Omega f_{K,2^V}$ (but not a homotopy homomorphism). The product of the maps i and j is the required homeomorphism. □

The following examples continue the theme of Examples 5.6 and 5.13. They refer to the second horizontal fibration of the diagram (6.6), which is homotopy equivalent to the third whenever $K = L$ is flag, by Proposition 6.1. The second examples also appeal to James's Theorem [19], which identifies the loop space ΩS^n with the free monoid $F^+(S^{n-1})$ for any $n > 1$.

Examples 6.8. *If K is the discrete flag complex V, then $\Omega U_{\mathbb{F}}(K)$ is homotopy equivalent to the commutator subgroup of the free product $*_{j=1}^m G_j$. If K is the non-flag complex $\partial(m)$, then $\Omega U_{\mathbb{F}}(K)$ is homotopy equivalent to $F^+(S^{m-2})$ for $G = \mathbb{Z}/2$, and $F^+(S^{2m-2})$ for $G = T$; the map i identifies the inclusion of the generating sphere with the higher Samelson product (of order m) in $\pi_*(\Omega DJ(K))$.*

Of course, both examples split topologically according to Lemma 6.7. The appearance of higher products in $\Omega DJ(\partial(m))$ shows that commutators alone cannot model $\Omega DJ(K)$ when K is not flag. More subtle structures are required, based on higher homotopy commutativity; they are related to Samelson and Whitehead products, as we explain elsewhere [27].

7. Homotopy colimits of topological monoids

We now turn to the loop space $\Omega DJ(K)$ for a general simplicial complex K, appealing to the theory of homotopy colimits. Although the resulting models are necessarily more complicated, they are homotopy equivalent to $\operatorname{colim}^{\text{TMG}} G^K$ when K is flag. The constructions depend fundamentally on the categorical ideas of Section 2, and apply to more general spaces than $DJ(K)$. We therefore work with an arbitrary diagram $D\colon \mathrm{A} \to \mathrm{TMG}$ for most of the section, and write $BD\colon \mathrm{A} \to \mathrm{TOP}_+$ for its classifying diagram. Our applications follow by substituting G^K for D.

We implement proposals of earlier authors (as in [36], for example) by forming the homotopy colimit $\operatorname{hocolim}^{\text{TMG}} D$ in TMG, rather than TOP_+. This is made possible by the observation of Section 2 that the categories TMG are T-cocomplete, and therefore have sufficient structure for the creation of internal homotopy colimits. We confirm that $\operatorname{hocolim}^{\text{TMG}} D$ is a model for the loop space $\Omega \operatorname{hocolim}^+ BD$ by proving that B commutes with homotopy colimits in the relevant sense. As usual, we work in TMG, but find it convenient to describe certain details in terms of topological monoids; whenever these monoids are topological groups, so is the output.

We recall the standard extension of the 2-sided bar construction to the based setting, with reference to (2.19). We write $B^+_\bullet(*, \mathrm{A}, D)$ for the diagram $\mathrm{B}^{op} \times \Delta^{op} \to \mathrm{TOP}_+$ given by

$$(b,(n)) \longmapsto \bigvee_{a_0,a_n} D(a_0,b) \wedge \mathrm{A}_n(a_0,a_n)_+,$$

where D is a diagram $\mathrm{A} \times \mathrm{B}^{op} \to \mathrm{TOP}_+$. Following Examples 2.21, we define the homotopy TOP_+-colimit as

$$\operatorname{hocolim}^+ D = B^+(*, \mathrm{A}, D),$$

and note the equivalent expressions $B^+(*, \mathrm{A}, A_+) \wedge_{\mathrm{A}} D \cong D \wedge_{\mathrm{A}^{op}} B^+(*, \mathrm{A}, A_+)$.

For TMG, we proceed by categorical analogy. We replace the TOP-coproduct in (2.19) by its counterpart in TMG, and the internal cartesian product in TOP by the tensored structure of TMG over TOP_+. For any diagram $D\colon \mathrm{A} \to \mathrm{TMG}$, the simplicial topological monoid $B^{\text{TMG}}_\bullet(*, \mathrm{A}, D)$ is therefore given by

$$(n) \longmapsto \mathop{*}_{a_0,a_n} D(a_0) \circledast \mathrm{A}_n(a_0,a_n)_+, \tag{7.1}$$

where $*$ denotes the free product of topological monoids. The face and degeneracy operators are defined as before, but are now homomorphisms. When A is of the form $\text{CAT}(K)$, the n-simplices (7.1) may be rewritten as the finite free product

$$B_n^{\text{TMG}}(*, \text{CAT}(K), D) = \mathop{*}_{\sigma_n \supseteq \cdots \supseteq \sigma_0} D(\sigma_0),$$

where there is one factor for each n-chain of simplices in K.

Definition 7.2. The *homotopy* TMG-*colimit* of D is given by

$$\operatorname{hocolim}^{\text{TMG}} D = |B_\bullet^{\text{TMG}}(*, \text{A}, D)|_{\text{TMG}}$$

in TMG, for any diagram $D\colon \text{A} \to \text{TMG}$. □

So $\operatorname{hocolim}^{\text{TMG}} D$ is an object of TMG. Following Construction 2.11, it may be described in terms of generators and relations as a quotient monoid of the form

$$\Big(\mathop{*}_{n\geq 0} \big(B_n(*, \text{A}, D) \circledast \Delta_+^n\big)\Big) \Big/ \Big\langle \big(d_n^i(b), s\big) = \big(b, \delta_n^i(s)\big),\ \big(s_n^i(b), t\big) = \big(b, \sigma_n^i(t)\big)\Big\rangle,$$

for all $b \in B_n(*, \text{A}, D)$, and all $s \in \Delta(n-1)$ and $t \in \Delta(n+1)$. Here δ_n^i and σ_n^i are the standard face and degeneracy maps of geometric simplices.

Example 7.3. *Suppose that* A *is the category* $\cdot \to \cdot$, *with a single non-identity. Then an* A-*diagram is a homomorphism* $M \to N$ *in* TMG, *and* $\operatorname{hocolim}^{\text{TMG}} D$ *is its* TMG *mapping cylinder. It may be identified with the* TMG-*pushout of the diagram*

$$M \circledast \Delta(1)_+ \xleftarrow{\ j\ } M \longrightarrow N,$$

where $j(m) = (m, 0)$ *in* $M \circledast \Delta(1)_+$ *for all* $m \in M$.

An alternative expression for the simplicial topological monoid $B_\bullet^{\text{TMG}}(*, \text{A}, D)$ arises by analogy with the equivalences (2.20).

Proposition 7.4. *There is an isomorphism* $D \circledast_{\text{A}^{op}} B_\bullet^+(*, \text{A}, A_+) \cong B_\bullet^{\text{TMG}}(*, \text{A}, D)$ *of simplicial topological monoids, for any diagram* $D\colon \text{A} \to \text{TMG}$.

Proof. By (2.17), the functor $D \circledast_{\text{A}^{op}}\ : [\text{A}^{op} \times \Delta^{op}, \text{TOP}_+] \to [\Delta^{op}, \text{TMG}]$ is left TOP_+-adjoint to $\text{TMG}(D,\ )$, and therefore preserves coproducts. So we may write

$$\begin{aligned} D \circledast_{\text{A}^{op}} B_\bullet^+(*, \text{A}, A_+) &\cong \mathop{*}_{a,b} D \circledast_{\text{A}^{op}} (\text{A}(\ , a)_+ \wedge \text{A}_\bullet(a, b)_+) \\ &\cong \mathop{*}_{a,b} D(a) \circledast \text{A}_\bullet(a, b)_+ \end{aligned}$$

as required, using the isomorphism $D \circledast_{\text{A}^{op}} \text{A}(\ , a) \cong D(a)$ of (2.16). □

We must decide when the simplicial topological monoids $B_\bullet^{\text{TMG}}(*, \text{A}, D)$ are proper simplicial spaces (in the sense of [25]) because we are interested in the homotopy type of their realisations. This is achieved in Proposition 7.8, and leads on to the analogue of the Homotopy Lemma for TMG. These are two of the more memorable of the following sequence of six preliminaries, which precede the proof of our main result. On several occasions we insist that objects of TMG are well pointed, and even that they have the homotopy type of a CW-complex. Such conditions certainly hold for our exponential diagrams, and do not affect the applications.

We consider families of monoids indexed by the elements s of an arbitrary set S.

Lemma 7.5. *Let $f_s\colon M_s \to N_s$ be a family of homomorphisms of well-pointed topological monoids, which are homotopy equivalences; then the coproduct homomorphism*

$$\mathop{*}_s f_s\colon \mathop{*}_s M_s \longrightarrow \mathop{*}_s N_s$$

is also a homotopy equivalence.

Proof. Let $f\colon M \to N$ denote the homomorphism in question, and write F_kM for the subspace of M of elements representable by words of length $\leq k$. Hence $F_0 = \{e\}$, and $F_{k+1}M$ is the pushout

$$\begin{array}{ccc} \bigvee_K W_K(M) & \xrightarrow{\subseteq} & \bigvee_K P_K(M) \\ \downarrow & & \downarrow \\ F_kM & \xrightarrow{j_k} & F_{k+1}M \end{array} \tag{7.6}$$

in TOP$_+$, where K runs through all $(k+1)$-tuples $(s_1, \ldots, s_{k+1}) \in S^{k+1}$ such that $s_{i+1} \neq s_i$, and $W_K(M) \subset P_K(M)$ is the fat wedge subspace of $M_{s_1} \times \cdots \times M_{s_{k+1}}$. Each M_s is well pointed, so $W_K(M) \subset P_K(M)$ is a closed cofibration, and therefore so is j_k. Since $M = \operatorname{colim}_k F_kM$ in TOP$_+$, it remains to confirm that the restriction $f_k\colon F_kM \to F_kN$ is a homotopy equivalence for all k. We proceed by induction, based on the trivial case $k = 0$.

The map f induces a homotopy equivalence $W_K(M) \to W_K(N)$ because M_s and N_s are well pointed, and a further homotopy equivalence $P_K(M) \to P_K(N)$ by construction. So the inductive hypothesis combines with Brown's Gluing Lemma [37, 2.4] to complete the proof. □

Lemma 7.7. *For any subset $R \subset S$, the inclusion $*_r M_r \to *_s M_s$ is a closed cofibration; in particular, $*_s M_s$ is well pointed.*

Proof. Let $B \to M$ be the inclusion in question, with F_kM as in the proof of Lemma 7.5 and $F'_kM = B \cup F_kM$. Then $F'_{k+1}M$ is obtained from F'_kM by attaching spaces $P_K(M)$, where K runs through all $(s_1, \ldots, s_{k+1})$ in $S^{k+1} \setminus R^{k+1}$ such that $s_{i+1} \neq s_i$. Thus $B \to F'_kM$ is a cofibration for all k, implying the result. □

Proposition 7.8. *Given any small category* A, *and any diagram $D\colon$ A $\to$ TMG of well-pointed topological monoids, the simplicial space $B^{\mathrm{TMG}}_\bullet(*, \mathrm{A}, D)$ is proper, and its realisation is well pointed.*

Proof. By Lemma 7.7, each degeneracy map $B^{\mathrm{TMG}}_n(*, \mathrm{A}, D) \to B^{\mathrm{TMG}}_{n+1}(*, \mathrm{A}, D)$ is a closed cofibration. The first result then follows from Lillig's Union Theorem [23] for cofibrations. So $B^{\mathrm{TMG}}_0(*, \mathrm{A}, D) \subset |B^{\mathrm{TMG}}_\bullet(*, \mathrm{A}, D)|$ is a closed cofibration and $B^{\mathrm{TMG}}_0(*, \mathrm{A}, D)$ is well pointed, yielding the second result. □

As described in Examples 2.14, every simplicial object $M_\bullet$ in TMG has two possible realisations. We now confirm that they agree, and identify their classifying space.

Lemma 7.9. *The realisations $|M_\bullet|_{\mathrm{TMG}}$ and $|M_\bullet|$ are naturally isomorphic objects of* TMG, *whose classifying space is naturally homeomorphic to $|B(M_\bullet)|$.*

Proof. We apply the techniques of [14, VII §3] and [26, §4] to the functors $|\ |_{\mathrm{TMG}}$ and the restriction of $|\ |$ to $[\Delta^{op},\mathrm{TMG}]$. Both are left TOP_+-adjoint to $Sin\colon \mathrm{TMG} \to [\Delta^{op},\mathrm{TMG}]$, and so are naturally equivalent. The homeomorphism $B|M_\bullet| \cong |B(M_\bullet)|$ arises by considering the bisimplicial object $(k,n) \mapsto (M_n)^k$ in TOP_+, and forming its realisation in either order. □

We may now establish our promised Homotopy Lemma.

Proposition 7.10. *Given diagrams D_1, $D_2\colon \mathrm{A} \to \mathrm{TMG}$ of well-pointed topological monoids, and a map $f\colon D_1 \to D_2$ such that $f(a)\colon D_1(a) \to D_2(a)$ is a homotopy equivalence of underlying spaces for each object a of* A, *the induced map*

$$\operatorname{hocolim}^{\mathrm{TMG}} D_1 \longrightarrow \operatorname{hocolim}^{\mathrm{TMG}} D_2$$

is a homotopy equivalence.

Proof. This follows directly from Lemmas 7.5 and 7.9, and Proposition 7.8. □

We need one more technical result concerning homotopy limits of simplicial objects. We work with diagrams $X_\bullet\colon \mathrm{A}\times\Delta^{op} \to \mathrm{TOP}_+$ of simplicial spaces, and $D_\bullet\colon \mathrm{A}\times\Delta^{op} \to \mathrm{TMG}$ of simplicial topological monoids.

Proposition 7.11. *With $X_\bullet$ and $D_\bullet$ as above, there are natural isomorphisms*

$$\operatorname{hocolim}^{+}|X_\bullet| \cong |\operatorname{hocolim}^{+} X_\bullet| \quad \textit{and} \quad \operatorname{hocolim}^{\mathrm{TMG}}|D_\bullet| \cong |\operatorname{hocolim}^{\mathrm{TMG}} D_\bullet|$$

in TOP_+ and TMG *respectively.*

Proof. The isomorphisms arise from realising the bisimplicial objects

$$(k,n) \longmapsto B_k^{+}(*,\mathrm{A},X_n) \qquad \text{and} \qquad (k,n) \longmapsto B_k^{\mathrm{TMG}}(*,\mathrm{A},D_n)$$

in either order. In the case of $D_\bullet$, we must also apply the first statement of Lemma 7.9. □

Parts of the proofs above may be rephrased using variants of the equivalences (2.16). They lead to our first general result, which states that the formation of classifying spaces commutes with homotopy colimits in an appropriate sense.

Theorem 7.12. *For any diagram $D\colon \mathrm{A} \to \mathrm{TMG}$ of well-pointed topological monoids with the homotopy types of CW-complexes, there is a natural map*

$$g_D\colon \operatorname{hocolim}^{+} BD \longrightarrow B\operatorname{hocolim}^{\mathrm{TMG}} D$$

which is a homotopy equivalence.

Proof. For each object a of A, let $D_\bullet(a)$ be the singular simplicial monoid of $D(a)$. The natural map $|D_\bullet(a)| \to D(a)$ is a homomorphism of well-pointed monoids and a homotopy equivalence, so it passes to a homotopy equivalence $B|D_\bullet(a)| \to BD(a)$ under the formation of classifying spaces. By Proposition 7.10 and the corresponding Homotopy Lemma for TOP_+, it therefore suffices to prove our result for diagrams of realisations of simplicial monoids.

Let $D_\bullet\colon \mathrm{A} \times \Delta^{op} \to \mathrm{TMG}$ be a diagram of simplicial monoids. By Lemma 7.9 and Proposition 7.11, we must exhibit a natural homotopy equivalence

$$|\operatorname{hocolim}^+ BD_\bullet| \longrightarrow |B \operatorname{hocolim}^{\mathrm{TMG}} D_\bullet|.$$

Since both simplicial spaces are proper, it suffices to find a natural map

$$\operatorname{hocolim}^+ BD_\bullet \longrightarrow B \operatorname{hocolim}^{\mathrm{TMG}} D_\bullet$$

which is a homotopy equivalence in each dimension n. But $\operatorname{hocolim}^+ BD_n$ is the realisation of the proper simplicial space $B_\bullet^+(*, \mathrm{A}, BD_n)$, and Lemma 7.9 confirms that $B \operatorname{hocolim}^{\mathrm{TMG}} D_n$ is naturally homeomorphic to the realisation of the proper simplicial space $B(B_\bullet^{\mathrm{TMG}}(*, \mathrm{A}, D_n))$; so g_D may be specified by a sequence of maps $B_k^+(*, \mathrm{A}, BD_n) \to B(B_k^{\mathrm{TMG}}(*, \mathrm{A}, D_n))$, where $k \geq 0$. They are most easily described as maps

$$\bigvee_{a_0 \to \cdots \to a_k} BD_n(a_0) \longrightarrow B\Big(\mathop{*}_{a_0 \to \cdots \to a_k} D_n(a_0)\Big), \tag{7.13}$$

and are induced by including each of the $D_n(a_0)$ into the free product. Since (7.13) is a homotopy equivalence by a theorem of Fiedorowicz [15, 4.1], the proof is complete. □

Various steps in the proof of Theorem 7.12 may be adapted to verify the following, which answers a natural question about tensored monoids.

Proposition 7.14. *For any well-pointed topological monoid M and based space Y, the natural map*

$$BM \wedge Y \longrightarrow B(M \circledast Y)$$

is a homotopy equivalence if M and Y have the homotopy type of CW-complexes.

Proof. As in Theorem 7.12, we need only work with the realisations $|M_\bullet|$ and $|Y_\bullet|$ of the total singular complexes. Since $B|M_\bullet| \wedge |Y_\bullet| \to B(|M_\bullet| \circledast |Y_\bullet|)$ is the realisation of the natural map $BM_n \wedge Y_n \to B(M_n \circledast Y_n)$, it suffices to assume that Y is discrete; in this case,

$$BM \wedge Y \longrightarrow B\big(\mathop{*}_{y} M_y\big)$$

is a homotopy equivalence by the same result of Fiedorowicz [15]. □

We apply Theorem 7.12 to construct our general model for $\Omega DJ(K)$, but require a commutative diagram to clarify its relationship with the special case h_K of Proposition 6.3. We deal with $\mathrm{A}^{op} \times \Delta^{op}$-diagrams $X_\bullet$ in TOP_+, and certain of their morphisms. These include $\theta\colon X_\bullet \to \mathrm{TOP}_+(BD, B(D \circledast_{\mathrm{A}^{op}} X_\bullet))$, defined for any $X_\bullet$ by $\theta(x) = B(d \mapsto d \circledast x)$, and the projection $\pi\colon B_\bullet^+ \to (*_+)_\bullet$, where

$B_\bullet^+$ and $(*_+)_\bullet$ denote $B_\bullet^+(*, \mathrm{A}, A)$ and the trivial diagram respectively. Under the homeomorphism

$$[\Delta^{op},\mathrm{TOP}_+]\big(BD \wedge_{\mathrm{A}^{op}} X_\bullet, B(D \circledast_{\mathrm{A}^{op}} X_\bullet)\big) \cong [\mathrm{A}^{op}\times\Delta^{op},\mathrm{TOP}_+]\big(X_\bullet, \mathrm{TOP}_+(BD, B(D \circledast_{\mathrm{A}^{op}} X_\bullet))\big)$$

of (2.17), θ corresponds to a map $\phi\colon BD \wedge_{\mathrm{A}^{op}} X_\bullet \to B(D \circledast_{\mathrm{A}^{op}} X_\bullet)$ of simplicial spaces.

Proposition 7.15. *For any diagram $D\colon \mathrm{A} \to \mathrm{TMG}$, there is a commutative square*

$$\begin{array}{ccc} \operatorname{hocolim}^+ BD & \xrightarrow{g_D} & B\operatorname{hocolim}^{\mathrm{TMG}} D \\ \big\downarrow{\scriptstyle p^+} & & \big\downarrow{\scriptstyle Bp^{\mathrm{TMG}}} \\ \operatorname{colim}^+ BD & \xrightarrow{f_D} & B\operatorname{colim}^{\mathrm{TMG}} D \end{array},$$

where p^+ and p^{TMG} are the natural projections.

Proof. By construction, the diagram

$$\begin{array}{ccc} B_\bullet^+ & \xrightarrow{\theta} & \mathrm{TOP}_+(BD, B(D \circledast_{\mathrm{A}^{op}} B_\bullet^+)) \\ \big\downarrow{\scriptstyle \pi} & & \big\downarrow{\scriptstyle B(1\circledast\pi)\cdot} \\ (*_+)_\bullet & \xrightarrow{\theta} & \mathrm{TOP}_+(BD, B(D \circledast_{\mathrm{A}^{op}} (*_+)_\bullet)) \end{array}$$

is commutative in $[\mathrm{A}^{op}\times\Delta^{op},\mathrm{TOP}_+]$, and has adjoint

$$\begin{array}{ccc} BD \wedge_{\mathrm{A}^{op}} B_\bullet^+ & \xrightarrow{\phi} & B(D \circledast_{\mathrm{A}^{op}} B_\bullet^+) \\ \big\downarrow{\scriptstyle 1\wedge\pi} & & \big\downarrow{\scriptstyle B(1\circledast\pi)} \\ BD \wedge_{\mathrm{A}^{op}} (*_+)_\bullet & \xrightarrow{\phi} & B(D \circledast_{\mathrm{A}^{op}} (*_+)_\bullet) \end{array} \tag{7.16}$$

in $[\Delta^{op},\mathrm{TOP}_+]$. By Proposition 7.4, the upper ϕ is the map $B_\bullet^+(*, \mathrm{A}, BD) \to B\left(B_\bullet^{\mathrm{TMG}}(*, \mathrm{A}, D)\right)$ obtained by applying the relevant map (7.13) in each dimension. By Examples 2.14, the lower ϕ is given by the canonical map $f_D\colon \operatorname{colim}^+ BD \to B\operatorname{colim}^{\mathrm{TMG}} D$ in each dimension. Since realisation commutes with B, the topological realisation of (7.16) is the diagram we seek; for Lemma 7.9 identifies the upper right-hand space with $B\operatorname{hocolim}^{\mathrm{TMG}} D$, and Examples 2.21 confirms that the vertical maps are the natural projections. □

Theorem 7.17. *There is a homotopy commutative square*

$$\begin{array}{ccc} \Omega\operatorname{hocolim}^+(BG)^K & \xrightarrow{\overline{h}_K} & \operatorname{hocolim}^{\mathrm{TMG}} G^K \\ \big\downarrow{\scriptstyle \Omega p_K} & & \big\downarrow{\scriptstyle q_K} \\ \Omega DJ(K) & \xrightarrow{h_K} & \operatorname{colim}^{\mathrm{TMG}} G^K \end{array}$$

of homotopy homomorphisms, where p_K and $\overline{h}_K$ are homotopy equivalences for any simplicial complex K.

Proof. We apply Proposition 7.15 with $D = G^K$, and loop the corresponding square; the projection $p_K\colon \operatorname{hocolim}^+(BG)^K \to DJ(K)$ is a homotopy equivalence, as explained in (5.3). The result follows by composing the horizontal maps with the canonical homotopy homomorphism $\Omega BH \to H$, where $H = \operatorname{hocolim}^{\mathrm{TMG}} G^K$ and $\operatorname{colim}^{\mathrm{TMG}} G^K$ respectively. □

It is an interesting challenge to describe good geometrical models for homotopy homomorphisms which are inverse to h_K and $\overline{h}_K$.

References

[1] Michael Barr and Charles Wells. *Toposes, Triples and Theories.* Number 278 in Grundlehren der mathematischen Wissenschaften. Springer-Verlag, 1985.

[2] Anders Björner. Subspace arrangements. In A. Joseph, F. Mignot, F. Murat, and B. Prum, editors, *Proceedings of the First European Congress of Mathematics (Paris 1992)*, volume 119 of *Progress in Mathematics*, pages 321–370. Birkhäuser, 1995.

[3] Francis Borceux. *Handbook of Categorical Algebra 2*, volume 51 of *Encyclopedia of Mathematics and its Applications.* Cambridge University Press, 1994.

[4] A.K. Bousfield and Daniel M. Kan. *Homotopy Limits, Completions and Localizations*, volume 304 of *Lecture Notes in Mathematics.* Springer Verlag, 1972.

[5] Michael Brinkmeier. Strongly homotopy-commutative monoids revisited. *Documenta Mathematica*, 5:613–624, 2000.

[6] Ronald Brown and J.P. Lewis Hardy. Topological groupoids 1: universal constructions. *Mathematische Nachrichten*, 71:273–286, 1976.

[7] Victor M. Buchstaber and Taras E. Panov. Torus actions, combinatorial topology, and homological algebra. *Russian Mathematical Surveys*, 55:825–921, 2000.

[8] Pierre Cartier and Dominique Foata. *Problèmes combinatoires de commutation et réarrangements*, volume 85 of *Lecture Notes in Mathematics.* Springer Verlag, 1969.

[9] Ruth Charney and Michael Davis. Finite $K(\pi, 1)$ for Artin groups. In *Prospects in Topology (Princeton NJ, 1994)*, volume 138 of *Annals of Mathematics Studies*, pages 110–124. Princeton University Press, 1995.

[10] Ian M. Chiswell. The Euler characteristic of graph products and of Coxeter groups. In William J. Harvey and Colin Maclachlan, editors, *Discrete Groups and Geometry (Birmingham, 1991)*, volume 173 of *London Mathematical Society Lecture Note Series*, pages 36–46. Cambridge University Press, 1992.

[11] Michael W. Davis. Groups generated by reflections and aspherical manifolds not covered by Euclidean space. *Annals of Mathematics*, 117:293–324, 1983.

[12] Michael W. Davis and Tadeusz Januszkiewicz. Convex polytopes, Coxeter orbifolds and torus actions. *Duke Mathematical Journal*, 62:417–451, 1991.

[13] Albrecht Dold and Richard Lashof. Principal quasifibrations and fibre homotopy equivalence of bundles. *Illinois Journal of Mathematics*, 3:285–305, 1959.

[14] Anthony D. Elmendorf, Igor Kriz, Michael P. Mandell, and J. Peter May. *Rings, Modules, and Algebras in Stable Homotopy Theory*, volume 47 of *Mathematical Surveys and Monographs.* American Mathematical Society, 1997.

[15] Zbigniew Fiedorowicz. Classifying spaces of topological monoids and categories. *American Journal of Mathematics*, 106:301–350, 1984.

[16] Tor Gulliksen and Gerson Levin. *Homology of local rings*, volume 20 of *Queen's papers in Pure and applied Mathematics*. Queen's University, Kingston, Ontario, 1969.

[17] Jørgen Herzog, Victor Reiner, and Volkmar Welker. Componentwise linear ideals and Golod rings. *Michigan Mathematical Journal*, 46:211–223, 1999.

[18] Jens Hollender and Rainer M. Vogt. Modules of topological spaces, applications to homotopy limits and E_∞ structures. *Archiv der Mathematik*, 59:115–129, 1992.

[19] Ioan M. James. Reduced product spaces. *Annals of Mathematics*, 62:259–280, 1955.

[20] S.A. Joni and G.-C. Rota. Coalgebras and bialgebras in combinatorics. *Studies in Applied Mathematics*, 61:93–139, 1979.

[21] G. Maxwell Kelly. *Basic Concepts of Enriched Category Theory*, volume 64 of *London Mathematical Society Lecture Note Series*. Cambridge University Press, 1982.

[22] Ki Hang Kim and Fred W. Roush. Homology of certain algebras defined by graphs. *Journal of Pure and Applied Algebra*, 17:179–186, 1980.

[23] Joachim Lillig. A union theorem for cofibrations. *Archiv der Mathematik*, 24:410–415, 1973.

[24] J. Peter May. *Simplicial Objects in Algebraic Topology*, volume 11 of *Van Nostrand Mathematical Studies*. Van Nostrand Reinhold, 1967.

[25] J. Peter May. E_∞-spaces, group completions and permutative categories. In Graeme Segal, editor, *New Developments in Topology*, volume 11 of *London Mathematical Society Lecture Notes Series*, pages 153–231. Cambridge University Press, 1974.

[26] James McClure, Roland Schwänzl, and Rainer Vogt. *THH(R)* $\cong R \otimes S^1$ for E_∞ ring spectra. *Journal of Pure and Applied Algebra*, 121:137–159, 1997.

[27] Taras Panov and Nigel Ray. The homology and homotopy theory of certain loop spaces. In preparation, University of Manchester.

[28] Irena Peeva, Victor Reiner, and Volkmar Welker. Cohomology of real diagonal subspace arrangments via resolutions. *Compositio Mathematica*, 117:99–115, 1999.

[29] Nigel Ray and William Schmitt. Combinatorial models for coalgebraic structures. *Advances in Mathematics*, 138:211–262, 1998.

[30] Roland Schwänzl and Rainer M Vogt. The categories of A_∞ and E_∞ monoids and ring spectra as closed simplicial and topological model categories. *Archiv der Mathematik*, 56:405–411, 1991.

[31] Peter Scott and Terry Wall. Topological methods in group theory. In C.T.C. Wall, editor, *Homological Group Theory (Proceedings of the Durham Symposium, 1977)*, volume 36 of *London Mathematical Society Lecture Note Series*, pages 137–203. Cambridge University Press, 1979.

[32] Graeme Segal. Classifying spaces and spectral sequences. *Institut des Hautes Études Scientifiques, Publications Mathématiques*, 34:105–112, 1968.

[33] Richard P. Stanley. *Combinatorics and Commutative Algebra, 2nd edition*, volume 41 of *Progress in Mathematics*. Birkhäuser, Boston, 1996.

[34] Masahiro Sugawara. On the homotopy-commutativity of groups and loop spaces. *Memoirs of the College of Science, University of Kyoto: Series A, Mathematics*, 33:257–269, 1960/61.

[35] Rainer M. Vogt. Convenient categories of topological spaces for homotopy theory. *Archiv der Mathematik*, 22:545–555, 1971.

[36] Rainer M. Vogt. Homotopy limits and colimits. *Mathematische Zeitschrift*, 134:11–52, 1973.

[37] Volkmar Welker, Günter M. Ziegler, and Rade T. Živaljević. Homotopy colimits – comparison lemmas for combinatorial applications. *Journal für die reine und angewandte Mathematik*, 509:117–149, 1999.

[38] Günter M. Ziegler and Rade T. Živaljević. Homotopy types of subspace arrangements via diagrams of spaces. *Mathematische Annalen*, 295:527–548, 1993.

Taras Panov
Department of Mathematics and Mechanics
Moscow State University
119899 Moscow, Russia
and
Institute for Theoretical and Experimental Physics
117259 Moscow, Russia
e-mail: tpanov@mech.math.msu.su

Nigel Ray
Department of Mathematics
University of Manchester
Manchester M13 9PL, England
e-mail: nige@ma.man.ac.uk

Rainer Vogt
Fachbereich Mathematik/Informatik
Universität Osnabrück
D-49069 Osnabrück, Germany
e-mail: rainer@mathematik.uni-osnabrueck.de

Progress in Mathematics, Vol. 215
Algebraic Topology: Categorical Decomposition Techniques, 293–302

Maps to Spaces in the Genus of Infinite Quaternionic Projective Space

Donald Yau

Abstract. Spaces in the genus of infinite quaternionic projective space which admit essential maps from infinite complex projective space are classified. In these cases the sets of homotopy classes of maps are described explicitly. These results strengthen the classical theorem of McGibbon and Rector on maximal torus admissibility for spaces in the genus of infinite quaternionic projective space. An interpretation of these results in the context of Adams-Wilkerson embedding in integral K-theory is also given.

1. Introduction and statement of results

In an attempt to understand Lie groups through their classifying spaces, Rector [9] classified the genus of $\mathbf{HP}^\infty$, the infinite projective space over the quaternions, considered as a model for the classifying space BS^3. The homotopy type of a space X is said to be in the genus of $\mathbf{HP}^\infty$, denoted $X \in \text{Genus}(\mathbf{HP}^\infty)$, if the p-localizations of X and $\mathbf{HP}^\infty$ are homotopy equivalent for each prime p. One often speaks of a space rather than its homotopy type when considering genus. Rector's classification [9] of the genus of $\mathbf{HP}^\infty$ is as follows.

Theorem 1.1 (Rector). *Let X be a space in the genus of $\mathbf{HP}^\infty$. Then for each prime p there exists a homotopy invariant $(X/p) \in \{\pm 1\}$ such that the following statements hold.*

1. *The (X/p) for p primes provide a complete list of homotopy classification invariants for the genus of $\mathbf{HP}^\infty$.*
2. *Any combination of values of the (X/p) can occur. In particular, the genus of $\mathbf{HP}^\infty$ is uncountable.*
3. *The invariant $(\mathbf{HP}^\infty/p)$ is 1 for all primes p.*
4. *The space X has a maximal torus if and only if X is homotopy equivalent to $\mathbf{HP}^\infty$.*

The invariant (X/p) is now known as the *Rector invariant at the prime p*. Actually, for the last statement about the maximal torus, Rector only proved it for the odd primes. That is, if X has a maximal torus, then (X/p) is equal to 1 for all odd primes p. Then McGibbon [6] proved it for the prime 2 as well. Here X is said to have a maximal torus if there exists a map from $\mathbf{CP}^\infty$, the infinite complex projective space, to X whose homotopy theoretic fiber has the homotopy type of a finite complex.

For a space X in the genus of $\mathbf{HP}^\infty$ which is not homotopy equivalent to $\mathbf{HP}^\infty$, the nonexistence of a maximal torus does not rule out the possibility that there could be some essential (that is, non-nullhomotopic) maps from $\mathbf{CP}^\infty$ to X. The main purposes of this paper are (1) to describe spaces in the genus of $\mathbf{HP}^\infty$ for which this can happen (Theorem 1.2) and (2) to compute the maps in these cases (Theorem 1.7).

The following is our first main result, which classifies spaces in the genus of $\mathbf{HP}^\infty$ which admit essential maps from $\mathbf{CP}^\infty$.

Theorem 1.2. *Let X be a space in the genus of $\mathbf{HP}^\infty$. Then the following statements are equivalent.*

1. *There exists an essential map from $\mathbf{CP}^\infty$ to X.*
2. *There exists a nonzero integer k such that $(X/p) = (k/p)$ for all but finitely many primes p.*
3. *There exists a cofinite set of primes L such that $\mathbf{HP}^\infty$ and X become homotopy equivalent after localization at L.*

Here (k/p) is the Legendre symbol of k, which is defined whenever p does not divide k. If p is odd and if p does not divide k, then $(k/p) = 1$ (resp. -1) if k is a quadratic residue (resp. non-residue) mod p. If $p = 2$ and if k is odd, then $(k/2) = 1$ (resp. -1) if k is a quadratic residue (resp. non-residue) mod 8.

Before discussing related issues, let us first record the following immediate consequence of Theorem 1.2.

Corollary 1.3. *There exist only countably many homotopically distinct spaces in the genus of $\mathbf{HP}^\infty$ which admit essential maps from $\mathbf{CP}^\infty$.*

Indeed, each nonzero integer k can determine only countably many homotopically distinct spaces X in the genus of $\mathbf{HP}^\infty$ satisfying the second condition in Theorem 1.2.

The second condition of Theorem 1.2 gives an arithmetic description of spaces in the genus of $\mathbf{HP}^\infty$ which occur as the target of essential maps from $\mathbf{CP}^\infty$. Since it involves Rector invariants, it is specific to the genus of $\mathbf{HP}^\infty$ and is not very convenient for generalizations. The last condition of Theorem 1.2, on the other hand, is geometric and is more suitable for possible generalizations of the theorem.

Having characterized spaces in the genus of $\mathbf{HP}^\infty$ which admit nontrivial maps from $\mathbf{CP}^\infty$, we proceed to compute the maps themselves. Now for any space X in the genus of $\mathbf{HP}^\infty$, the K-theory $K(X)$ of X, as a filtered ring, is a power series ring $\mathbf{Z}[[b^2u_X]]$ (see Proposition 2.1), where u_X is some element in $K^4(X)$ and b is the Bott element in $K^{-2}(\mathrm{pt})$. Here, and throughout the rest of the paper, $K(-)$ denotes complex K-theory with coefficients over the integers $\mathbf{Z}$. So if $f\colon \mathbf{CP}^\infty \to X$ is any map, then its induced map in K-theory defines an integer $\deg(f)$, called the *degree of f*, by the equation

$$f^*(b^2u_X) = \deg(f)(b\xi)^2 + \text{higher terms in } b\xi, \tag{1.4}$$

where $b\xi$ is the ring generator in the power series ring $K(\mathbf{CP}^\infty) = \mathbf{Z}[[b\xi]]$. Note that $\deg(f)$ is, up to a sign, simply the degree of the induced map of f in integral homology in dimension 4. According to a result of Dehon and Lannes [4], the homotopy class of such a map f is determined by its degree. We will therefore identify such a map with its degree in the sequel. The degrees of a self-map of X, a self-map of the p-localization $\mathbf{HP}^\infty_{(p)}$ of $\mathbf{HP}^\infty$, or a map from $\mathbf{CP}^\infty$ to $\mathbf{HP}^\infty_{(p)}$ can be defined similarly.

To describe the maps from $\mathbf{CP}^\infty$ to $X \in \mathrm{Genus}(\mathbf{HP}^\infty)$ up to homotopy, we need only describe the possible degrees of such maps. Let's first consider the classical case. There is a maximal torus inclusion $i\colon \mathbf{CP}^\infty \to \mathbf{HP}^\infty$ of degree 1, and any other map $f\colon \mathbf{CP}^\infty \to \mathbf{HP}^\infty$ factors through i up to homotopy. A special case of a theorem of Ishiguro, Møller, and Notbohm [5, Thm. 1] says that for any space X in the genus of $\mathbf{HP}^\infty$, the degrees of essential self-maps of X consist of precisely the squares of odd numbers. For the classical case, $X = \mathbf{HP}^\infty$, this result is due to Sullivan [10, p. 58-59]. Therefore, the degrees of essential maps from $\mathbf{CP}^\infty$ to $\mathbf{HP}^\infty$ also consist of precisely the odd squares.

The situation in general is quite similar. Recall that any space X in the genus of $\mathbf{HP}^\infty$ can be obtained as a homotopy inverse limit [3]

$$X = \mathrm{holim}_q \left\{ \mathbf{HP}^\infty_{(q)} \xrightarrow{r_q} \mathbf{HP}^\infty_{(0)} \xrightarrow{n_q} \mathbf{HP}^\infty_{(0)} \right\}. \tag{1.5}$$

Here q runs through all primes, r_q is the natural map from the q-localization to the rationalization of $\mathbf{HP}^\infty$, and n_q is an integer relatively prime to q, satisfying $(n_q/q) = (X/q)$. The integer n_2 also satisfies $n_2 \equiv 1 \pmod 4$.

Now if X admits an essential map from $\mathbf{CP}^\infty$, and thus satisfies the condition in Theorem 1.2 for some nonzero integer k, then the integers n_q can be chosen so that the set $\{n_q\colon q \text{ primes}\}$ contains only finitely many distinct integers. So it makes sense to talk about the least common multiple of the integers n_q, denoted $\mathrm{LCM}(n_q)$. Now we define an integer T_X as

$$T_X = \min\{\mathrm{LCM}(n_q) : X = \mathrm{holim}_q\,(n_q \circ r_q)\}. \tag{1.6}$$

That is, choose the integers n_q as in (1.5) so as to minimize their least common multiple, and T_X is defined to be $\mathrm{LCM}(n_q)$.

We are now ready for the second main result of this paper, which describes the maps from $\mathbf{CP}^\infty$ to $X \in \mathrm{Genus}(\mathbf{HP}^\infty)$.

Theorem 1.7. *Let X be a space in the genus of $\mathbf{HP}^\infty$ which admits an essential map from $\mathbf{CP}^\infty$. Then the following statements hold.*

1. *There exists a map $i_X\colon \mathbf{CP}^\infty \to X$ of degree T_X.*
2. *The map $i_{\mathbf{HP}^\infty}\colon \mathbf{CP}^\infty \to \mathbf{HP}^\infty$ is the maximal torus inclusion.*
3. *Given any map $f\colon \mathbf{CP}^\infty \to X$, there exists a self-map g of X such that f is homotopic to $g \circ i_X$.*
4. *The degrees of essential maps from $\mathbf{CP}^\infty$ to X are precisely the odd squares multiples of T_X.*

It should be noted that the integer T_X does **not** determine the homotopy type of X. For example, consider the spaces X and Y in the genus of $\mathbf{HP}^\infty$ with Rector invariants

$$(1.8) \qquad (X/p) = \begin{cases} 1 & \text{if } p \neq 3 \\ -1 & \text{if } p = 3 \end{cases}, \qquad (Y/p) = \begin{cases} 1 & \text{if } p \neq 5 \\ -1 & \text{if } p = 5. \end{cases}$$

Then, of course, X is not homotopy equivalent to Y because their Rector invariants at the prime 3 are distinct. But it is easy to see that $T_X = 2 = T_Y$.

Theorems 1.2 and 1.7 are closely related to the (non)existence of Adams-Wilkerson type embeddings of finite H-spaces in integral K-theory. As mentioned before, a map $f\colon \mathbf{CP}^\infty \to X \in \mathrm{Genus}(\mathbf{HP}^\infty)$ is essential if and only if $\deg(f)$ is nonzero. Thus, if there exists an essential map from $\mathbf{CP}^\infty$ to $X \in \mathrm{Genus}(\mathbf{HP}^\infty)$, then $K(X)$ can be embedded into $K(\mathbf{CP}^\infty)$ as a sub-λ-ring. The converse is also true. Indeed, a theorem of Notbohm and Smith [8, Thm. 5.2] says that the function

$$\alpha\colon [\mathbf{CP}^\infty, X] \to \mathrm{Hom}_\lambda(K(X), K(\mathbf{CP}^\infty))$$

which sends (the homotopy class of) a map to its induced map in K-theory, is a bijection. (Here $[-,-]$ and $\mathrm{Hom}_\lambda(-,-)$ denote, respectively, sets of homotopy classes of maps between spaces and of λ-ring homomorphisms.) So a λ-ring embedding $K(X) \to K(\mathbf{CP}^\infty)$ must be induced by an essential map from $\mathbf{CP}^\infty$ to X. Therefore, Theorem 1.2 and Corollary 1.3 can be restated in this context as follows.

Theorem 1.9. *Let X be a space in the genus of $\mathbf{HP}^\infty$. Then $K(X)$ can be embedded into $K(\mathbf{CP}^\infty)$ as a sub-λ-ring if, and only if, there exists a nonzero integer k such that $(X/p) = (k/p)$ for all but finitely many primes p. This is true if, and only if, there exists a cofinite set of primes L such that $\mathbf{HP}^\infty$ and X become homotopy equivalent after localization at L.*

In particular, there exist only countably many homotopically distinct spaces X in the genus of $\mathbf{HP}^\infty$ whose K-theory λ-rings can be embedded into that of $\mathbf{CP}^\infty$ as a sub-λ-ring.

Before Theorem 1.9, there is at least one space in the genus of $\mathbf{HP}^\infty$ whose K-theory λ-ring was known to be non-embedable into the K-theory λ-ring of $\mathbf{CP}^\infty$. This example was due to Adams [1, p. 79].

Remark 1.10. Theorems 1.2 and 1.7 can also be regarded as an attempt to understand the set of homotopy classes of maps from X to Y, where $X \in \mathrm{Genus}(BG)$ and $Y \in \mathrm{Genus}(BK)$ with G and K some connected compact Lie groups. This problem, especially the case $G = K = S^3 \times \cdots \times S^3$, was studied extensively by Ishiguro, Møller, and Notbohm [5].

This finishes the presentation of our main results. The rest of this paper is organized as follows. Theorems 1.2 and 1.7 are proved in §3 and §4, respectively. For use in §3, we recall in §2 some results from [12], with sketches of proofs, about K-theory filtered rings.

Acknowledgment

The author would like to thank Dietrich Notbohm for his interest and comments in this work, and to the referee whose suggestions, especially the last condition of Theorem 1.2, are very helpful. The author would like to express his deepest gratitude to his thesis advisor Haynes Miller for many hours of stimulating conversations and encouragement.

2. K-theory filtered ring

In preparation for the proof of Theorem 1.2 in §3, a result from [12] about K-theory filtered ring is reviewed in this section.

To begin with, a *filtered ring* is a pair $(R, \{I_n^R\})$ consisting of:

1. A commutative ring R with unit;
2. A decreasing filtration $R = I_0^R \supset I_1^R \supset \cdots$ of ideals of R such that $I_i^R I_j^R \subset I_{i+j}^R$ for all $i, j \geq 0$.

A map between two filtered rings is a ring homomorphism which preserves the filtrations. With these maps as morphisms, the filtered rings form a category.

Every space Z of the homotopy type of a CW complex gives rise naturally to an object $(K(Z), \{K_n(Z)\})$, which is usually abbreviated to $K(Z)$, in this category. Here $K(Z)$ and $K_n(Z)$ denote, respectively, the complex K-theory ring of Z and the kernel of the restriction map $K(Z) \to K(Z_{n-1})$, where Z_{n-1} denotes the $(n-1)$-skeleton of Z. Using a different CW structure of Z will not change the filtered ring isomorphism type of $K(Z)$, as can be easily seen by using the cellular approximation theorem. The symbol $K_s^r(Z)$ denotes the subgroup of $K^r(Z)$ consisting of elements whose restrictions to $K^r(Z_{s-1})$ are equal to 0.

The following result, which is proved in [12], will be needed in §3. For the reader's convenience we include here a sketch of the proof. In what follows $b \in K^{-2}(\mathrm{pt})$ will denote the Bott element.

Proposition 2.1. *Let X be a space in the genus of* $\mathbf{HP}^\infty$. *Then the following statements hold.*

1. *There exists an element $u_X \in K_4^4(X)$ such that $K(X) = \mathbf{Z}[[b^2 u_X]]$ as a filtered ring.*
2. *For any odd prime p, the Adams operation ψ^p satisfies*

$$\psi^p\left(b^2 u_X\right) = \left(b^2 u_X\right)^p + 2\left(X/p\right) p \left(b^2 u_X\right)^{(p+1)/2} + pw + p^2 z, \tag{2.2}$$

where w and z are some elements in $K_{2p+3}^0(X)$ and $K_4^0(X)$, respectively. In particular, we have

$$\psi^p\left(b^2 u_X\right) = 2\left(X/p\right) p \left(b^2 u_X\right)^{(p+1)/2} \quad \left(\textit{mod } K_{2p+3}^0(X) \textit{ and } p^2\right). \tag{2.3}$$

Sketch of the proof of Proposition 2.1. For the first assertion, Wilkerson's proof of the classification theorem [11, Thm. I] of spaces of the same n-type for all n can be easily adapted to show the following. There is a bijection between the following two pointed sets:

1. The pointed set of isomorphism classes of filtered rings $(R, \{I_n^R\})$ with the properties:
   (a) The natural map $R \to \varprojlim_n R/I_n^R$ is an isomorphism, and
   (b) R/I_n^R and $K(\mathbf{HP}^\infty)/K_n(\mathbf{HP}^\infty)$ are isomorphic as filtered rings for all $n > 0$.
2. The pointed set $\varprojlim_n^1 \operatorname{Aut}(K(\mathbf{HP}^\infty)/K_n(\mathbf{HP}^\infty))$.

Here Aut$(-)$ denotes the group of filtered ring automorphisms, and the $\varprojlim^1$ of a tower of not-necessarily abelian groups is as defined in [3]. It is not difficult to check that the hypothesis on X implies that the filtered ring $K(X)$ has the above two properties, (a) and (b). Moreover, by analyzing the subquotients $K_n(\mathbf{HP}^\infty)/K_{n+1}(\mathbf{HP}^\infty)$, one can show that the map

$$\operatorname{Aut}(K(\mathbf{HP}^\infty)/K_{n+1}(\mathbf{HP}^\infty)) \to \operatorname{Aut}(K(\mathbf{HP}^\infty)/K_n(\mathbf{HP}^\infty))$$

is surjective for each integer n greater than 4.

The point is that any automorphism of $K(\mathbf{HP}^\infty)/K_n(\mathbf{HP}^\infty)$ can be lifted to an endomorphism of $K(\mathbf{HP}^\infty)/K_{n+1}(\mathbf{HP}^\infty)$ without any difficulty. Then, since the quotient $K(\mathbf{HP}^\infty)/K_{n+1}(\mathbf{HP}^\infty)$ is a finitely generated abelian group, one only has to observe that the chosen lift is surjective. Thus, the above $\varprojlim^1$ is the one-point set, and hence $K(X)$ and $K(\mathbf{HP}^\infty)$ are isomorphic as filtered rings. This establishes the first assertion.

The second assertion concerning the Adams operations ψ^p is an easy consequence of the first assertion, Atiyah's theorem [2, Prop. 5.6], and the definition of the Rector invariants (X/p) [9]. □

3. Proof of Theorem 1.2

In this section the proof of Theorem 1.2 is given.

Recall that the complex K-theory of $\mathbf{CP}^\infty$ as a filtered λ-ring is given by $K(\mathbf{CP}^\infty) = \mathbf{Z}[[b\xi]]$ for some $\xi \in \tilde{K}_2^2(\mathbf{CP}^\infty)$, where $b \in K^{-2}(\mathrm{pt})$ is the Bott element. The Adams operations on the generator are given by

$$\psi^r(b\xi) = (1 + b\xi)^r - 1 \quad (r = 1, 2, \ldots). \tag{3.1}$$

Fix a space X in the genus of $\mathbf{HP}^\infty$ and write $\mathbf{Z}[[b^2 u_X]]$ for its K-theory filtered ring (cf. Proposition 2.1).

We will prove Theorem 1.2 by proving the implications (1) ⇒ (2) ⇒ (3) ⇒ (1). Each implication is contained in one subsection below.

3.1. Proof of (1) implies (2)

This part of Theorem 1.2 is contained in the next Lemma.

Lemma 3.2. *Let p be an odd prime and k be a nonzero integer relatively prime to p. If there exists an essential map $f\colon \mathbf{CP}^\infty \to X$ of degree k, then $(X/p) = (k/p)$.*

Proof. We will compare the coefficients of $(b\xi)^{p+1}$ in the equation

$$(3.3)\qquad f^*\psi^p\left(b^2u_X\right) = \psi^p f^*\left(b^2u_X\right) \quad \left(\text{mod } K^0_{2p+3}(\mathbf{CP}^\infty) \text{ and } p^2\right).$$

Working modulo $K^0_{2p+3}(\mathbf{CP}^\infty)$ and p^2, it follows from (1.4) and (2.3) that

$$\begin{aligned} f^*\psi^p\left(b^2u_X\right) &= 2\,(X/p)\,p\left(kb^2\xi^2\right)^{(p+1)/2} \\ &= 2\,(X/p)\,p\,k^{(p+1)/2}(b\xi)^{p+1}. \end{aligned}$$

Similarly, still working modulo $K^0_{2p+3}(\mathbf{CP}^\infty)$ and p^2, it follows from (1.4) and (3.1) that

$$\begin{aligned} \psi^p f^*\left(b^2u_X\right) &= k\psi^p(b^2\xi^2) \\ &= k\psi^p(b\xi)^2 \\ &= 2pk(b\xi)^{p+1}. \end{aligned}$$

Thus, we obtain the congruence relation

$$(3.4)\qquad 2\,(X/p)\,p\,k^{(p+1)/2} \equiv 2pk \pmod{p^2}.$$

Since (k/p) is congruent to $k^{(p-1)/2}$ (mod p) (see, for example, [7, Thm. 3.12]) and since p is odd and relatively prime to k, (3.4) is equivalent to the congruence relation

$$(3.5)\qquad (X/p)\,(k/p) \equiv 1 \pmod{p}.$$

Hence $(X/p) = (k/p)$, as desired.

This finishes the proof of Lemma 3.2. □

This shows that (1) implies (2) in Theorem 1.2.

3.2. Proof of (2) implies (3)

Suppose that there exists a nonzero integer k such that $(X/p) = (k/p)$ for all primes p, except possibly $p_1, \ldots, p_s$. The prime factors of k are among the p_i. Let L be the set consisting of all primes except the p_i, $1 \le i \le s$. We will show that $\mathbf{HP}^\infty$ and X become homotopy equivalent after localization at L.

First note that for any space Y in the genus of $\mathbf{HP}^\infty$ and for any subset I of primes, the I-localization of Y can be obtained as

$$(3.6)\qquad Y_{(I)} = \operatorname{holim}_{q\in I}\left\{\mathbf{HP}^\infty_{(q)} \xrightarrow{n_q\circ r_q} \mathbf{HP}^\infty_{(0)}\right\}.$$

In particular, we have

$$(3.7)\qquad X_{(L)} = \operatorname{holim}_{q\in L}\left\{\mathbf{HP}^\infty_{(q)} \xrightarrow{k\circ r_q} \mathbf{HP}^\infty_{(0)}\right\}$$

and

$$(3.8)\qquad \mathbf{HP}^\infty_{(L)} = \operatorname{holim}_{q\in L}\left\{\mathbf{HP}^\infty_{(q)} \xrightarrow{r_q} \mathbf{HP}^\infty_{(0)}\right\}.$$

Now for each prime $q \in L$, let f_q be a self-map of $\mathbf{HP}^\infty_{(q)}$ of degree k^{-1}. Since k is a q-local unit (because q does not divide k), it is easy to see that each f_q is a homotopy equivalence. Moreover, the two maps

$$r_q,\, k \circ r_q \circ f_q \colon \mathbf{HP}^\infty_{(q)} \to \mathbf{HP}^\infty_{(0)} \tag{3.9}$$

coincide. Therefore, the maps f_q $(q \in L)$ glue together to yield a map

$$f \colon \mathbf{HP}^\infty_{(L)} \to X_{(L)} \tag{3.10}$$

which is a homotopy equivalence, since each f_q is.

This shows that (2) implies (3) in Theorem 1.2.

3.3. Proof of (3) implies (1)

Suppose that there exists a cofinite set of primes L such that $\mathbf{HP}^\infty_{(L)}$ and $X_{(L)}$ are homotopy equivalent. Write $p_1, \ldots, p_s$ for the primes not in L, and write r_L for the natural map from $X_{(L)}$ to $\mathbf{HP}^\infty_{(0)}$.

To construct an essential map from $\mathbf{CP}^\infty$ to X, first note that X can be constructed as the homotopy inverse limit of the diagram

$$\mathbf{HP}^\infty_{(p_i)} \xrightarrow{r_{p_i}} \mathbf{HP}^\infty_{(0)} \xrightarrow{n_{p_i}} \mathbf{HP}^\infty_{(0)} \xleftarrow{r_L} X_{(L)} \tag{3.11}$$

in which i runs from 1 to s.

For each i, $1 \le i \le s$, let f_{p_i} be a map from $\mathbf{CP}^\infty$ to $\mathbf{HP}^\infty_{(p_i)}$ of degree M/n_{p_i}, where $M = \prod_{i=1}^{s} n_{p_i}$. Also, let f_L denote a map from $\mathbf{CP}^\infty$ to $X_{(L)}$ of degree M, which exists because $X_{(L)}$ has the same homotopy type as $\mathbf{HP}^\infty_{(L)}$. It is then easy to see that the two maps

$$r_L \circ f_L,\, n_{p_i} \circ r_{p_i} \circ f_{p_i} \colon \mathbf{CP}^\infty \to \mathbf{HP}^\infty_{(0)} \tag{3.12}$$

coincide for any $1 \le i \le s$. Therefore, the maps f_{p_i} $(1 \le i \le s)$ and f_L glue together to yield an essential map

$$f \colon \mathbf{CP}^\infty \to X \tag{3.13}$$

through which all the maps f_{p_i} and f_L factor.

This shows that (3) implies (1) in Theorem 1.2.

The proof of Theorem 1.2 is complete.

4. Proof of Theorem 1.7

In this section we prove Theorem 1.7.

Fix a space X in the genus of $\mathbf{HP}^\infty$ which admits an essential map from $\mathbf{CP}^\infty$.

First we note that part (2) follows from the discussion preceding Theorem 1.7, since it is obvious that the integer $T_{\mathbf{HP}^\infty}$ is 1.

Since any essential self-map of X is a rational equivalence, part (4) is an immediate consequence of parts (1) and (3) and a result of Ishiguro, Møller, and Notbohm [5, Thm. 1] which says that the degrees of essential self-maps of X are precisely the odd squares.

Now we consider part (1). Suppose that the integers n_q as in (1.5) are chosen so that there are only finitely many distinct integers in the set $\{n_q\colon q \text{ primes}\}$ and that T_X is their least common multiple (see (1.6) for the definition of T_X). Denote by $l_q\colon \mathbf{HP}^\infty \to \mathbf{HP}^\infty_{(q)}$ the q-localization map and by $i\colon \mathbf{CP}^\infty \to \mathbf{HP}^\infty$ the maximal torus inclusion. A self-map of $\mathbf{CP}^\infty$ of degree m on $H^2(\mathbf{CP}^\infty;\mathbf{Z})$ is simply denoted by m. Now for each prime q define a map $f_q\colon \mathbf{CP}^\infty \to \mathbf{HP}^\infty_{(q)}$ to be the composition

$$\mathbf{CP}^\infty \xrightarrow{M/n_q} \mathbf{CP}^\infty \xrightarrow{i} \mathbf{HP}^\infty \xrightarrow{l_q} \mathbf{HP}^\infty_{(q)}. \tag{4.1}$$

It is then easy to see that the two maps

$$n_q \circ r_q \circ f_q,\ n_{q'} \circ r_{q'} \circ f_{q'}\colon \mathbf{CP}^\infty \to \mathbf{HP}^\infty_{(0)} \tag{4.2}$$

coincide for any two primes q and q'. Therefore, the maps f_q glue together to yield an essential map

$$f\colon \mathbf{CP}^\infty \to X \tag{4.3}$$

through which every map f_q factors. The map f has degree T_X because its induced map in rational cohomology in dimension 4 does.

Finally, for part (3), suppose that $f\colon \mathbf{CP}^\infty \to X$ is a map. Write $f_p\colon \mathbf{CP}^\infty \to \mathbf{HP}^\infty_{(p)}$ for the component map of f corresponding to the prime p. That is, f_p is the composition

$$\mathbf{CP}^\infty \xrightarrow{f} X \to \mathbf{HP}^\infty_{(p)} \tag{4.4}$$

where the second map is the natural map arising from the construction of X. Then for any prime p we have the equality

$$\deg(f) \;=\; n_p \deg(f_p). \tag{4.5}$$

Since each n_p divides $\deg(f)$, so does their least common multiple T_X. Moreover, by writing $(i_X)_p$ for the component map of i_X corresponding to the prime p, (4.5) implies that for any prime p we have the equalities

$$\frac{\deg(f_p)}{\deg(i_X)_p} = \frac{\deg(f)/n_p}{T_X/n_p} = \frac{\deg(f)}{T_X}. \tag{4.6}$$

Since there are self-maps of $\mathbf{HP}^\infty_{(q)}$ (q any prime) and $\mathbf{HP}^\infty_{(0)}$ of degree $\deg(f)/T_X$, one can construct a self-map g of X such that $\deg(g)$ is equal to $\deg(f)/T_X$ and that f is homotopic to $g \circ i_X$. This proves part (3).

The proof of Theorem 1.7 is complete.

References

[1] J.F. Adams, *Infinite loop spaces*, Annals of Math. Studies **90**, Princeton Univ. Press, 1978.

[2] M.F. Atiyah, *Power operations in K-theory*, Quart. J. Math. Oxford. **17** (1966), 165–193.

[3] A.K. Bousfield and D.M. Kan, *Homotopy limits, completions and localizations*, Springer Lecture Notes in Math. **304**, 1972.

[4] F.-X. Dehon and J. Lannes, *Sur les espaces fonctionnels dont la source est le classifiant d'un groupe de Lie compact commutatif*, Inst. Hautes Etudes Sci. Publ. Math. **89** (1999), 127–177.

[5] K. Ishiguro, J. Møller, and D. Notbohm, *Rational self-equivalences of spaces in the genus of a product of quaternionic projective spaces*, J. Math. Soc. Japan **51** (1999), 45–61.

[6] C.A. McGibbon, *Which group structures on S^3 have a maximal torus?*, Springer Lecture Notes in Math. **657** (1978), 353–360.

[7] M.B. Nathanson, *Elementary Methods in Number Theory*, Graduate Texts in Math., vol. 195, Springer-Verlag, New York, 2000.

[8] D. Notbohm and L. Smith, *Fake Lie groups and maximal tori I*, Math. Ann. **288** (1990), 637–661.

[9] D.L. Rector, *Loop structures on the homotopy type of S^3*, Springer Lecture Notes in Math. **249** (1971), 99–105.

[10] D. Sullivan, *Genetics of homotopy theory and the Adams conjecture*, Ann. Math. **100** (1974), 1–79.

[11] C. Wilkerson, *Classification of spaces of the same n-type for all n*, Proc. Amer. Math. Soc. **60** (1976), 279–285.

[12] D. Yau, *On adic genus, Postnikov conjugates, and lambda-rings*, preprint. `http://www.math.uiuc.edu/~dyau/papers.html`

Donald Yau
Department of Mathematics
University of Illinois at Urbana-Champaign
1409 W. Green Street
Urbana, IL 61801, USA
e-mail: `dyau@math.uiuc.edu`

PM 212: **Lubotzky, A. / Segal, D.** Subgroup Growth. (2003) ISBN 3-7643-6989-2

PM 211: **Hernández-Lerma, O. / Lasserre, J.B.** Markov Chains and Invariant Probabilities. (2003) ISBN 3-7643-7000-9

PM 210: **Joseph, A. / Melnikov, A. / Rentschler, R.** (Eds.) Studies in Memory of Issai Schur (2003) ISBN 0-8176-4208-0

PM 209: **Unterberger, A.** Automorphic Pseudo-differential Analysis and Higher-Level Weyl Calculi (2002) ISBN 3-7643-6909-4

PM 208: **Turaev, V.** Torsions of 3-dimensional Manifolds (2002) ISBN 3-7643-6911-6

PM 207: **Long, Y.** Index Theory for Symlectic Paths with Applications (2002) ISBN 3-7643-6647-8

PM 206: **Snaith, V. P.** Algebraic K-Groups as Galois Modules (2002) ISBN 3-7643-6717-2

PM 205: **Bouwknegt, P. / Wu, S.** Geometric Analysis and Applications to Quantum Field Theory (2002) ISBN 0-8176-4287-0

PM 204: **Kumar, S.** Kac-Moody Groups, their Flag Varieties and Representation Theory (2002) ISBN 0-8176-4227-7

PM 203: **Blair, D.E.** Riemannian Geometry of Contact and Symplectic Manifolds (2001) ISBN 0-8176-4261-7

PM 201/202: **Casacuberta, C. / Miró-Roig, R.M. / Verdera, J. / Xambó-Descamps, S.** European Congress of Mathematics. Barcelona, July 10-14, 2000, Volume I (2001) ISBN 3-7643-6417-3

Volume II (2001) ISBN 3-7643-6418-1

PM 200: **Golubitsky, M. / Stewart, I.** The Symmetry Perspective. From Equilibrium to Chaos in Phase Space and Physical Space (2001) ISBN 3-7643-6609-5

PM 199: **Peyre, E. / Tschinkel, Y.** Rational Points on Algebraic Varieties (2001) ISBN 3-7643-6612-5

PM 198: **Landsman, N.P. / Pflaum, M., / Schlichenmaier, M.** Quantization of Singular Symplectic Quotients (2001) ISBN 3-7643-6608-7

PM 197: **Cherix, P.-A. / Cowling, M. / Jolissaint, P. / Julg, P.** Groups with the Haagerup Property. Gromov's a-T-menability (2001) ISBN 3-7643-6598-6

Edited by
Hyman Bass, University of Michigan, USA
Joseph Oesterlé, Université Paris VI, France
Alan Weinstein, University of California, Berkeley, USA

Progress in Mathematics

Progress in Mathematics is a series of books intended for professional mathematicians and scientists, encompassing all areas of pure mathematics. This distinguished series, which began in 1979, includes research level monographs, polished notes arising from seminars or lecture series, graduate level textbooks, and proceedings of focused and refereed conferences. It is designed as a vehicle for reporting ongoing research as well as expositions of particular subject areas.

For orders originating from all over the world except USA and Canada:

Birkhäuser Verlag AG
c/o Springer GmbH & Co
Haberstrasse 7
D-69126 Heidelberg
Fax: +49 / 6221 / 345 4229
e-mail: birkhauser@springer.de

For orders originating in the USA and Canada:

Birkhäuser
333 Meadowland Parkway
USA-Secaucus
NJ 07094-2491
Fax: +1 201 348 4033
e-mail: orders@birkhauser.com

Birkhäuser

Progress in Mathematics

Previously published titles

PM 196: **Aguadé, J. / Broto, C. / Casacuberta, C.** (Eds.)
Cohomological Methods in Homotopy Theory (2001) ISBN 3-7643-6588-9

PM 195: **Faber, C. / van der Geer, G. / Oort, F.** (Eds.)
Moduli of Abelian Varieties (2001)
ISBN 3-7643-6517-X

PM 194: **Juhl, A.**
Cohomological Theory of Dynamical Zeta Functions (2001) ISBN 3-7643-6405-X

PM 193: **Knudson, K.P.**
Homology of Linear Groups (2001)
ISBN 3-7643-6415-7

PM 192: **Dehornoy, P.**
Braids and Self-Distributivity (2000)
ISBN 3-7643-6343-6

PM 191: **Kashiwara, M. / Miwa, T.** (Eds.)
Physical Combinatorics (2000)
ISBN 0-8176-4175-0

PM 190: **Essen van den, A.**
Polynomial Automorphisms and the Jacobian Conjecture (2000)
ISBN 3-7643-6350-9

PM 189: **André, Y. / Baldassarri, F.**
De Rham Cohomology of Differential Modules on Algebraic Varieties (2001)
ISBN 3-7643-6348-7

PM 188: **Dolbeault, P. / Iordan, A. / Henkin, G. / Skoda, H.** (Eds.)
Complex Analysis and Geometry. International Conference in Honor of Pierre Lelong (2000) ISBN 3-7643-6352-5

PM 187: **Bost, J.-B. / Loeser, F. / Raynaud, M.** (Eds.)
Courbes semi-stables et groupe fondamental en géométrie algébrique. Luminy, Décembre 1998 (2000)
ISBN 3-7643-6308-8

PM 186: **Kleinert, E.**
Units in Skew Fields (2000)
ISBN 3-7643-6293-6

PM 185: **Faraut, J. / Kaneyuki, S. /** Korányi, A. / **Lu, Q.**
Analysis and Geometry on Complex Homogeneous Domains (2000)
ISBN 0-8176-4138-6

PM 184: **du Sautoy, M. / Segal, D. / Shalev, A.** (Eds.)
New Horizons in pro-p Groups ((p kursiv)) (2000) ISBN 0-8176-4171-8

PM 183: **Kapovich, M.**
Hyperbolic Manifolds and Discrete Groups (2000) ISBN 0-8176-3904-7

PM 182: **Billey, S. / Lakshmibai, V.**
Singular Loci of Schubert Varieties (2000)
ISBN 0-8176-4092-4

PM 181: **Hauser, H. / Lipmann, J. / Oort, F. / Quirós, A.** (Eds.)
Resolution of Singularities. A research textbook in tribute to Oscar Zariski Based on the courses given at the Working Week in Obergurgl, Austria, September 7-14, 1997 (2000)
ISBN 3-7643-6178-6

PM 180: **Paternain, G.P.**
Geodesic Flows (1999)
ISBN 0-8176-4144-0

PM179: **Morales Ruiz, J.J.**
Differential Galois Theory and Non-Integrability of Hamiltonian Systems. (Sunyer Prize 1998) (1999)
ISBN 3-7643-6078-X

PM 178: **Puig, L.**
On the Local Structure of Morita and Rickard Equivalences between Brauer Blocks (1999) ISBN 3-7643-6156-5

PM 177: **Birkenhake, C. / Lange, H.**
Complex Tori (1999)
ISBN 0-8176-4103-3

PM 176: **Bass, H. / Lubotsky, A.**
Tree Lattices (2000)
ISBN 0-8176-4120-3

PM 175: **Bañuelos, R. / Moore, C.N.**
Probabilistic Behavior of Harmonic Functions (1999)
ISBN 3-7643-6062-3

PM 174: **Goerss, P.G. / Jardine, J.F.**
Simplicial Homotopy Theory (1999)
ISBN 3-7643-6064-X

PM 173: **Dräxler, P. / Michler, G. / Ringel, C.M.** (Eds.)
Computational Methods for Representations of Groups and Algebras. Euroconference in Essen (Germany), April 1-5, 1997 (1999) ISBN 3-7643-6063-1

PM 172: **Brylinski, J.-L. / Brylinski, R. / Nistor, V. / Tsygan, B.** et al. (Eds.)
Advances in Geometry (1998)
ISBN 0-8176-4044-4

PM 171: **Reznikov, A. / Schappacher, N.** (Eds.) Regulators in Analysis, Geometry and Number Theory (1999)
ISBN 0-8176-4115-7

PM 170: **Paterson, A.**
Groupoids, Inverse Semigroups and their Operator Algebras (1998)
ISBN 0-8176-4051-7

PM 169: **Balog, A. / Katona, G.O.H. / Sza'sz, D. / Recski, A.** (Eds.)
European Congress of Mathematics, Budapest, July 22-26, 1996. Volume II (1998) ISBN 3-7643-5498-4

PM 168: **Balog, A. / Katona, G.O.H. / Sza'sz, D. / Recski, A.** (Eds.)
European Congress of Mathematics, Budapest, July 22-26, 1996. Volume I (1998) ISBN 3-7643-5497-6

PM 167: **Facchini, A.**
Module Theory. Endomorphism Rings and Direct Sum Decompositions in some Classes of Modules (1998)
ISBN 3-7643-5908-0

PM 166: **Elias, J. / Giral, J.M. / Miró-Roig, R.M. / Zarzuela, S.** (Eds.)
Six Lectures on Commutative Algebra (1998) ISBN 3-7643-5951-X

PM 165: **Migliore, J.C.**
Introduction to Liaison Theory and Deficiency Modules (1998)
ISBN 0-8176-4027-4

PM 164: **Roussarie, R.**
Bifurcations of Planar Vector Fields and Hilbert's Sixteenth Problem (1998)
ISBN 3-7643-5900-5

PM 163: **Berndt, R. / Schmidt, R.**
Elements of the Representation Theory of the Jacobi Group (1998)
ISBN 3-7643-5922-6

PM 162: **Arnold, V.I. / Greuel, G.M. / Steenbrink, J.H.M.** (Eds.)
Singularities. The Brieskorn Anniversary Volume (1998)
ISBN 3-7643-5913-7

PM 161: **Sagan, B.E. / Stanley, R.P.**
Mathematical Essays in Honor of Gian-Carlo Rota (1998) ISBN 0-8176-3872-5

PM 160: **Kashiwara, M. / Matsuo, A. / Saito, K. / Satake, I.** (Eds.)
Topological Field Theory, Primitive Forms and Related Topics (1998)
ISBN 0-8176-3975-6

PM 159: **Thangavelu, S.**
Harmonic Analysis on the Heisenberg Group (1998) ISBN 0-8176-4050-9

PM 158: **Tirao, J. / Vogan, D.A. / Wolf, J.A.**
Geometry and Representation. Theory of Real and p-Adic Groups (1997)
ISBN 0-8176-3931-4

PM 157: **Murty, M.R. / Murty, V.K.**
Non-vanishing of L-Functions and Applications. (Sunyer pize 1996) (1997)
ISBN 3-7643-5801-7

PM 156: **Guivarc'h, Y. / Ji, L. / Taylor, J.C.**
Compactifications of Symmetric Spaces (1998) ISBN 0-8176-3899-7

PM 155: **Dragomir, S. / Ornea, L.**
Locally Conformal Kähler Geometry (1998)
ISBN 0-8176-4020-7

PM 154: **Böttcher, A. / Karlovich, Y.I.**
Carleson Curves, Muckenhoupt Weights, and Toeplitz Operators. Sunyer prize 1997 (1997) ISBN 3-7643-5796-7

PM 153 : **Buescu, J.**
Exotic Attractors. From Liapunov Stability to Riddled Basins (1997)
ISBN 3-7643-5793-2

PM 152: **Gromov, M. / Bates, S.M.**
Metric Structures for Riemannian and Non-Riemannian Spaces (2001)
ISBN 3-7643-3898-9

PM 151: **Hummel, C.**
Gromov's Compactness Theorem for Pseudo-holomorphic Curves (1997)
ISBN 3-7643-5735-5

PM 150: **Shiota, M.**
Geometry of Subanalytic and Semialgebraic Sets (1997)
ISBN 0-8176-4000-2

PM 149: **Souriau, J.-M.**
A Symplectic View of Physics. Structure of Dynamical Systems (1997)
ISBN 0-8176-3695-1

PM 148: **Huang, Y.-Z.**
Two-Dimensional Conformal Geometry and Vertex Operator Algebras (1997)
ISBN 0-8176-3829-6

PM 147: **Di Biase, F.**
Fatou Type Theorems. Maximal Functions and Approach Regions (1998)
ISBN 0-8176-3976-4

PM 146: **Jardine, J.F.**
Generalized Etale Cohomology Theories (1997) ISBN 3-7643-5494-1

PM 145: **Albert, C. / Brouzet, R. / Dufour, J.P.**
Integrable Systems and Foliations. Feuilletages et Systèmes Intégrables (1997)
ISBN 0-8176-3894-6

PM 144: **Bellaïche, A. / Risler, J.J.** (Eds.)
Sub-Riemannian Geometry (1996)
ISBN 3-7643-5476-3

PM 143: **González-Vega, L. / Recio, T.**
Algorithms in Algebraic Geometry and Applications (1996)
ISBN 3-7643-5274-4

PM 142: **Monk, J.D.**
Cardinal Invariants on Boolean Algebras (1996) ISBN 3-7643-5402-X

PM 141: **Cabanes, M.** (Ed.)
Finite Reductive Groups: Related Structures and Representations. Proceedings of an International conference held in Luminy, France (1996)
ISBN 0-8176-3885-7

PM 140: **Knapp, A.W.**
Lie Groups Beyond an Introduction (1996)
ISBN 0-8176-3926-8

MIX
Papier aus verantwortungsvollen Quellen
Paper from responsible sources
FSC® C105338

If you have any concerns about our products,
you can contact us on
ProductSafety@springernature.com

In case Publisher is established outside the EU,
the EU authorized representative is:
Springer Nature Customer Service Center GmbH
Europaplatz 3, 69115 Heidelberg, Germany

Printed by Libri Plureos GmbH
in Hamburg, Germany